Mechanics of fracture

VOLUME 3

Plates and shells with cracks

Mechanics of fracture

edited by GEORGE C. SIH

VOLUME 1

Methods of analysis and solutions of crack problems

VOLUME 2

Three-dimensional crack problems

VOLUME 3

Plates and shells with cracks

Mechanics of fracture   3

Plates and shells with cracks

A collection of stress intensity factor solutions for cracks in plates and shells

Edited by

G. C. SIH
Professor of Mechanics and
Director of the Institute of
Fracture and Solid Mechanics

Lehigh University
Bethlehem, Pennsylvania

NOORDHOFF INTERNATIONAL PUBLISHING
LEYDEN

ISBN-13:978-94-010-1294-2   e-ISBN-13:978-94-010-1292-8
DOI: 10.1007/978-94-010-1292-8

VAN DER LOEFF / DRUKKERS BV – ENSCHEDE

Contents

Chapter 3

Through cracks in multilayered plates
R. Badaliance, G. C. Sih and E. P. Chen

Chapter 4

Asymptotic approximations to crack problems in shells
E. S. Folias

Chapter 5

Crack problems in cylindrical and spherical shells
F. Erdogan

Chapter 6

On cracks in shells with shear deformation
G. C. Sih and H. C. Hagendorf

Chapter 7

Dynamic analysis of cracked plates in bending and extension
G. C. Sih and E. P. Chen

Chapter 8

A specialized finite element approach for three-dimensional crack problems   *P. D. Hilton*

Editor's preface

This third volume of a series on Mechanics of Fracture deals with cracks in plates and shells. It was noted in Volume 2 on three-dimensional crack problems that additional free surfaces can lead to substantial mathematical complexities, often making the analysis unmanageable. The theory of plates and shells forms a part of the theory of elasticity in which certain physical assumptions are made on the basis that the distance between two bounded surfaces, either flat or curved, is small in comparison with the overall dimensions of the body. In modern times, the broad and frequent applications of plate- and shell-like structural members have acted as a stimulus to which engineers and researchers in the field of fracture mechanics have responded with a wide variety of solutions of technical importance. These contributions are covered in this book so that the reader may gain an understanding of how analytical treatments of plates and shells containing initial imperfections in the form of cracks are carried out.

The development of plate and shell theories has involved long standing controversy on the consistency of omitting certain small terms and at the same time retaining others of the same order of magnitude. This deficiency depends on the ratio of the plate or shell thickness, h, to other characteristic dimensions and cannot be completely resolved in view of the approximations inherent in the transverse dependence of the extensional and bending stresses. The presence of a crack tends to further complicate the situation and requires additional classification of the basic assumptions. First of all, the conventional subdivision of thin and thick plates or shells is no longer adequate. Suppose that a thin shell satisfies the usual requirement of $(h/R)_{max} \leq 1/20$ with R being the smallest radius of curvature of the shell. At small distances (as compared with h) from the crack, the shell would appear to be thick and the assumptions* of the classical thin shell theory would lead to an inaccu-

---

* Points near the crack edge in a plate or shell, which lie on the normal to the middle surface before deformation, cannot be assumed to remain on the normal after deformation. This in effect implies that shear forces cannot be disregarded.

rate description of the local stress state. In particular, the replacement of boundary conditions on twisting moment and shearing force by one of equivalent or Kirchhoff shear results in the satisfaction of only four boundary conditions instead of five as required on the crack surface. A priori, the qualitative character of the crack front stresses in a plate or shell should be the same as that derived from the three-dimensional theory of elasticity. The approximations mentioned earlier should only influence the results in a quantitative sense through the stress intensity factors.

Another inadequacy of the classical approach is that the dependence of the stresses on the transverse coordinate is assumed arbitrarily. Along the crack front, the normal stresses in most of the interior region except in a layer near the free surface should satisfy the plane strain condition, $\sigma_z = v(\sigma_x + \sigma_y)$, in which σ_z and σ_x, σ_y are the transverse and in-plane normal stresses while v stands for the Poisson's ratio. The stress solution in the surface layer can be constructed separately by adopting the concept of a boundary layer within which the stress intensity factors decrease sharply and tend to zero* on the free surface. This desirable feature of the solution can serve as a useful guideline in selecting the appropriate plate or shell theory.

In structural applications, nonalignment of loads with crack orientation is a common practice. As a rule, the stress analysis of plates or shells involves more than one stress intensity factor. For example, misalignment in a plane involves k_1 and k_2 while k_2 and k_3 arise simultaneously in plates or shells that are twisted or sheared. Combined loadings often lead to all three stress intensity factors k_j ($j = 1, 2, 3$). The calculation of allowable load in fracture mechanics will depend on some combination of k_1, k_2 and k_3 reaching some critical condition. This requires the knowledge of a suitable fracture criterion that can predict non-coplanar** crack extension. The theory of strain

---

* The three-dimensional theory of elasticity yields a stress solution different from $1/\sqrt{r}$ on the free surface where r is a radial distance. This implies the vanishing of stress intensity factors. Since a hypothetical surface of zero thickness cannot be conceived physically, the necessity of admitting a boundary layer in engineering applications is apparent.

** The classical energy release rate concept based on cracks propagating in a self-similar manner cannot be extended to the general case by adding G_1, G_2 and G_3 as associated with k_1, k_2 and k_3:

$$G = [\pi(1 + v)/E] [(1 - v) (k_1^2 + k_2^2) + k_3^2]$$

This is simply because cracks do not propagate straight ahead (as this equation requires) in combined loading. Any attempts made to extend the energy release concept to the bent

energy density or S-theory provides a simple means of evaluating allowable stress in plates and shells subjected to complex loadings. Since this is a relatively new concept, a few introductory remarks are in order.

One important feature of the S-theory, which is fundamentally different from all existing failure theories, is that it simultaneously accounts for S_v, energy factor associated with volume change (dilatation) and S_d, energy factor associated with shape change (distortion). The proportion of S_v and S_d in $S = S_v + S_d$ is determined at locations of the stationary values of S. Depending on the material constants, the relative minimums, S_{min}, occur at locations where S_v is normally greater than S_d and the relative maximums, S_{max}, correspond to locations where $S_d > S_v$. The basic assumption is that brittle fracture takes place at locations of S_{min} (excessive dilatation) while yielding occurs at locations of S_{max} (excessive distortion). In general, it is not justified to neglect S_v against S_d or vice versa in a material element unless they differ from one another by one order or magnitude or more which is seldom the case. The application of S-theory to analyze cracks in plates is presented in the Introductory Chapter. For brittle fracture, the onset of crack propagation is assumed to occur when S_{min} reaches a critical value, S_c, characteristic of the material. This value is then related to the load and geometry of a plate or shell for determining the sub-critical or critical size of cracks as a function of the allowable load or vice versa.

The first chapter deals with the stress intensity factor solutions for a variety of crack geometries by using the Poisson-Kirchhoff plate bending theory. Among the problems treated are collinear cracks, parallel cracks, cracks inclined at arbitrary positions, etc. The versatility of the complex variable technique is again demonstrated for solving two-dimensional crack problems involving rather complicated crack geometries. Although the theory does not satisfy the physical crack surface boundary conditions, useful information on stress intensity factors can be assessed for complex problems which may not be so easily obtained from the higher order plate theories. Chapter Two contains a thorough review of existing plate theories in connection with the mixed boundary value crack problem. Assumptions made in each of the theories as developed from the minimum principles in variational calculus and how they would affect the end results are discussed. Numerical solutions computed from integral equations for several crack

or kinked crack configuration would be unfeasible as the uncertainties encountered in the mathematics cannot be easily resolved.

problems show the interaction of plate thickness with crack length. A formulation of the bending and/or extension of layered plates containing a through crack is presented in Chapter Three. Each layer can have different elastic properties and thickness. The three-layered plate problem with a through crack subjected to a tensile load is solved as an example. Lamination is found to have a significant influence on the intensity of load transmitted to the crack tip region. Graphical plots of stress intensity factors for various material and geometric parameters are displayed.

In Chapter Four, the classical theory of thin shallow shells is applied to analyze the stress distribution around cracks in shells. The shallowness assumption justifies the projection of the cracked portion of the shell onto a flat plane such that the problem is formulated in terms of rectangular coordinates in two dimensions. Solutions based on singular integral equations are provided for cracks in shells of different shapes such as spherical, cylindrical, conical, toroidal, etc. The presence of curvature in a shell generates deviation from the behavior of flat plates, in that stretching loads induce both extensional and bending stresses, while bending loads also lead to both types of stresses. Chapter Five considers a method of solution for cracked shells possessing anisotropy. The method is illustrated by solving the problem of an orthotropic cylindrical shell with an axial crack. Both symmetric and skew-symmetric loadings are considered. Other examples dealing with the effect of Poisson's ratio and interaction of two collinear cracks in shells of an isotropic nature are also provided. Developed in Chapter Six is a higher order thin shell theory in which transverse shear deformation is included. A tenth order system differential equation is obtained such that satisfaction of the five natural boundary conditions on the crack surface is made possible. This is an improvement over the classical theory which considers only four conditions. Solutions showing the influence of shell curvature, shell thickness, crack length, etc., on the stress intensity factor are given and they can differ appreciably from those obtained by the classical theory corresponding to an eighth order system of equations.

The Seventh Chapter is concerned with dynamic loading of flat plates with cracks where inertia effects can no longer be neglected. Two types of dynamic-load sources are treated: namely, vibratory and impact. The classical and Mindlin theories are used for crack problems in plate bending and the Kane-Mindlin theory is applied to solve the elastodynamic plate stretching problems. Dynamic stress intensity factors are defined and found to vary as functions of time. They tend to the static values as time becomes increasingly large for the case of impact loading.

Chapter Eight is devoted to numerical analysis of the three-dimensional plate problem with cracks. The solution is considered to be approximate in that the equilibrium equations of elasticity are satisfied only in an average sense over discrete regions of the plate medium. The numerical procedure involves three-dimensional isoparametric finite elements specialized for solving crack problems where the crack edge singularity is assumed to be known and embedded into the analysis. This is done by including in the crack tip elements the asymptotic solution for the displacement field as well as the usual polynomial approximation. Values for the stress intensity factors at nodes along the crack edge are treated as unconstrained degrees of freedom or generalized nodal displacements to be determined through the finite element procedure. Studies are made on the stress distribution in a finite thickness plate with a through crack subject to tensile and bending loads.

Much of the material in this volume is the result of research efforts initiated in the 1960's at the California Institute of Technology and Lehigh University. The fruitful discussions and interchange of ideas among the researchers associated with these two institutions over the years have made possible significant progress in the understanding of fracture behavior of plates and shells weakened by cracks or initial imperfections. My gratitude goes to colleagues and students who have contributed in this field. The many valuable hours the authors have spent in preparing their manuscripts are deeply appreciated. I am also thankful to Mrs. Barbara DeLazaro for neatly typing portions of the book and Mrs. Constance Weaver for willingly providing constant secretarial assistance.

Lehigh University   G. C. SIH
Bethlehem, Pennsylvania

March, 1976

Contributing authors

R. BADALIANCE
Foster-Wheeler Corporation, Livingston, New Jersey

E. P. CHEN
Institute of Fracture and Solid Mechanics, Lehigh University, Bethlehem, Pennsylvania

F. ERDOGAN
Institute of Fracture and Solid Mechanics, Lehigh University, Bethlehem, Pennsylvania

E. S. FOLIAS
Department of Civil Engineering, University of Utah, Salt Lake City, Utah

H. C. HAGENDORF
Fatigue and Fracture Consultants, Hawthorne, California

R. J. HARTRANFT
Institute of Fracture and Solid Mechanics, Lehigh University, Bethlehem, Pennsylvania

P. D. HILTON
Institute of Fracture and Solid Mechanics, Lehigh University, Bethlehem, Pennsylvania

M. ISIDA
School of Engineering, Kyushu University, Fukuoka 812, Japan

G. C. SIH
Institute of Fracture and Solid Mechanics, Lehigh University, Bethlehem, Pennsylvania

G. C. Sih

Introductory chapter:

Strain energy density theory applied to plate bending problems

I. Introductory remarks

When a slender member is stretched gradually with consideration given only to the principal stress in the axial direction, then failure in the global sense is said to occur by yielding if this stress reaches the elastic limit or yield point and by fracturing if the ultimate strength of the material is reached. Material elements within this member, however, are in a multiaxial stress state. Hence, yielding and/or fracture in the local sense or at a point cannot be adequately described by considering just one of the six stress or strain components even if the remote loading is uniaxial. The distinction between localized and global yielding or fracture must be made before selecting a quantity whose critical value will be assigned to determine the load carrying capacity of the member. In principle, this quantity should involve some combination of the stresses or strains.

The *energy* quantity has been frequently associated with the development of failure theories. Two of the commonly used theories are the total energy or Beltrami-Haigh theory and the distortion energy or Huber-Von Mises-Hencky theory. According to the former, failure by yielding in the material occurs when the total strain energy per unit volume

$$
\frac{\mathrm{d}W}{\mathrm{d}V} = \frac{1+v}{2E}\left[\sigma_x^2 + \sigma_y^2 + \sigma_z^2 - \frac{v}{1+v}(\sigma_x + \sigma_y + \sigma_z)^2 \right.
$$
$$
\left. + 2(\tau_{xy}^2 + \tau_{yz}^2 + \tau_{zx}^2)\right] \tag{1}
$$

absorbed by the material is equal to the energy per unit volume stored in the material upon reaching the yield point under a uniaxial stress state, i.e., $\mathrm{d}W/\mathrm{d}V = \sigma_{yp}^2/2E$. The latter theory can be stated in the same way by replacing the total strain energy density, dW/dV, with the distortional strain energy density:

$$\left(\frac{dW}{dV}\right)_d = \frac{1+v}{6E}[(\sigma_x - \sigma_y)^2 + (\sigma_y - \sigma_z)^2 + (\sigma_z - \sigma_x)^2$$
$$+ 6(\tau_{xy}^2 + \tau_{yz}^2 + \tau_{zx}^2)] \tag{2}$$

where E is the Young's modulus of elasticity and v the Poisson's ratio. Equation (2) is obtained by decomposing dW/dV in equation (1) into two parts

$$\frac{dW}{dV} = \left(\frac{dW}{dV}\right)_v + \left(\frac{dW}{dV}\right)_d \tag{3}$$

in which

$$\left(\frac{dW}{dV}\right)_v = \frac{1-2v}{6E}(\sigma_x + \sigma_y + \sigma_z)^2 \tag{4}$$

represents that part of the strain energy density associated with volume change. It has been implicitly assumed in the two foregoing theories that the failure mode by yielding for an element in a multiaxial stress state is the same as that in a uniaxial stress state.

The Huber-Von Mises-Hencky theory associates failure by yielding with $(dW/dV)_d$ only while $(dW/dV)_v$ is neglected*. Except for the extreme situations of a hydrostatic stress state, where $(dW/dV)_d = 0$, and a pure shear stress state, where $(dW/dV)_v = 0$, the failure of a unit volume of material ranging from incipient yielding to brittle fracture involves the release of both energy due to changing in volume and shape. For instance, equations (2) and (4) indicate that for a uniaxial stress state both $(dW/dV)_d$ and $(dW/dV)_v$ are present:

$$\left(\frac{dW}{dV}\right)_d = \frac{1+v}{3E}\sigma_x^2, \left(\frac{dW}{dV}\right)_v = \frac{1-2v}{6E}\sigma_x^2 \tag{5}$$

Note that the distortional energy is only four times greater than the dilata-

---

* This assumption grew out of the test results of Bridgman [1] who showed that many materials did not exhibit yielding when subjected to very high hydrostatic pressure. In this case, the material elements experience volume change only with $(dW/dV)_d = 0$. It would be inaccurate to conclude in general that the energy absorbed in changing volume has no effect in causing failure by yielding for other stress states as well.

tional energy if $v = 0.2$. This difference, being less than one order of magnitude, cannot justify neglecting $(\mathrm{d}W/\mathrm{d}V)_v$ in comparison with $(\mathrm{d}W/\mathrm{d}V)_d{}^\star$.

A more general situation prevails in Figure 1 where the material element is in a triaxial stress state. Let r_0 be the radius of a core region centered

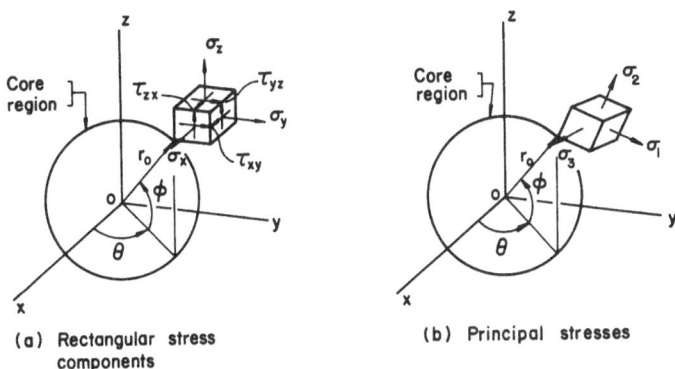

(a) Rectangular stress components

(b) Principal stresses

Figure 1. Stress element outside of core region

around a convenient reference point, the origin. The element under consideration is located outside of this core region. In order to focus attention on the location of failure, Sih [2] has defined a strain energy density factor $S = S_v + S_d$ such that

$$S_v = r_0\left(\frac{\mathrm{d}W}{\mathrm{d}V}\right)_v \tag{6}$$

is the factor associated with volume change and

$$S_d = r_0\left(\frac{\mathrm{d}W}{\mathrm{d}V}\right)_d \tag{7}$$

is the factor associated with shape change. With reference to the principal stresses σ_1, σ_2 and σ_3 in Figure 1(b), equations (4) and (6) may be written as

$$S_v = r_0\left(\frac{1-2v}{6E}\right)(\sigma_1 + \sigma_2 + \sigma_3)^2 \tag{8}$$

and equations (2) and (7) yield

$\star$ An accurate description of the uniaxial extension of metal bars with v between 0.2 and 0.3 should involve both changes in volume and shape.

$$S_d = r_0 \left(\frac{1 + v}{6E} \right) [(\sigma_1 - \sigma_2)^2 + (\sigma_2 - \sigma_3)^2 + (\sigma_3 - \sigma_1)^2] \tag{9}$$

For a state of plane strain where

$$\sigma_3 = v(\sigma_1 + \sigma_2) \tag{10}$$

the ratio S_v/S_d obtained from equations (8) to (10) takes the form

$$\frac{S_v}{S_d} = \frac{(1 - 2v)(1 + v)[(\sigma_1/\sigma_2) + 1]^2}{[(\sigma_1/\sigma_2) - 1]^2 + [(1 - v) - v(\sigma_1/\sigma_2)]^2 + [(1 - v)(\sigma_1/\sigma_2) - v]^2} \tag{11}$$

Figure 2 displays a plot of S_v/S_d versus the ratio of the principal stresses σ_1/σ_2 in equation (11) for $v = 0.0, 0.1, 0.2, 0.3$, and 0.4. The greatest volume change takes place when $\sigma_1 = \sigma_2$ corresponding to a two-dimensional hydrostatic stress state. For most metals* with v ranging from 0.2 to 0.3, S_v/S_d varies from 4 to 6.5. The relative magnitude of S_v and S_d shows that both of them should be accounted for in considering the failure of material elements.

More recently, Sih [3, 4] has proposed a failure theory based on the stationary value of the strain energy density factor

$$S = S_v + S_d \tag{12}$$

reaching some critical value. Unlike that of Beltrami-Haigh, Sih's theory possesses the additional feature of being able to locate the element initiating yielding or brittle fracture. This is accomplished by obtaining the stationary values of S with respect to the angles θ and ϕ shown in Figure 1(a) or 1(b). The point (θ_0, ϕ_0) making S a minimum ($= S_{min}$) gives the position of the element that initiates fracture while the location of S_{max} determines the onset of yielding. The initiation of yielding and/or fracture usually occurs in the vicinity of a reentrant corner, a crack front or a notch border from which r_0, the radius of the core region in Figures 1, is measured. The radius r_0 serves as a limiting distance within which the microstructural effect of the material comes into play. In contrast to the classical theory of fracture mechanics [5] which utilizes the concept of a critical stress intensity factor k_{1c} or an energy release rate quantity G_{1c}, the strain energy density factor theory, or S-theory, is not restricted to loading symmetric with respect to

---

* Keep in mind that the present discussion holds only if yielding or fracture occurs under small strain. It does not apply to rubber-like or polymer materials undergoing large deformation.

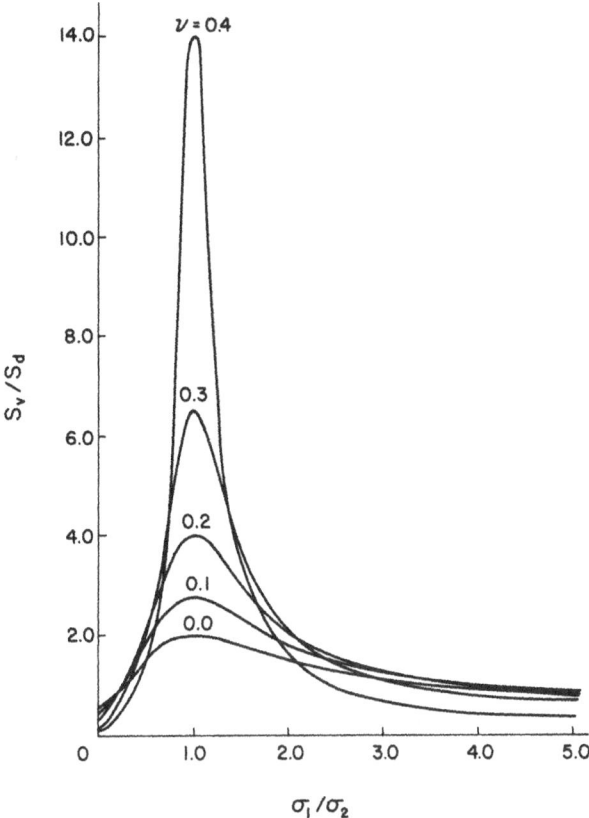

Figure 2. Ratio of dilatational and distortional energy for a state of plane strain

the crack plane and does not require the material to possess an initial crack. In this introductory chapter, the *S*-theory will be further employed in plate bending problems involving the interaction of crack size with plate thickness.

II. Strain energy density factor theory

A rational procedure for the design of structural members requires that the most likely of several possible failure modes of a member in service be determined and that a quantity be assigned to predict failure. Fracture mechanics deals with the conditions under which a member attains uncontrollable fracture by crack propagation. The critical value of the selected quantity which limits the allowable load must not be sensitive to changes

in the nature of loading, the type of geometry, etc., and should be a characteristic of the material. The k_{1c} value in the classical theory of fracture
mechanics [5] is one such example, where $k_{1c} = \sigma\sqrt{a}$ for a given material
should remain constant at fracture while σ, the applied stress, and a, the
half crack length, can vary individually at crack instability. Suitable experiments may be designed to test a given theory by using two different loading
conditions and comparing the experimental results with those predicted by
the theory. Erdogan and Sih [6] have measured the critical tensile stress σ_c
in uniaxial loading and critical shear stress τ_c in pure shear loading for plexiglass plate specimens containing a central crack. Their measured values of
σ_c and τ_c yield an average of 0.915 for τ_c/σ_c and agree favorably with Sih's
prediction [2] of $\tau_c/\sigma_c = 0.905$ for $v = 0.33$. The theoretical prediction based
on the strain energy density theory was obtained by using the same critical
value of S or S_c for the two different stress states. Detailed calculations of
σ_c and τ_c and additional comments will follow subsequently.

Before proceeding further, it is necessary to state the failure mode to be
treated in this discussion. In situations where the phenomenon of material
separation occurs suddenly without noticeable warning such as slow crack
growth, the fracture process is classified as *brittle* or simply as brittle fracture*. The application of the S-theory to ductile fracture is discussed elsewhere and will not be dealt with here. Within the framework of brittle
fracture where failure initiates from an element nearby the crack border,
it suffices to compute the strain energy density S from the local stresses,
$\sigma_x, \sigma_y, \text{---}, \tau_{zx}$ as shown in Figure 3. Without loss of generality, the crack
surface R bounded by an arbitrary contour C is assumed to coincide with
the xz-plane. As the point 0 travels around the crack border, the x-axis is
directed normal to C, the y-axis is perpendicular to R and the z-axis is tangent
to C. The element under consideration is located at (r, θ, ω) which is related
to (x, y, z) through the relationships

---

* Brittle fracture includes those failures which occur without warning and give no evidence of ductility or stretching prior to rapid crack propagation but does not exclusively
imply that the material must be inherently brittle in the metallurgical sense. Materials
with the same chemical composition and microstructure can behave in a brittle fashion
under one set of conditions (i.e., temperature, rate of loading, mechanical constraint, etc.)
and in a ductile fashion under other conditions. Since the continuum mechanics model
of the S-theory excludes metallurgical effects, it would not be accurate to say that the
theory is restricted to brittle materials but rather it deals with an instability phenomenon
that exhibits *brittle behavior*.

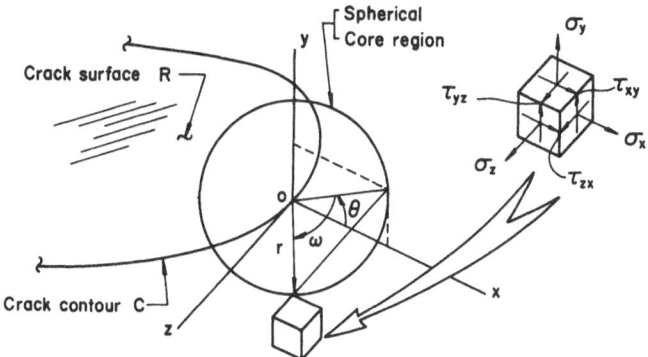

Figure 3. A special spherical coordinate system

$$x = r \cos \theta \cos \omega \tag{13a}$$

$$y = r \sin \theta \cos \omega \tag{13b}$$

$$z = r \sin \omega \tag{13c}$$

where $-\pi/2 \leq \omega \leq \pi/2$ and $-\pi \leq \theta \leq \pi$. The choice of spherical coordinates (r, θ, ω) in Figure 3 yields a simple form* for the local stresses

$$\sigma_x = \frac{k_1}{(2r \cos \omega)^{\frac{1}{2}}} \cos \frac{\theta}{2} \left(1 - \sin \frac{\theta}{2} \sin \frac{3\theta}{2} \right)$$

$$- \frac{k_2}{(2r \cos \omega)^{\frac{1}{2}}} \sin \frac{\theta}{2} \left(2 + \cos \frac{\theta}{2} \cos \frac{3\theta}{2} \right) + O(1) \tag{14a}$$

$$\sigma_y = \frac{k_1}{(2r \cos \omega)^{\frac{1}{2}}} \cos \frac{\theta}{2} \left(1 + \sin \frac{\theta}{2} \sin \frac{3\theta}{2} \right)$$

$$+ \frac{k_2}{(2r \cos \omega)^{\frac{1}{2}}} \sin \frac{\theta}{2} \cos \frac{\theta}{2} \cos \frac{3\theta}{2} + O(1) \tag{14b}$$

$$\sigma_z = 2v \left(\frac{k_1}{(2r \cos \omega)^{\frac{1}{2}}} \cos \frac{\theta}{2} - \frac{k_2}{(2r \cos \omega)^{\frac{1}{2}}} \sin \frac{\theta}{2} \right) + O(1) \tag{14c}$$

* An alternate choice of spherical coordinates (r, θ, φ) such as those shown in Figures 1 lead to more complicated stress expressions [7] which are equivalent to equations (14). An error in the asymptotic expression for one of the ellipsoidal coordinates in [7] needs to be corrected.

$$\tau_{xy} = \frac{k_1}{(2r \cos \omega)^{\frac{1}{3}}} \sin \frac{\theta}{2} \cos \frac{\theta}{2} \cos \frac{3\theta}{2}$$

$$+ \frac{k_2}{(2r \cos \omega)^{\frac{1}{3}}} \cos \frac{\theta}{2} \left(1 - \sin \frac{\theta}{2} \sin \frac{3\theta}{2}\right) + O(1) \tag{14d}$$

$$\tau_{xz} = -\frac{k_3}{(2r \cos \omega)^{\frac{1}{3}}} \sin \frac{\theta}{2} + O(1) \tag{14e}$$

$$\tau_{yz} = \frac{k_3}{(2r \cos \omega)^{\frac{1}{3}}} \cos \frac{\theta}{2} + O(1) \tag{14f}$$

Substituting equations (14) into (1) and using the relation $dW/dV = S/r$, it is found that

$$S = a_{11}k_1^2 + 2a_{12}k_1k_2 + a_{22}k_2^2 + a_{33}k_3^2 \tag{15}$$

The coefficients a_{ij} ($i, j = 1,2$) depend on the material constants and the angles (θ, ω) as given by

$$16\mu \cos \omega \, a_{11} = (3 - 4v - \cos \theta)(1 + \cos \theta) \tag{16a}$$

$$16\mu \cos \omega \, a_{12} = 2 \sin \theta (\cos \theta - 1 + 2v) \tag{16b}$$

$$16\mu \cos \omega \, a_{22} = 4(1 - v)(1 - \cos \theta) + (3 \cos \theta - 1)(1 + \cos \theta) \tag{16c}$$

$$16\mu \cos \omega \, a_{33} = 4 \tag{16d}$$

in which μ stands for the shear modulus of elasticity.

On physical grounds, material elements are capable of storing energy by experiencing volume change (dilatation) and shape change (distortion). Excessive dilatation may result in brittle fracture while excessive distortion may tend to yield the material. The proportion of energy absorbed in dilatation and distortion can be determined for various material properties at locations of the stationary values of the strain energy density factor. In fracture mechanics, interest is centered on the conditions that trigger crack instability. The assumptions made in the strain energy density theory [3, 4] may be stated as

(1) *Crack growth is directed along the line from the center of the sperical core (Figure 3) to the point on the spherical surface with the minimum strain energy density factor S_{min}.*

(2) *Growth along this direction begins when S_{min} reaches the maximum critical value S_c which the material will tolerate.*

The values of (θ_0, ω_0) corresponding to S_{min} are obtained by taking $\partial S/\partial \theta = 0$ and $\partial S/\partial \omega = 0$. At these locations, the elements tend to experience more volume change ΔV than those located at S_{max} where yielding is assumed to take place. The path along which S is a relative minimum has coincided accurately with the experimentally observed trajectory of brittle fracture [8, 9].

Experience indicates that failure can seldom be associated solely with volume change or shape change. It normally involves some of each depending on the stress state. The amount of dilatation and distortion for each material element can be determined by decomposing equation (15) into

$$S_v = b_{11}k_1^2 + 2b_{12}k_1k_2 + b_{22}k_2^2 + b_{33}k_3^2 \tag{17}$$

for dilatation and

$$S_d = c_{11}k_1^2 + 2c_{12}k_1k_2 + c_{22}k_2^2 + c_{33}k_3^2 \tag{18}$$

for distortion. The sum of the coefficients b_{ij} and c_{ij} $(i, j = 1, 2, 3)$ in equations (17) and (18) equal a_{ij} given by equations (16). Without going into details, it can be shown that

$$12\mu \cos \omega \, b_{11} = (1 - 2v)(1 + v)(1 + \cos \theta) \tag{19a}$$

$$12\mu \cos \omega \, b_{12} = (1 - 2v)(1 + v) \sin \theta \tag{19b}$$

$$12\mu \cos \omega \, b_{22} = (1 - 2v)(1 + v)(1 - \cos \theta) \tag{19c}$$

$$12\mu \cos \omega \, b_{33} = 0 \tag{19d}$$

and the result of subtracting equations (19) from (16) yields

$$16\mu \cos \omega \, c_{11} = (1 + \cos \theta)\left[\tfrac{2}{3}(1 - 2v)^2 + 1 - \cos \theta\right] \tag{20a}$$

$$16\mu \cos \omega \, c_{12} = 2 \sin \theta \left[\cos \theta - \tfrac{1}{3}(1 - 2v)^2\right] \tag{20b}$$

$$16\mu \cos \omega \, c_{22} = \tfrac{2}{3}(1 - 2v)^2 (1 - \cos \theta) + 4 - 3 \sin^2 \theta \tag{20c}$$

$$16\mu \cos \omega \, c_{33} = 4 \tag{20d}$$

Once the location (θ_0, ω_0) for a given element is known, S_v and S_d can be computed from equations (17) and (18) for determining the effect of dilatation and distortion on the onset of fracture or yielding.

Now, if failure is confined to the plane normal to the crack front, then $\omega = 0$ and the core region in Figure 3 becomes a circle of radius r centered around the crack edge. Moreover, if the load is applied such that the same stress state prevails along the crack border, then the problem is said to be two-dimensional. The assumption of plane strain leads to additional simplification since in this case k_3 drops out of equations (14) and (15). Consider the plane strain specimens in Figures 4(a) and 4(b) for the uniaxial tension

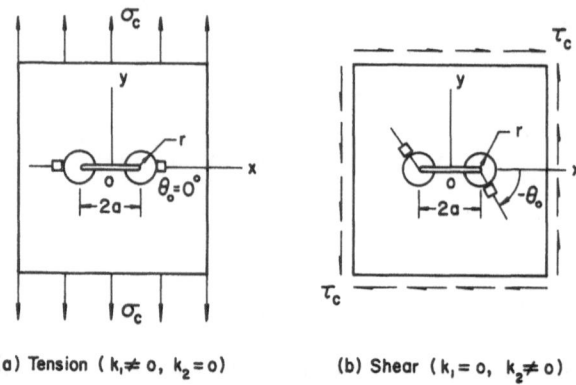

(a) Tension ($k_1 \neq 0$, $k_2 = 0$)   (b) Shear ($k_1 = 0$, $k_2 \neq 0$)

Figure 4. Plane strain specimens

and in-plane shear of a solid containing a crack of length 2a. It will be shown subsequently that, according to the strain energy density theory, the ratio of the critical stresses τ_c/σ_c at incipient fracture is

$$\frac{\tau_c}{\sigma_c} = \left(\frac{3(1-2v)}{2(1-v)-v^2}\right)^{\frac{1}{2}} \tag{21}$$

For $v = 0.33$, this ratio gives $\tau_c/\sigma_c = 0.905$ which agrees well with the experimentally measured value of 0.915 for cracked plexiglass plates [6] as discussed earlier.

Uniaxial tension test. When a tensile load is applied normal to the crack plane, only k_1 is nonzero and equation (15) reduces to

$$S = \frac{1}{16\mu}(3 - 4v - \cos\theta)(1 + \cos\theta)k_1^2 \tag{22}$$

in which $k_1 = \sigma\sqrt{a}$ for a large plate. Referring to Figure 4(a), σ_c denotes the critical value of σ at which crack propagation begins. The direction of crack initiation is found from the condition $\partial S/\partial \theta = 0$ that makes S a relative minimum. This corresponds to $\theta_0 = 0$ implying that the crack grows in a self-similar manner and

$$S_{\min} = \frac{1 - 2v}{4\mu} k_1^2 \tag{23}$$

At the point of crack instability, $S_{\min} \rightarrow S_c$ and equation (23) becomes

$$S_c = \frac{(1 + v)(1 - 2v)}{2E} \sigma_c \sqrt{a} \tag{24}$$

where S_c represents the fracture toughness of the material. The measured values of S_c for a number of metal alloys can be found in [10]. With the help of equations (17) and (18), $S_{\min}$ in equation (23) can be decomposed into

$$S_v = \frac{(1 + v)(1 - 2v)}{6\mu} \sigma^2 a, \; S_d = \frac{(1 - 2v)^2}{12\mu} \sigma^2 a \tag{25}$$

The ratio

$$\frac{S_v}{S_d} = \frac{2(1 + v)}{1 - 2v} \tag{26}$$

shows that S_v is always greater than S_d for $\theta_0 = 0$. Hence, the element assumed to trigger fracture undergoes more volume change than shape change.

The relative maximum of S can be found in the same way. It occurs at $\theta_0 = \cos^{-1}(1 - 2v)$ which may be inserted into equation (22) to give

$$S_{\max} = \frac{(1 - v)^2}{4\mu} \sigma^2 a \tag{27}$$

Separating equation (27) into S_d and S_v renders

$$S_d = \frac{(1 - v)(1 - v + 4v^2)}{12\mu} \sigma^2 a, \; S_v = \frac{(1 - v^2)(1 - 2v)}{6\mu} \sigma^2 a \tag{28}$$

It is easily seen from

$$\frac{S_d}{S_v} = \frac{2(1 - v + 4v^2)}{(1 + v)(1 - 2v)} \tag{29}$$

that $S_d > S_v$ corresponds to elements located at $\theta_0 = \cos^{-1}(1 - 2v)$ where yielding is assumed to take place.

Pure shear test. The specimen in Figure 4(b) is subjected to a state of in-plane shear where $k_1 = 0$ and $k_2 = \tau \sqrt{a}$. With this information, equation

(15) can be written as

$$S = \frac{1}{16\mu} [4(1-v)(1-\cos\theta) + (1+\cos\theta)(3\cos\theta - 1)]\tau^2 a \qquad (30)$$

Equation (30) may be minimized with respect to θ giving

$$\theta_0 = \cos^{-1}\left(\frac{1-2v}{3}\right) \qquad (31)$$

as the direction of crack initiation*, Figure 4(b). Values of θ_0 for different Poisson's ration are given in Table 1. Putting equation (31) into (30) renders

$$S_{\min} = \frac{2(1-v)-v^2}{12\mu}\tau^2 a \qquad (32)$$

TABLE I

Fracture angles

v	0.0	0.1	0.2	0.3	0.4	0.5
$-\theta_0$	70.5⁰	74.5⁰	78.5⁰	82.3⁰	86.2⁰	90⁰

Crack instability starts when $S_{\min} = S_c$ or

$$S_c = \frac{(1+v)[2(1-v)-v^2]}{6E}\tau_c^2 a \qquad (33)$$

If the specimens in Figures 4(a) and 4(b) are made of the same material, then S_c in equations (24) and (33) may be eliminated to give the ratio τ_c/σ_c as shown in equation (21). As in the case of tensile loading, equation (32) may also be separated into S_v and S_d. Under in-plane shear load, the critical element prior to fracture undergoes more distortion than the element situated along the line of symmetry in Figure 4(a).

---

* It should be cautioned that equation (31) applies to a perfectly sharp crack of zero radius of curvature, ρ. Furthermore, the distance r_0 at which θ_0 is measured has not been defined. In a more refined analysis, Sih and Kipp [11] have shown that small values of ρ and r_0 can greatly affect the calculated and measured values of the fracture angle θ_0. Comparison of theory and experiment on θ_0 without considering the effect of ρ and r_0 could lead to premature conclusions.

III.   Bending and twisting of cracked plates

The accurate translation of physical phenomena into mathematical terms often presents many difficulties. Too much importance cannot be attached to the initial stages of sifting the evidence to enable a useful analytical treatment. Some of these difficulties and complexities are brought out in Volume 2 of this series on Mechanics of Fracture dealing with three-dimensional crack problems. The strict requirement of satisfying all the governing differential equations and boundary conditions in the theory of three-dimensional theory of elasticity cannot always be met. Assumptions and/or approximations are frequently made to relax either the governing differential equations or boundary conditions or both so as to make the problem manageable. Since the discipline of fracture mechanics requires an accurate description of the stress state near the crack tip region, it is of primary concern to preserve the exact nature of the crack surface boundary conditions in formulating approximate theories. More specifically, the basic character of the local stresses in equations (14) as determined from the exact three-dimensional theory of elasticity should be retained. Such a preference implies that approximations will be made only in a quantitative sense through stress intensity factors k_1, k_2 and k_3.

Character of local moments and shear forces.   Let the thickness h of an elastic body with a crack be small when compared with the other dimensions such as crack length, width of the body, etc. Then the body is said to be a plate. The conventional plate theories* assume that the stress variations in terms of the thickness coordinate are known as a priori. For thin plates, the transverse normal stress component σ_z is assumed to be zero throughout the plate. The middle plane bisecting the plate thickness as shown in Figure 5 is usually taken as reference. With respect to this plane $z = 0$, the in-plane stresses σ_x, σ_y and τ_{xy} on the element located at (x, y, z) are assumed to vary linearly in z, i.e.,

$$\sigma_x = \frac{12z}{h^3} M_x, \, \sigma_y = \frac{12z}{h^3} M_y, \, \tau_{xy} = \frac{12z}{h^3} M_{xy} \tag{34}$$

where M_x and M_y are the bending moments and M_{xy} the twisting moment

---

* The theory of Hartranft and Sih [12] does not preassume the stress distribution across the plate thickness. They determine this distribution from the plane strain condition ahead of the crack. More detailed information can be found in Chapter 2 of this volume.

per unit length acting on any section of the plate parallel to the xz and yz planes. They can be obtained by integration as

$$M_x(x, y) = \int_{-h/2}^{h/2} \sigma_x(x, y, z)z \, dz \tag{35a}$$

$$M_y(x, y) = \int_{-h/2}^{h/2} \sigma_y(x, y, z)z \, dz \tag{35b}$$

$$M_{xy}(x, y) = \int_{-h/2}^{h/2} \tau_{xy}(x, y, z)z \, dz \tag{35c}$$

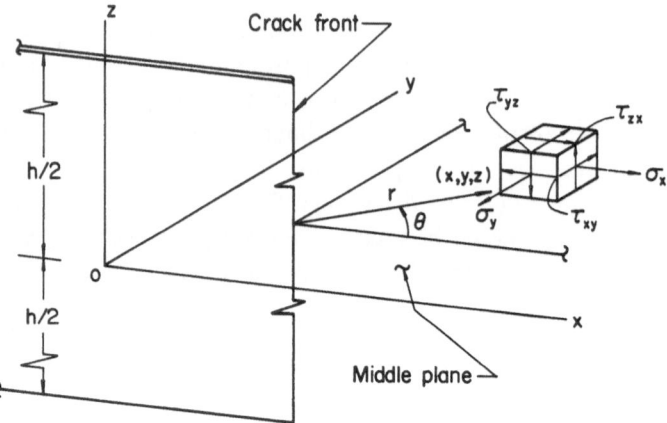

Figure 5. Stresses on an element ahead of crack in a thin plate

With σ_z already assumed to vanish everywhere, the free surface conditions at $z = \pm h/2$ can be satisfied by further requiring that

$$\tau_{xz}(x, y, \pm h/2) = \tau_{yz}(x, y, \pm h/2) = 0 \tag{36}$$

It follows that the equations of equilibrium require both τ_{xz} and τ_{yz} to vary parabolically in z as

$$\tau_{xz} = \frac{3}{2h}\left[1 - \left(\frac{2z}{h}\right)^2\right]V_x \tag{37a}$$

$$\tau_{yz} = \frac{3}{2h}\left[1 - \left(\frac{2z}{h}\right)^2\right]V_y \tag{37b}$$

In equations (37), V_x and V_y are the shearing forces per unit length on sections parallel to the y and x axes and they are given by

$$V_x(x, y) = \int_{-h/2}^{h/2} \tau_{xz}(x, y, z)\,dz \tag{38a}$$

$$V_y(x, y) = \int_{-h/2}^{h/2} \tau_{yz}(x, y, z)\,dz \tag{38b}$$

Since the z-dependence of the stress components is assumed to be known, the plate bending problem has essentially been reduced to two dimensions and the boundary conditions will be specified in terms of the moments and shear forces.

For a crack whose surfaces are free from tractions, the bending moment, twisting moment and shear force must all vanish individually. If the crack is placed on the x-axis, then the conditions*

$$M_y(x, 0) = M_{xy}(x, 0) = V_y(x, 0) = 0 \tag{39}$$

must prevail on the segment of the x-axis where the crack is located. To this end, the Reissner sixth order plate bending theory [13] is used to satisfy equations (39) and yields the desired character of the crack tip moments and shear forces:

$$M_x = \frac{K_1}{(2r)^{\frac{1}{2}}} \cos\frac{\theta}{2}\left(1 - \sin\frac{\theta}{2}\sin\frac{3\theta}{2}\right)$$

$$- \frac{K_2}{(2r)^{\frac{1}{2}}} \sin\frac{\theta}{2}\left(2 + \cos\frac{\theta}{2}\cos\frac{3\theta}{2}\right) + O(1) \tag{40a}$$

$$M_y = \frac{K_1}{(2r)^{\frac{1}{2}}} \cos\frac{\theta}{2}\left(1 + \sin\frac{\theta}{2}\sin\frac{3\theta}{2}\right)$$

$$+ \frac{K_2}{(2r)^{\frac{1}{2}}} \sin\frac{\theta}{2}\cos\frac{\theta}{2}\cos\frac{3\theta}{2} + O(1) \tag{40b}$$

* These conditions cannot be completely satisfied by the Poisson-Kirchhoff theory of plate bending which neglects the effect of transverse shear deformation. As a result, the two conditions on M_{xy} and V_y in equation (39) are replaced by a single one on $V_y + \partial M_{xy}/\partial x$, referred to as the equivalent or Kirchhoff shear. The consequences are that the angular variations of the local stresses become dependent on the Poisson's ratio and differ from those in equations (14) and the transverse shear stresses τ_{xz} and τ_{yz} attain a singularity of the order $r^{-\frac{3}{2}}$. This is in contrast to the exact three-dimensional solution of $r^{-\frac{1}{2}}$-singularity for τ_{xz} and τ_{yz}.

$$M_{xy} = \frac{K_1}{(2r)^{\frac{1}{2}}} \sin \frac{\theta}{2} \cos \frac{\theta}{2} \cos \frac{3\theta}{2}$$

$$+ \frac{K_2}{(2r)^{\frac{1}{2}}} \cos \frac{\theta}{2} \left(1 - \sin \frac{\theta}{2} \sin \frac{3\theta}{2} \right) + O(1) \tag{40c}$$

$$V_x = - \frac{K_3}{(2r)^{\frac{1}{2}}} \sin \frac{\theta}{2} + O(1) \tag{40d}$$

$$V_y = \frac{K_3}{(2r)^{\frac{1}{2}}} \cos \frac{\theta}{2} + O(1) \tag{40e}$$

in which K_1 and K_2 are the moment intensity factors and K_3 the shear force intensity factor. Note that the $1/\sqrt{r}$ singularity and the θ-dependence in equations (40) are the same as those shown in equations (14).

Moment and shear force intensity factors. Consider the problem of a through crack in a bent plate. The crack is tilted at an angle β with reference to the plane about which a bending moment of magnitude M is applied. This loading as shown in Figure 6(a) is equivalent to the application of bending

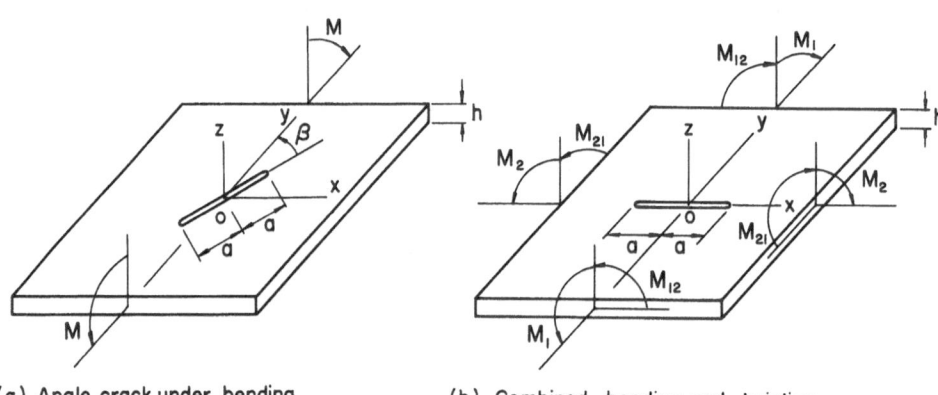

(a) Angle crack under bending   (b) Combined bending and twisting

Figure 6. Loading configuration for bending and twisting of a cracked plate

moments M_1, M_2 and twisting moment M_{12} in Figure 6(b) through the transformation properties of the stresses:

$$M_1 = M \sin^2 \beta \tag{41a}$$

$$M_2 = M \cos^2 \beta \tag{41b}$$

$$M_{12} = M \sin \beta \cos \beta \tag{41c}$$

For this problem, Hartranft and Sih [12] and Wang [14] have provided the K_j ($j = 1, 2, 3$) expressions and they are

$$K_1 = \Phi(1) M_1 \sqrt{a} \tag{42a}$$

$$K_2 = \Psi(1) M_{12} \sqrt{a} \tag{42b}$$

$$K_3 = -\frac{\sqrt{10}}{(1 + v)h} \Omega(1) M_{12} \sqrt{a} \tag{42c}$$

Equations (42) show that only M_1 and M_{12} affect the moment intensity near the crack tip while M_2 has no contribution. The separate effects of M_1, M_2 and M_{12} can be best illustrated by Figures 7(a) to 7(c) where normal

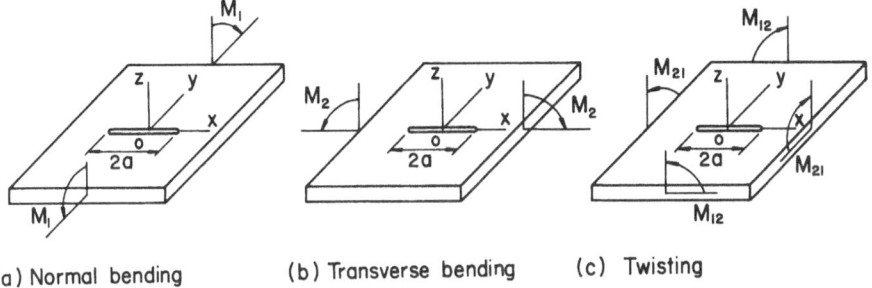

(a) Normal bending   (b) Transverse bending   (c) Twisting

Figure 7. Three separate problems of plate bending

bending produces K_1, parallel bending gives rise to no moment singularity or intensity and pure twisting leads to both K_2 and K_3. It is apparent from equations (41) and (42) that

$$K_1 = \Phi(1) M \sqrt{a} \sin^2 \beta \tag{43a}$$

$$K_2 = \Psi(1) M \sqrt{a} \sin \beta \cos \beta \tag{43b}$$

$$K_3 = -\frac{\sqrt{10}}{(1 + v)h} \Omega(1) M \sqrt{a} \sin \beta \cos \beta \tag{43c}$$

where the functions $\Phi(1)$, $\Psi(1)$ and $\Omega(1)$ are computed numerically from integral equations [12, 14]. Their values as a function of $h/a\sqrt{10}$ for different Poisson's ratio can be found in Figures 8 to 10.

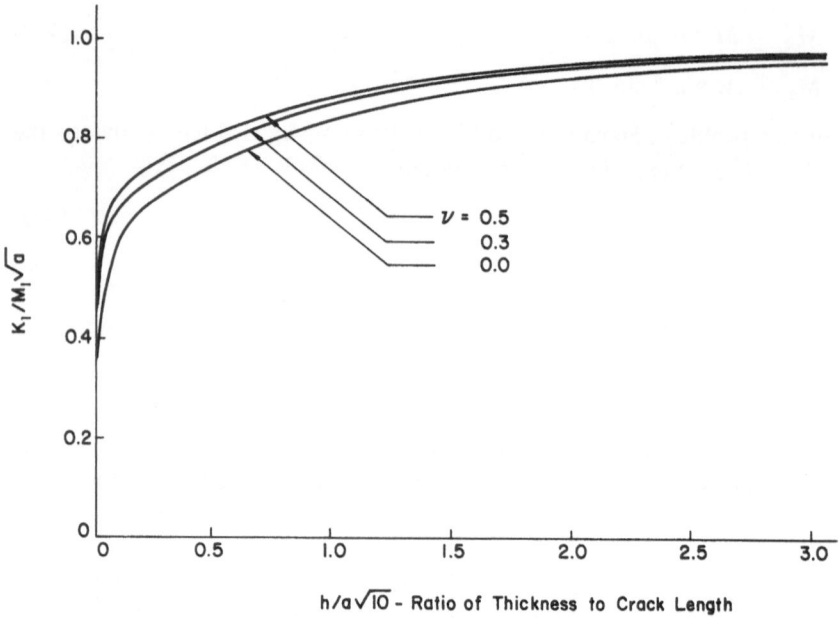

Figure 8. Bending-moment-intensity factor versus ratio of thickness to crack length

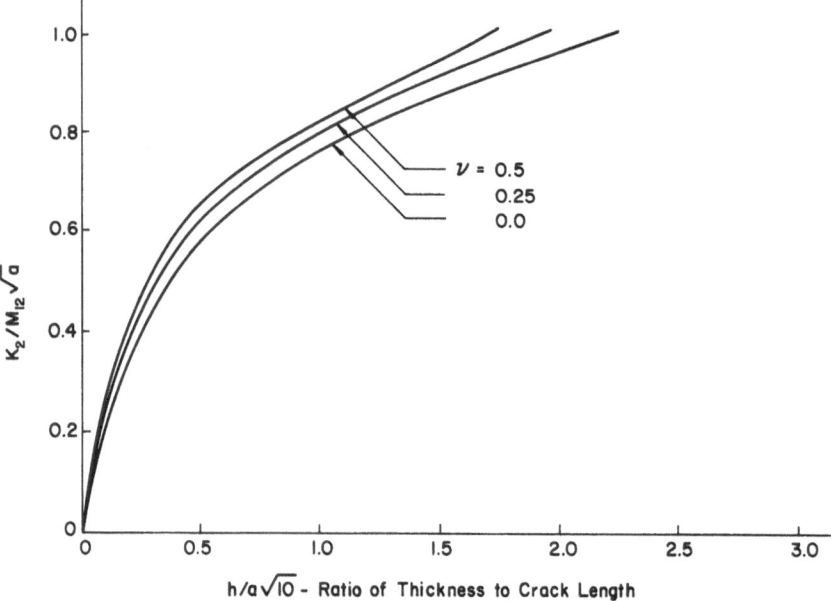

Figure 9. Twisting moment-intensity factor versus ratio of thickness to crack length

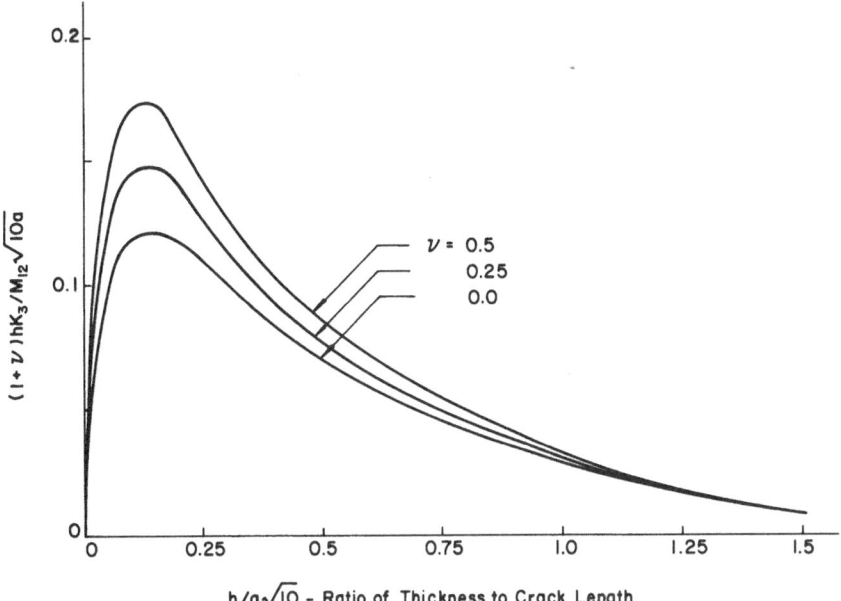

Figure 10. Shear intensity factor versus ratio of thickness to crack length

IV. Direction of crack growth

In the absence of symmetry, the path of crack propagation is no longer obvious since it may not coincide with its initial plane. The direction of crack initiation as predicted from the strain energy density theory depends on the intensity of the $1/r$ energy field near the crack tip. This intensity measured by S varies as a function of r and θ through the stresses on the element shown in Figure 5. With $\sigma_z = 0$, equation (1) simplifies to

$$\frac{dW}{dV} = \frac{1}{2E} \left[\sigma_x^2 + 2\tau_{xy}^2 + \sigma_y^2 + 2v(\tau_{xy}^2 - \sigma_x\sigma_y) + 2(1 + v)(\tau_{xz}^2 + \tau_{yz}^2) \right] \qquad (44)$$

If the radius of the core region is kept sufficiently small, then consideration need only be given to the singular stresses. Substitution of equations (40) into (34) and (37) yields

$$\sigma_x = \frac{k_1}{(2r)^{\frac{1}{2}}} \cos\frac{\theta}{2}\left(1 - \sin\frac{\theta}{2}\sin\frac{3\theta}{2}\right)$$

$$-\frac{k_2}{(2r)^{\frac{1}{2}}}\sin\frac{\theta}{2}\left(2+\cos\frac{\theta}{2}\cos\frac{3\theta}{2}\right)+O(1) \tag{45a}$$

$$\sigma_y = \frac{k_1}{(2r)^{\frac{1}{2}}}\cos\frac{\theta}{2}\left(1+\sin\frac{\theta}{2}\sin\frac{3\theta}{2}\right)$$

$$+\frac{k_2}{(2r)^{\frac{1}{2}}}\sin\frac{\theta}{2}\cos\frac{\theta}{2}\cos\frac{3\theta}{2}+O(1) \tag{45b}$$

$$\tau_{xy} = \frac{k_1}{(2r)^{\frac{1}{2}}}\sin\frac{\theta}{2}\cos\frac{\theta}{2}\cos\frac{3\theta}{2}$$

$$+\frac{k_2}{(2r)^{\frac{1}{2}}}\cos\frac{\theta}{2}\left(1-\sin\frac{\theta}{2}\sin\frac{3\theta}{2}\right)+O(1) \tag{45c}$$

$$\tau_{xz} = -\frac{k_3}{(2r)^{\frac{1}{2}}}\sin\frac{\theta}{2}+O(1) \tag{45d}$$

$$\tau_{yz} = \frac{k_3}{(2r)^{\frac{1}{2}}}\cos\frac{\theta}{2}+O(1) \tag{45e}$$

The coefficients k_j ($j = 1, 2, 3$) are the stress intensity factors* which vary along the crack front as a function of z, i.e.,

$$k_1 = \frac{12z}{h^3}\Phi(1)\,M\,\sqrt{a}\,\sin^2\beta \tag{46a}$$

$$k_2 = \frac{12z}{h^3}\Psi(1)\,M\,\sqrt{a}\,\sin\beta\cos\beta \tag{46b}$$

$$k_3 = -\frac{3\sqrt{10}}{2(1+v)h^2}\left[1-\left(\frac{2z}{h}\right)^2\right]\Omega(1)\,M\,\sqrt{a}\,\sin\beta\cos\beta \tag{46c}$$

Inserting equations (45) into (44), the result may be cast into the form $dW/dV = S/r$ where the strain energy density factor S is given by

$$S = A_{11}k_1^2 + 2A_{12}k_1k_2 + A_{22}k_2^2 + A_{33}k_3^2 \tag{47}$$

The quantities A_{ij} ($i, j = 1, 2, 3$) stand for

---

* Note from equations (43) and (46) that k_j ($j = 1, 2, 3$) are related to K_j ($j = 1, 2, 3$) as

$k_1 = (12z/h^3)\,K_1,\ k_2 = (12z/h^3)\,K_2,\ k_3 = (3/2h)\,[1-(2z/h)^2]K_3$

$$A_{11} = \frac{1}{8E} (1 + \cos \theta) [3 - v - (1 + v) \cos \theta] \tag{48a}$$

$$A_{12} = \frac{1}{4E} \sin \theta [(1 + v) \cos \theta - (1 - v)] \tag{48b}$$

$$A_{22} = \frac{1}{8E} [4(1 - \cos \theta) + (1 + v)(3 \cos \theta - 1)(1 + \cos \theta)] \tag{48c}$$

$$A_{33} = \frac{1 + v}{2E} \tag{48d}$$

For the plate bending problem at hand, equation (47) becomes

$$S = \frac{2a}{E} \left[\frac{3\Phi(1)M \sin \beta}{h^2} \right]^2 \left\{ \left(\frac{z}{h} \right)^2 F(\theta, \beta) + \left[1 - \left(\frac{2z}{h} \right)^2 \right]^2 G(\beta) \right\} \tag{49}$$

with the functions $F(\theta, \beta)$ and $G(\beta)$ being defined as

$$F(\theta, \beta) = B_{11} \sin^2 \beta + 2\lambda B_{12} \sin \beta \cos \beta + \lambda^2 B_{22} \cos^2 \beta \tag{50a}$$

$$G(\beta) = \frac{5}{8(1 + v)} \gamma^2 \cos^2 \beta \tag{50b}$$

where $B_{ij} = 8E A_{ij}$ $(i, j = 1, 2, 3)$. The parameters λ and γ are

$$\lambda = \frac{\Psi(1)}{\Phi(1)}, \gamma = \frac{\Omega(1)}{\Phi(1)} \tag{51}$$

The numerical values of λ and γ can be obtained from the graphs in Figures 8 to 10.

The strain energy density theory [3, 4] assumes that the crack grows in a direction corresponding to $\theta = \theta_0$ which makes S in equation (49) a minimum:

$$\frac{\partial S}{\partial \theta} = 0; \frac{\partial^2 S}{\partial \theta^2} > 0 \text{ at } \theta = \theta_0 \tag{52}$$

The vanishing of the first derivative of S with respect to θ gives

$$(1 + v)\{2(1 - 3\lambda^2) \sin 2\theta_0 - (1 - \lambda)(1 - 3\lambda) \sin [2(\theta_0 + \beta)]$$

$$- (1 + \lambda)(1 + 3\lambda) \sin [2(\theta_0 - \beta)]\} - 2(1 - v)[2(1 - \lambda^2) \sin \theta_0$$

$$- (1 - \lambda)^2 \sin (\theta_0 + 2\beta) - (1 + \lambda)^2 \sin (\theta_0 - 2\beta)] = 0 \tag{53}$$

There exist two minimum values of S. One corresponds to the negative angles θ_0 on the tension side of the plate and the other corresponds to the positive angles on the compression side. If the crack is always assumed to initiate from the tension side, then only negative fracture angles will be considered. Figure 11 gives a plot of $-\theta_0$ in equation (53) as a function of

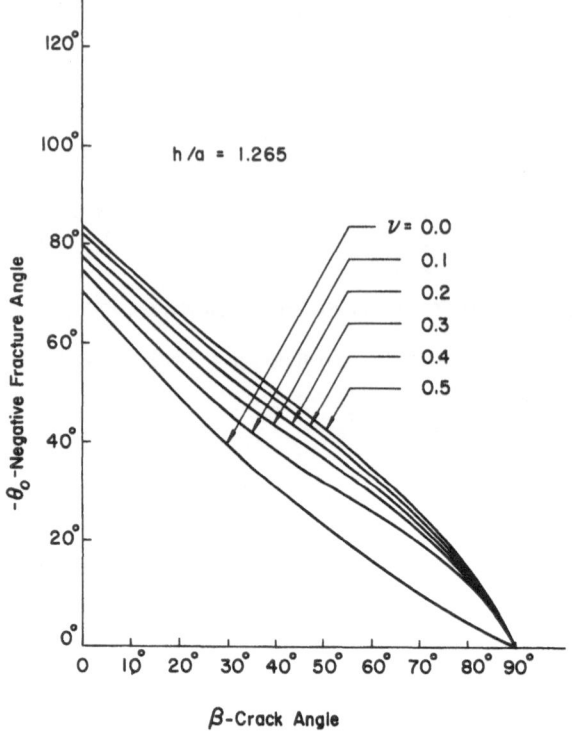

Figure 11. Variations of fracture angle with crack angle for constant h/a

the crack angle β for a fixed plate thickness to half crack length ratio of $h/a = 1.265$. Under normal bending $\beta = 90°$, the predicted fracture angle is $\theta_0 = 0$ which implies that the crack grows in a self-similar manner. As the crack is tilted into the direction of bending by decreasing the angle β, the direction of crack initiation deviates more and more away from its initial plane as the fracture angle $-\theta_0$ increases. The values of $-\theta_0$ are also seen to be affected by the Poisson's ratio of the material. The variation appears to be more pronounced for crack angle β in the middle range. Unlike the plane theory of elasticity, the plate thickness to crack length ratio exerts an

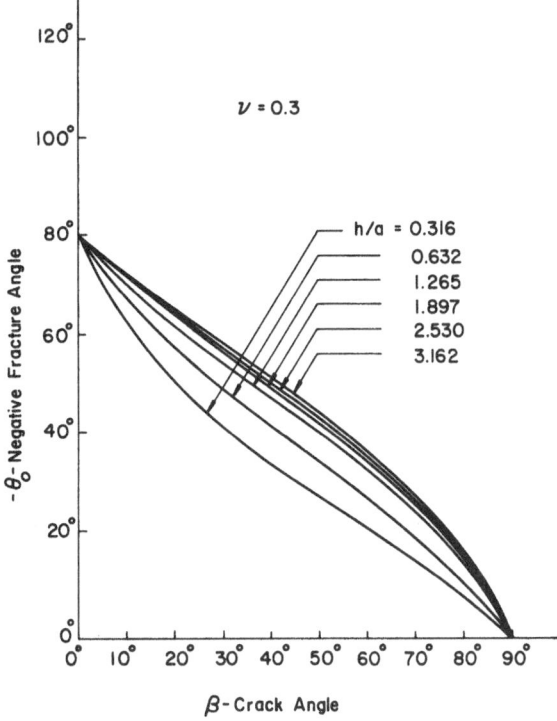

Figure 12. Variations of fracture angle with crack angle for constant v

influence on the direction of crack initiation. For a fixed crack length and angle β, Figure 12 shows that the fracture angle $-\theta_0$ decreases with increasing plate thickness. Such an influence tends to diminish for β close to $0°$ and $90°$. Hence, changes in the h/a ratio* should be accounted for when comparing experimental data with theory.

V. Minimum strain energy density factors and allowable bending moments

Once the fracture angles $-\theta_0$ are known, they may be inserted into equation (49) to obtain the minimum S values or $S_{\min}$. Since $F(\theta, \beta)$ is the only

---

* Variations of fracture angle $-\theta_0$ with h/a also occurs in the stretching of plates. The higher order theory of Hartranft and Sih [12] when applied to the mixed mode crack problem in plane extension can predict such an effect.

function in equation (49) that depends on θ, its minimum values $F_{\min}(\theta_0, \beta)$ are displayed graphically in Figure 13 for $h/a = 1.265$. The general trend of

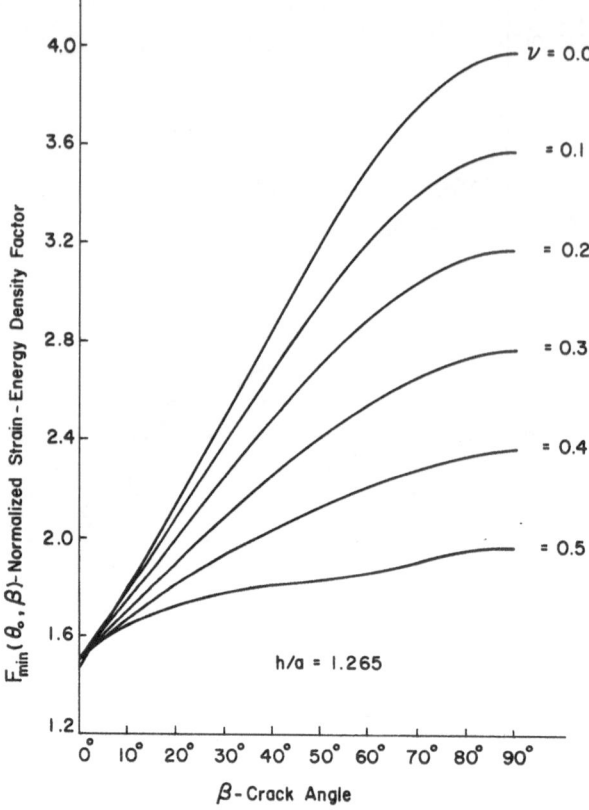

Figure 13. Normalized strain-energy density factor versus crack angle for constant h/a

the curves for $F_{\min}$ or $S_{\min}$ tends to increase with the crack angle β and the rate of this increase becomes greater as the Poisson's ratio is decreased. A similar plot for $F_{\min}(\theta_0, \beta)$ versus β with $v = 0.3$ is given in Figure 14. Here, the ratio h/a is varied from 0.316 to 3.162. For small values of β, significant change of $F_{\min}$ are observed as h/a is varied. This change diminishes as β approaches 90° where all the curves converge to the same value.

In order to estimate the allowable moment, M_c, that a cracked plate can sustain without causing fracture, a normalized moment quantity $M_c/(Eh^3 S_{\min})^{\frac{1}{2}}$ is defined by using equation (49):

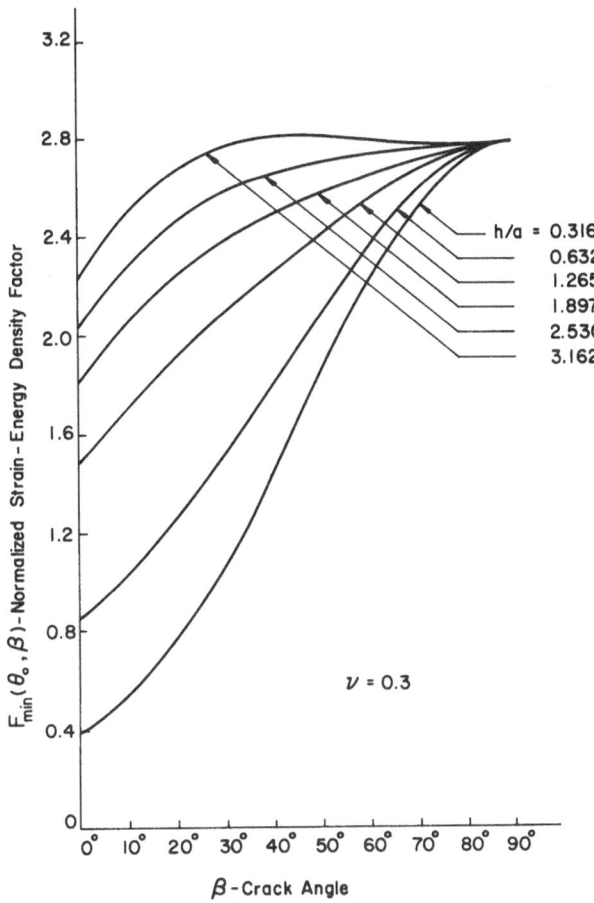

Figure 14. Normalized strain-energy density factor versus crack angle for constant v

$$\frac{M_c}{(Eh^3 S_{\min})^{\frac{1}{4}}} = \frac{\sqrt{h/a}}{3 \sin \beta \{2\Phi(1)[(z/h)^2 F_{\min}(\theta_0, \beta) + [1 - (2z/h)^2]G(\beta)\}^{\frac{1}{4}}} \quad (54)$$

where M_c is undefined for $\beta = 0$. Assuming that both v and h/a take the respective values of 0.3 and 1.265, the normalized moment can then be plotted against β for different values of z/h, say from 0.1 to 0.5 as illustrated in Figure 15. The curve for $z/h = 0$ is outside the scale range of the graph. For a fixed $S_{\min}$, the moment M is seen to decrease monotonically with β. The minimum value of M_c occurs at a position where the bending moment is applied normal to the crack plane, i.e., $\beta = 90°$. Each value of z/h corre-

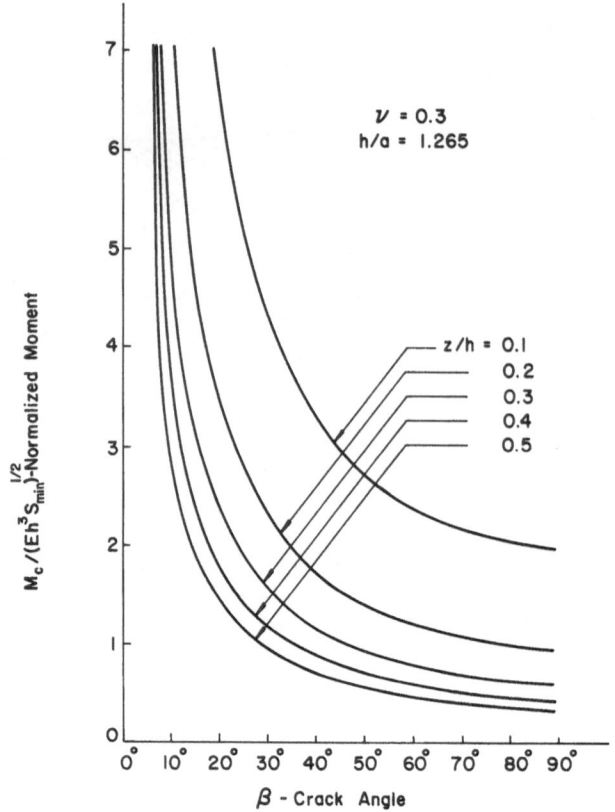

Figure 15. Normalized moment versus crack angle for constant v and h/a.

sponds to a layer of material in the plate with $z = 0$ being the middle layer and $z = 0.5h$ the outer or surface layer. According to the S-theory [3, 4], unstable crack propagation commences when $S_{min} \to S_c$. The critical value S_c is a constant representing the fracture toughness of the material. When $S_{min} = S_c$ in equation (54), M becomes M_c or the critical moment at incipient fracture. With this understanding, the results in Figure 15 indicate that fracture always starts in the surface layer of the plate ($z/h = 0.5$) which corresponds to the lowest critical moment M_c for all crack angles β.

With attention focused on the surface layer ($z/h = 0.5$) where fracture initiates, Figures 16 and 17 study the influence of Poisson's ratio and h/a ratio. The changes in M_c as v is varied from 0.0 to 0.5 are not appreciable. This can be seen from Figure 16. Increasing plate thickness or h/a, say with

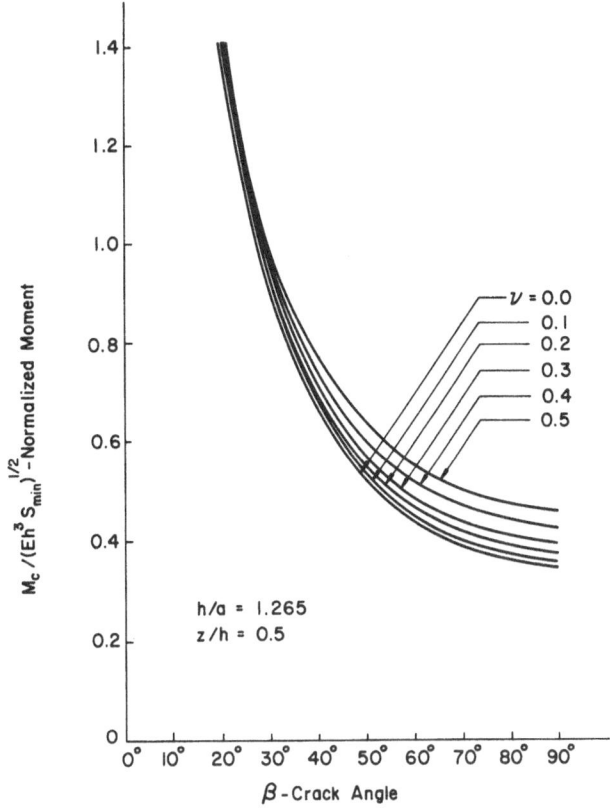

Figure 16. Normalized moment versus crack angle for constant h/a and z/h

a being constant, tends to lower the normalized moment that can be applied to a plate without causing fracture. This tendency is clearly shown in Figure 17.

VI. Conclusions

When plate-like structural members are subjected to bending loads, the in-plane stresses are required to change sign in the thickness direction and the problem acquires a three-dimensional character. Since the through crack geometry* prohibits any exact three-dimensional solutions. the elas-

---

* The major drawback arises from an incomplete knowledge of the stress singularity on the plate surface where it intersects with the crack. Further discussions on this subject can be found in [15, 16].

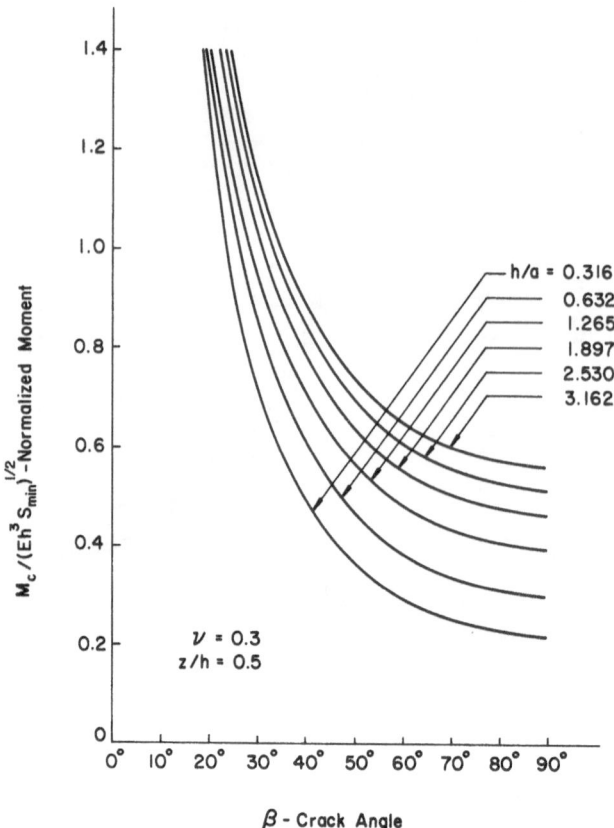

Figure 17. Normalized moment versus crack angle for constant v and z/h

ticity theory must be approximated by appealing to minimum principles in the calculus of variations. A variety of approximations have been made and led to various formulations known as plate theories. For instance, the most widely used theory of Poisson-Kirchhoff in plate bending involves alteration of the crack surface boundary conditions. Such a change has a drastic influence on the qualitative character of the stresses near the crack tip and will seriously affect the direction of crack initiation predicted from any assumed fracture criterion. As an example, the maximum principal stress criterion* states that crack growth takes place in a direction normal to the plane on which the tangential normal stress σ_θ is a maximum or the shear stress $\tau_{r\theta}$

* This criterion yields sufficiently accurate results only when one of the principal stresses dominate.

vanishes. Consider the pure twisting case shown in Figure 7(c). The $\tau_{r\theta} = 0$ plane based on the Poisson-Kirchhoff theory can be found from Sih's work [15] as

$$\theta_0 = -2 \cos^{-1} \left(\frac{2(2 + v)}{5 + 3v} \right)^{\frac{1}{2}} \tag{55}$$

which gives fracture angles of θ_0 from $-53.2°$ to $-57.4°$ for v ranging from 0.0 to 0.5. This differs from the value of θ_0 obtained from the Reissner's theory in which the crack surface boundary conditions are satisfied. Equation (45c) when expressed in cylindrical polar coordinates (r, θ) and $\tau_{r\theta}$ set equal to zero gives a fracture angle*

$$\theta_0 = -\cos^{-1} (\tfrac{1}{3}) = -70.5° \tag{56}$$

The difference in θ_0 as obtained from equations (55) and (56) is significant and cannot be ignored in failure analsyis. A similar discrepancy is found if the strain energy density factor theory is used.

Another important feature of the plate bending problem is that the material elements in the thickness direction may not fail simultaneously as assumed in the theory of plane elasticity. This is because the stress intensity factors k_1, k_2 and k_3 no longer remain constant along the crack front. The location of the critical element that triggers fracture must depend on some combination of k_j ($j = 1, 2, 3$) reaching a critical value. Referring to the cylindrical polar coordinate system (r, θ, z) in Figure 5, the position of the critical element will be denoted by (r_0, θ_0, z_0), where r_0 is the radius of core region, θ_0 the fracture angle obtained from the condition of S reaching a minimum and z_0 corresponds to the greatest value of S_{min} along the crack front. The bending of an angle crack as shown in Figure 6(a) was analyzed. Predictions of the S-theory showed that the fracture angle θ_0 not only depended on the crack position relative to the direction of bending but also on the Poisson's ratio of the material and the plate thickness to crack length ratio. These factors should not be ignored in comparing experimental data with theoretical results.

For the sake of simplicity, the in-plane stresses have been assumed to vary linearly through the plate thickness and the transverse shear stresses are parabolic functions of z. Higher order plate theories [12, 17] involving more

---

* The experimental result of Erdogan and Sih [6] on the twisting of cracked plates gives values of θ_0 closer to 70.5^0 than those predicted by equation (55).

generalized functions of z are available. Sih [17] has proposed to expand the z-dependance, $f'(2z/h)$, for the transverse shear stresses τ_{xz} and τ_{yz} in a trigonometric series*

$$f'(2z/h) = b_0 - \sum_{n=1}^{m} (-1)^n b_n \cos \left[(2\pi n z/h) \right] \tag{57}$$

which satisfies the conditions that $\tau_{xz} = \tau_{yz} = 0$ on the plate surfaces $z = \pm h/2$, i.e., $f'(\pm 1) = 0$. The constants b_0 and b_n are given by

$$b_0 = \sum_{n=1}^{m} (-1)^n b_n, \; b_n = \frac{1}{1 + (an/h)^2} \tag{58}$$

where a is the half crack length. The derivative of $f'(2z/h)$ with respect to z or $f''(2z/h)$ gives the variations of the in-plane stresses σ_x, σ_y and τ_{xy} with z. A series of experiments on the bending of plexiglass plates [18] gave data checking remarkably well with the predictions made by Sih [17]. The stress intensity factor solution for this problem may also be used in conjunction with the strain energy density factor theory to predict the fracture behavior of plates under bending and/or twisting. The theoretical results will differ but not significantly from those based on the Reissner theory.

Another shortcoming of the conventional plate theory is that σ_z is neglected throughout the plate and hence the plane strain condition

$$\sigma_z = \nu(\sigma_x + \sigma_y) \tag{59}$$

ahead of the crack cannot be realized. Hartranft and Sih [12] have constructed a theory of plate stretching or bending in which the function $f(2z/h)$ in

$$\sigma_z = f(2z/h) \, Z_z(x, y) \tag{60}$$

will not be assigned arbitrarily but be determined from equation (59), a condition derived from the exact three-dimensional theory of elasticity [19]. Villarreal, Sih and Hartranft [20] have used a frozen stress photoelastic technique to measure the stress variation through the thickness of a thick plate containing a crack and the results were in excellent agreement with those postulated by the theory of Hartranft and Sih.

---

* Reference can also be made to equation (2.82) in Chapter 2.

References

[1] Bridgman, P. W., *Studies in Large Plastic Flow and Fracture with Special Emphasis on the Effects of Hydrostatic Pressure*, First edition, McGraw-Hill Book Co., New York (1952).

[2] Sih, G. C., Strain energy density factor applied to mixed mode crack problems, *International Journal of Fracture*, 10, pp. 305–321 (1974).

[3] Sih, G. C., A special theory of crack propagation: methods of analysis and solutions of crack problems, *Mechanics of Fracture I*, edited by G. C. Sih, Noordhoff International Publishing, Leyden, pp. 21–45 (1973).

[4] Sih, G. C., A three-dimensional strain energy density factor theory of crack propagation: three-dimensional crack problems, *Mechanics of Fracture II*, edited by G. C. Sih, Noordhoff International Publishing, Leyden, pp. 15–53 (1975).

[5] *Linear Fracture Mechanics*, edited by G. C. Sih, R. P. Wei and F. Erdogan, Envo Publishing Co., Inc. Lehigh Valley, Pennsylvania (1976).

[6] Erdogan, F. and Sih, G. C., On the crack extension in plates under plane loading and transverse shear, *Journal of Basic Engineering*, 85, pp. 519–527 (1963).

[7] Sih, G. C. and Cha, B. C. K., A fracture criterion for three-dimensional crack problems, *Journal of Engineering Fracture Mechanics*, pp. 699–723 (1974).

[8] Kipp, M. E. and Sih, G. C., The strain energy density failure criterion applied to notched elastic solids, *International Journal of Solids and Structures*, 2, pp. 153–173 (1975).

[9] Sih, G. C., Discussion on some observations on Sih's strain energy density approach for fracture prediction by I. Finnie and H. O. Weiss, *International Journal of Fracture*, 10, pp. 279–283 (1974).

[10] Sih, G. C. and Macdonald, B., Fracture mechanics applied to engineering problems —strain energy density fracture criterion, *Journal of Engineering Fracture Mechanics*, 6, pp. 361–386 (1974).

[11] Sih, G. C. and Kipp, M. E., Discussion on fracture under comples stress—the angle crack problem by J. G. Williams and P. D. Ewing, *International Journal of Fracture*, 10, pp. 261–265 (1974).

[12] Hartranft, R. J. and Sih, G. C., An approximate three-dimensional theory of plates with application to crack problems, *International Journal of Engineering Science*, 8, pp. 711–729 (1970).

[13] Reissner, E., On bending of elastic plates, *Quarterly of Applied Mathematics*, 5, pp. 55–68 (1947).

[14] Wang, N. M., Twisting of an elastic plate containing a crack, *International Journal of Fracture Mechanics*, 6, pp. 367–378 (1970).

[15] Sih, G. C., A review of the three-dimensional stress problem for a cracked plate, *International of Fracture Mechanics*, 7, pp. 39–61 (1971).

[16] Benthem, J. P., *Three-Dimensional State of Stress at the Vertex of a Quarter-Infinite Crack in a Half-Space*, Delft University Report No. 563 (1975).

[17] Sih, G. C., Bending of a cracked plate with an arbitrary stress distribution across the thickness, *Journal of Engineering for Industry*, 92, pp. 350–356 (1970).

[18] Rubayi, N. A. and Ved, R., Photoelastic analysis of a thick square plate containing

central crack and loaded by pure bending, *International Journal of Fracture*, 12, pp. 435–451 (1976).

[19] Hartranft, R. J. and Sih, G. C., The use of eigenfunction expansions in the general solution of three-dimensional crack problems, *Journal of Mathematics and Mechanics*, 19, pp. 123–138 (1969).

[20] Villarreal, G., Sih, G. C. and Hartranft, R. J., Photoelastic investigation of a thick plate with a transverse crack, *Journal of Applied Mechanics*, 42, pp. 9–14 (1975).

M. Isida

1 | Interaction of arbitrary array of cracks in wide plates under classical bending

1.1 Introduction

This paper presents a general method of analysis of arbitrary arrays of cracks in wide plates by means of the classical theory of plate bending. This theory is based on the Kirchhoff boundary conditions, which lead to approximate satisfaction of the free surface conditions on the crack. In spite of this approximation, useful stress intensity factor solutions for complicated crack geometries can be obtained.

The analysis is similar to that in the author's previous papers on two-dimensional [1] and antiplane shear problems [2], and is based on the concept of the Laurent series expansions of the complex potentials proposed by S. Moriguti [3]. In the numerical calculations a perturbation procedure is employed, and the crack tip stress intensity factors are given in power series of the relative crack length. Typical problems of practical importance are analysed and the numerical results are discussed.

1.2 Basic relations

Consider the axes x, y in the middle plane of the plate with thickness h, and the third axis t as shown in Figure 1.1. The moments M_x, M_y, M_{xy} and shearing forces Q_x, Q_y are defined as follows:

$$M_x = \int_{-h/2}^{h/2} t\sigma_x dt, \ M_y = \int_{-h/2}^{h/2} t\sigma_y dt, \ M_{xy} = \int_{-h/2}^{h/2} t\tau_{xy} dt \tag{1.1a}$$

$$Q_x = \int_{-h/2}^{h/2} \tau_{xt} dt, \ Q_y = \int_{-h/2}^{h/2} \tau_{yt} dt \tag{1.1b}$$

In small deflection of thin plates, the bending stresses are assumed to be proportional to the distance from the middle plane, and the first two expressions in equations (1.1a) give

$$\sigma_x = \frac{M_x t}{3h^3}, \sigma_y = \frac{M_y t}{3h^3}$$

(1.2)

and

$$\sigma_{B,x} = \frac{M_x}{6h^2}, \sigma_{B,y} = \frac{M_y}{6h^2}$$

(1.3)

are the extreme outer fiber stresses.

Equations of equilibrium of a small plate element are

$$\frac{\partial Q_x}{\partial x} + \frac{\partial Q_y}{\partial y} + p(x, y) = 0$$

(1.4a)

$$\frac{\partial M_x}{\partial x} + \frac{\partial M_{xy}}{\partial y} = Q_x$$

(1.4b)

$$\frac{\partial M_{xy}}{\partial x} + \frac{\partial M_y}{\partial y} = Q_y$$

(1.4c)

where $p(x, y)$ is the density of the transverse loading acting on the plate surface.

The moment components are related to the deflection w as follows:

$$M_x = -D\left(\frac{\partial^2 w}{\partial x^2} + v\frac{\partial^2 w}{\partial y^2}\right)$$

(1.5a)

$$M_y = -D\left(\frac{\partial^2 w}{\partial y^2} + v\frac{\partial^2 w}{\partial x^2}\right)$$

(1.5b)

$$M_{xy} = -D(1 - v)\frac{\partial^2 w}{\partial x \partial y}$$

(1.5c)

where D is the bending rigidity of the plate defined as

$$D = \frac{Eh^3}{12(1 - v^2)}$$

(1.6)

and E and v are, respectively, the Young's modulus and Poisson's ratio. Substituting equations (1.5) into (1.4), the governing equation for the deflection is obtained as follows:

$$\frac{\partial^4 w}{\partial x^4} + 2\frac{\partial^4 w}{\partial x^2 \partial y^2} + \frac{\partial^4 w}{\partial y^4} = \frac{p(x, y)}{D}$$

(1.7)

If the plate is free from transverse loading a biharmonic equation is obtained:

$$\frac{\partial^4 w}{\partial x^4} + 2\frac{\partial^4 w}{\partial x^2 \partial y^2} + \frac{\partial^4 w}{\partial y^4} = 0 \qquad (1.8)$$

Thus the displacement is a biharmonic function, and can be expressed by two complex functions $\phi(z)$ and $\psi(z)$ as follows:

$$Dw = \text{Re}\left[\bar{z}\phi(z) + \psi(z)\right] \qquad (1.9)$$

The moments and shearing forces which are written in terms of w as shown in equations (1.5) and (1.4) can also be expressed in terms of the derivatives of $\phi(z)$ and $\psi(z)$:

$$M_x + M_y = -4(1+v)\,\text{Re}[\phi'(z)] \qquad (1.10a)$$

$$M_x - M_y - 2iM_{xy} = -2(1-v)\left[\bar{z}\phi''(z) + \psi''(z)\right] \qquad (1.10b)$$

$$Q_x - iQ_y = -4\phi''(z) \qquad (1.10c)$$

A few comments will be made on the Kirchhoff boundary conditions in the classical plate bending. It is sufficient, without loss of generality, to consider the plate boundary parallel to x-axis. The physical boundary conditions require that M_y, M_{xy} and Q_y take some assigned values, but only two conditions are allowed on the plate edge as required in the classical plate bending theory. In this connection, the quantities M_{xy} and Q_y are replaced by single quantity referred to as the equivalent shearing force $V_y = Q_y + \partial M_{xy}/\partial x$. A traction free boundary would then require M_y and V_y to vanish which in terms of the complex functions may be written as

$$\phi'(z) - \mu\bar{\phi}'(\bar{z}) + \bar{z}\phi''(z) + \psi''(z) = \text{an imaginary constant} \qquad (1.11)$$

where

$$\mu = \frac{3+v}{1-v} \qquad (1.12)$$

Now, consider the local stress field close to the crack tip. Express the stress components obtained by the classical bending theory in the polar coordinates as shown in Figure 1.1, and expand them in ascending power series of r. The first terms which are predominant in the vicinity of the crack tip become [4]:

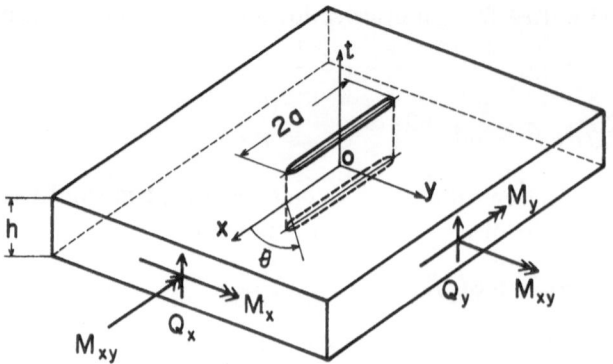

Figure 1.1. Bending of a plate containing a crack

$$\sigma_x = \frac{k_1}{(2r)^{\frac{1}{2}}}\frac{t}{2\mu h}\left(-3\cos\frac{\theta}{2} - \cos\frac{5\theta}{2}\right) + \frac{k_2}{(2r)^{\frac{1}{2}}}\frac{t}{2\mu h}\left(-\frac{9+7v}{1-v}\sin\frac{\theta}{2}\right.$$

$$\left. + \sin\frac{5\theta}{2}\right) \tag{1.13a}$$

$$\sigma_y = \frac{k_1}{(2r)^{\frac{1}{2}}}\frac{t}{2\mu h}\left(\frac{11+5v}{1-v}\cos\frac{\theta}{2} + \cos\frac{5\theta}{2}\right) + \frac{k_2}{(2r)^{\frac{1}{2}}}\frac{t}{2\mu h}\left(\sin\frac{\theta}{2} - \sin\frac{5\theta}{2}\right) \tag{1.13b}$$

$$\tau_{xy} = \frac{k_1}{(2r)^{\frac{1}{2}}}\frac{t}{2\mu h}\left(-\frac{7+v}{1-v}\sin\frac{\theta}{2} - \sin\frac{5\theta}{2}\right) + \frac{k_2}{(2r)^{\frac{1}{2}}}\frac{t}{2\mu h}\left(\frac{5+3v}{1-v}\cos\frac{\theta}{2}\right.$$

$$\left. - \cos\frac{5\theta}{2}\right) \tag{1.13c}$$

$$\tau_{xt} = \frac{f(t)}{(2r)^{\frac{1}{2}}}\left(-k_1\cos\frac{3\theta}{2} + k_2\sin\frac{3\theta}{2}\right) \tag{1.13d}$$

$$\tau_{yt} = -\frac{f(t)}{(2r)^{\frac{1}{2}}}\left(k_1\sin\frac{3\theta}{2} + k_2\cos\frac{3\theta}{2}\right) \tag{1.13e}$$

where

$$f(t) = \frac{h^2 - 4t^2}{4t(3+v)} \tag{1.14}$$

Coefficients k_1 and k_2 in equations (1.13) are the stress intensity factors in the classical plate bending theory. In general, the above expansion procedure may be skipped to get the stress intensity factors. They can be obtained directly from the complex potential $\phi(z)$ by the relation [5]

$$k_1 - ik_2 = -\frac{12\sqrt{2}(3+v)}{h^2}\lim_{z\to a}\left[(z-a)^{\frac{1}{2}}\phi'(z)\right] \qquad (1.15)$$

1.3 Complex potentials for traction free cracks

Consider a crack of length $2a$ in the physical plane and introduce the mapping function

$$z = \frac{a}{2}\left(\zeta + \frac{1}{\zeta}\right) \qquad (1.16)$$

which transforms the crack edge in z-plane and its outer region into a unit circle and its outer region in ζ-plane as shown in Figure 1.2.

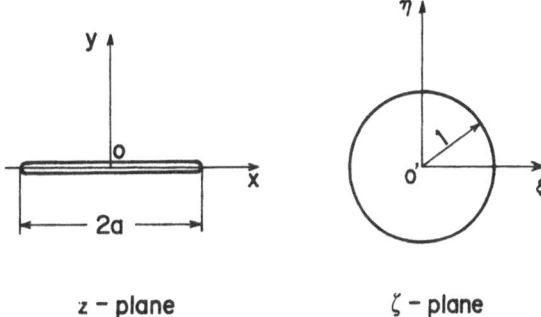

z – plane   ζ – plane

Figure 1.2. Conformal mapping

The inverse of equation (1.16) yields

$$\zeta = \frac{z}{a}\left[1 + \left(1 - \left(\frac{a}{z}\right)^2\right)^{\frac{1}{2}}\right] \qquad (1.17a)$$

$$\zeta^{-1} = \frac{z}{a}\left[1 - \left(1 - \left(\frac{a}{z}\right)^2\right)^{\frac{1}{2}}\right] \qquad (1.17b)$$

which together with equation (1.16) determine the points at infinity in the both planes.

General expressions of the complex potentials for traction free cracks will be developed in series forms. Assuming that the components of the shearing force and moment have no singularities within a region enclosing the crack as it is usually the case, the complex potentials $\phi'(z)$ and $\psi''(z)$ must be analytic functions within that region as it can be seen from equations (1.10). It follows from a well known theorem that they can be expressed in Laurent series of z convergent in a region bounded by two concentric circles. But here, for convenience of calculation, let us begin with the following expansions in ζ:

$$\phi'(z) = \alpha_0 + \sum_{m=1}^{\infty} (\alpha_m \zeta^m + \beta_m \zeta^{-m}) \tag{1.18a}$$

$$\frac{d\psi'(z)}{d\zeta} = \frac{a}{2}\left[\gamma_0 + \sum_{m=1}^{\infty} (\gamma_m \zeta^m + \delta_m \zeta^{-m})\right] \tag{1.18b}$$

where the coefficients in the righthand sides are generally complex. Now, let

$$\alpha_m = \overset{\bullet}{\alpha}_m + i\alpha'_m, \ \beta_m = \overset{\bullet}{\beta}_m + i\beta'_m$$
$$\gamma_m = \overset{\bullet}{\gamma}_m + i\gamma'_m, \ \delta_m = \overset{\bullet}{\delta}_m + i\delta'_m \tag{1.19}$$

such that dots and primes denote real and imaginary parts of the corresponding quantities.

The complex functions should also satisfy the single-valuedness condition of displacement. Derive the expressions for $\phi(z)$ and $\psi(z)$ by integrating equations (1.18) and making use of the relations

$$\int \zeta^m dz = \int \zeta^m \frac{dz}{d\zeta} \, d\zeta = \begin{cases} \dfrac{a}{2}\left(\dfrac{\zeta^{m+1}}{m+1} - \dfrac{\zeta^{m-1}}{m-1}\right) (m \neq \pm 1) \\[2ex] \dfrac{a}{2}\left(\log \zeta + \dfrac{1}{2\zeta^2}\right) (m = -1) \\[2ex] \dfrac{a}{2}\left(\dfrac{\zeta^2}{2} - \log \zeta\right) (m = 1) \end{cases} \tag{1.20}$$

results are then substituted into equation (1.9). The terms in the imaginary part of $\log \zeta$ cause multivalued displacement, and by taking their coefficients to be zero the following relations are obtained:

$$\alpha_1 - \beta_1 + \bar{\delta}_1 = 0, \ \gamma'_0 + \delta'_2 = 0 \tag{1.21}$$

Now, the traction free relation in equation (1.11) is rewritten with the help of the complex potentials in equation (1.18) as follows:

$$(1 - \bar{\zeta}^{-2})\left[\alpha_0 + \sum_{m=1}^{\infty}(\alpha_m\zeta^m + \beta_m\bar{\zeta}^m) - \mu\left\{\bar{\alpha}_0 + \sum_{m=1}^{\infty}(\bar{\alpha}_m\zeta^m + \bar{\beta}_m\bar{\zeta}^{-m})\right\}\right]$$

$$+ (\bar{\zeta} + \bar{\zeta}^{-1})\sum_{m=1}^{\infty}m(\alpha_m\zeta^{m-1} - \beta_m\zeta^{-(m+1)}) + \gamma_0$$

$$+ \sum_{m=1}^{\infty}(\gamma_m\zeta^m + \delta_m\zeta^{-m}) = iC(1 - \bar{\zeta}^{-2}) \tag{1.22}$$

where C is a pure imaginary constant. Replace $\bar{\zeta}$ by ζ^{-1} for the crack edge, rearrange both sides of equation (1.22) in ascending power series of ζ, equate the coefficients of the same powers, and the following relations are obtained:

$$\mu\beta_2 = -iC + \mu\alpha_0 - \bar{\alpha}_0 - \bar{\alpha}_2 - \bar{\gamma}_0 \tag{1.23a}$$

$$\mu(\beta_{m+2} - \beta_m) = -(m+1)(\bar{\alpha}_m + \bar{\alpha}_{m+2}) - \bar{\gamma}_m \ (m \geq 1) \tag{1.23b}$$

$$\delta_1 = \mu(\bar{\alpha}_1 - \bar{\beta}_1) \tag{1.23c}$$

$$\delta_m = (-iC + \alpha_0 - \mu\bar{\alpha}_0)\Delta_m^2 - \mu(\bar{\alpha}_{m-2} - \bar{\alpha}_m)$$
$$+ (m-1)(\beta_{m-2} + \beta_m)(m \geq 2) \tag{1.23d}$$

Equations (1.23), together with (1.21) can be used to express β_m and δ_m in terms of α_m and γ_m. Hence, the expressions in equations (1.18), (1.21) and (1.23) give the general forms of the complex potentials in terms of ζ which are suitable for solving traction free crack problems.

Finally, the results can be transformed into the z-plane, by using equations (1.17). The algebraic calculations are tedious and are similar to those in the plane problem [6]. The final results of the general Laurent series expansions for the complex potentials of stress free cracks are given by

$$Dw = \text{Re}\left[\bar{z}\phi(z) + \psi(z)\right] \tag{1.24a}$$

$$\phi(z) = \sum_{n=0}^{\infty}[(F_n^{\cdot} + iF_n')z^{-(n+1)} + (L_n^{\cdot} + iL_n')z^{n+1}] \tag{1.24b}$$

$$\psi(z) = -D_0^{\cdot}\log z + \sum_{n=1}^{\infty}(D_n^{\cdot} + iD_n')z^{-n} + \sum_{n=1}^{\infty}(K_n^{\cdot} + iK_n')z^{n+2} \tag{1.24c}$$

where

$$D_{2n}^{\cdot} = \sum_{p=0}^{\infty}a^{2n+2p+2}(P_{2p}^{2n}K_{2p}^{\cdot} + R_{2p}^{2n}L_{2p}^{\cdot}) \tag{1.25a}$$

$$F_{2n}^{\cdot} = \sum_{p=0}^{\infty} a^{2n+2p+2} (Q_{2p}^{2n} K_{2p}^{\cdot} + S_{2p}^{2n} L_{2p}^{\cdot}) \tag{1.25b}$$

$$D_{2n+1}^{\cdot} = \sum_{p=0}^{\infty} a^{2n+2p+4} (P_{2p+1}^{2n+1} K_{2p+1}^{\cdot} + R_{2p+1}^{2n+1} L_{2p+1}^{\cdot}) \tag{1.25c}$$

$$F_{2n+1}^{\cdot} = \sum_{p=0}^{\infty} a^{2n+2p+4} (Q_{2p+1}^{2n+1} K_{2p+1}^{\cdot} + S_{2p+1}^{2n+1} L_{2p+1}^{\cdot}) \tag{1.25d}$$

$$D_{2n}' = - \sum_{p=0}^{\infty} a^{2n+2p+2} (T_{2p}^{2n} K_{2p}' + V_{2p}^{2n} L_{2p}') \tag{1.25e}$$

$$F_{2n}' = - \sum_{p=0}^{\infty} a^{2n+2p+2} (U_{2p}^{2n} K_{2p}' + W_{2p}^{2n} L_{2p}') \tag{1.25f}$$

$$D_{2n+1}' = - \sum_{p=0}^{\infty} a^{2n+2p+4} (T_{2p+1}^{2n+1} K_{2p+1}' + V_{2p+1}^{2n+1} L_{2p+1}') \tag{1.25g}$$

$$F_{2n+1}' = - \sum_{p=0}^{\infty} a^{2n+2p+4} (U_{2p+1}^{2n+1} K_{2p+1}' + W_{2p+1}^{2n+1} L_{2p+1}') \tag{1.25h}$$

The coefficients P_{2p}^{2n} to W_{2p+1}^{2n+1} in the above expressions are defined as

$$P_{2p}^{0} = \left(1 - \frac{1}{\mu}\right) \frac{2p+2}{2^{2p+2}} \binom{2p+1}{p} \tag{1.26a}$$

$$\begin{Bmatrix} P_{2p}^{2n} \\ T_{2p}^{2n} \end{Bmatrix} = \frac{2p+2}{2^{2p+1}} \left[\frac{1}{2n} \sum_{m=\{{}^{0}_{1}\}}^{n,\,p} \left(\pm 1 - \frac{2m+1}{\mu}\right) \binom{2p+1}{p-m} A_{n-m,\,2m+1} \right.$$

$$\left. - \frac{1}{\mu} \sum_{m=\{{}^{0}_{1}\}}^{n-1,\,p} \frac{2m+1}{2m+2} \binom{2p+1}{p-m} A_{n-m-1,\,2m+2} \right] \quad (n \geqq 1 \text{ for } P_{2p}^{2n}) \tag{1.26b}$$

$$\begin{Bmatrix} Q_{2p}^{2n} \\ U_{2p}^{2n} \end{Bmatrix} = \frac{2p+2}{\mu \cdot 2^{2p+1}} \sum_{m=\{{}^{0}_{1}\}}^{n,\,p} \binom{2p+1}{p-m} A_{n-m,\,2m+1} \tag{1.26c}$$

$$R_{2p}^{0} = \left(1 - \frac{1}{\mu}\right) \frac{2p+1}{2^{2p+2}} \binom{2p}{p} \left(2 - \frac{\mu+1}{p+1}\right) \tag{1.26d}$$

$$R_{2p}^{2n} = \frac{2p+1}{2^{2p+2}} \left[\left(1 - \frac{1}{\mu}\right) \binom{2p}{p} A_{n-1,\,2} - \frac{4}{\mu} \sum_{m=0}^{n-1,\,p} \frac{2m}{2m+2} \binom{2p}{p-m} A_{n-m-1,\,2m+2} \right.$$

$$+ \frac{1}{n}\left\{\left(\mu - \frac{1}{\mu}\right)\binom{2p}{p}A_{n,1} + \sum_{m=0}^{n,p-1}\left(\frac{\mu}{2m+1} - \frac{2m+1}{\mu}\right)\binom{2p}{p-m-1}A_{n-m,2m+1}\right.$$

$$\left. - \sum_{m=1}^{n,p}\left(\frac{\mu}{2m+1} + \frac{6m-1}{\mu}\right)\binom{2p}{p-m}A_{n-m,2m+1}\right\}\Bigg], \quad (n \geq 1) \qquad (1.26e)$$

$$S_{2p}^{2n} = \frac{2p+1}{2^{2p+1}}\left[\left(\frac{1}{\mu}-1\right)\binom{2p}{p}A_{n,1} + \frac{1}{\mu}\sum_{m=0}^{n,p}\frac{4m}{2m+1}\binom{2p}{p-m}A_{n-m,2m+1}\right.$$

$$\left. + \frac{2}{2n+1}\sum_{m=1}^{n+1,p}\left(1 + \frac{2m-1}{\mu}\right)\binom{2p}{p-m}A_{n-m+1,2m}\right] \qquad (1.26f)$$

$$V_{2p}^{2n} = \frac{2p+1}{2^{2p+2}}\left[\left(\frac{1}{\mu}-1\right)\frac{p}{p+1}\binom{2p}{p}A_{n-1,2}\right.$$

$$- \frac{4}{\mu}\sum_{m=0}^{n-1,p}\frac{2m}{2m+2}\binom{2p}{p-m}A_{n-m-1,2m+2}$$

$$+ \frac{1}{n}\left\{\sum_{m=1}^{n,p-1}\left(\frac{\mu}{2m+1} - \frac{2m+1}{\mu}\right)\binom{2p}{p-m-1}A_{n-m,2m+1}\right.$$

$$\left.\left. - \sum_{m=1}^{n,p}\left(\frac{\mu}{2m+1} + \frac{6m-1}{\mu}\right)\binom{2p}{p-m}A_{n-m,2m+1}\right\}\right] \qquad (1.26g)$$

$$W_{2p}^{2n} = \frac{2p+1}{2^{2p+1}}\left[\left(1 - \frac{1}{\mu}\right)\frac{p}{p+1}\binom{2p}{p}A_{n,1}\right.$$

$$+ \frac{1}{\mu}\sum_{m=0}^{n,p}\frac{4m}{2m+1}\binom{2p}{p-m}A_{n-m,2m+1}$$

$$\left. - \frac{2}{2n+1}\sum_{m=1}^{n+1,p}\left(1 - \frac{2m-1}{\mu}\right)\binom{2p}{p-m}A_{n-m+1,2m}\right] \qquad (1.26h)$$

$$\begin{Bmatrix}P_{2p+1}^{2n+1}\\T_{2p+1}^{2n+1}\end{Bmatrix} = \frac{2p+3}{2^{2p+2}}\left[-\frac{1}{\mu}\sum_{m=0}^{n-1,p}\frac{2m+2}{2m+3}\binom{2p+2}{p-m}A_{n-m-1,2m+3}\right.$$

$$\left. - \frac{1}{2n+1}\sum_{m=0}^{n,p}\left(\frac{2m+2}{\mu}\mp 1\right)\binom{2p+2}{p-m}A_{n-m,2m+2}\right] \qquad (1.26i)$$

$$\begin{Bmatrix}Q_{2p+1}^{2n+1}\\U_{2p+1}^{2n+1}\end{Bmatrix} = \frac{2p+3}{\mu \cdot 2^{2p+2}}\sum_{m=0}^{n,p}\binom{2p+2}{p-m}A_{n-m,2m+2} \qquad (1.26j)$$

$$\begin{Bmatrix} R_{2p+1}^{2n+1} \\ V_{2p+1}^{2n+1} \end{Bmatrix} = \frac{2p+2}{2^{2p+2}} \left[\pm \binom{2p+1}{p} A_{n,1} \right.$$

$$- \frac{2}{\mu} \sum_{m=0}^{n-1,\,p} \frac{2m+1}{2m+3} \binom{2p+1}{p-m} A_{n-m-1,\,2m+3}$$

$$+ \frac{1}{2n+1} \left\{ \sum_{m=0}^{n,\,p-1} \left(\frac{\mu}{2m+2} - \frac{2m+2}{\mu} \right) \binom{2p+1}{p-m-1} A_{n-m,\,2m+2} \right.$$

$$\left. - \sum_{m=0}^{n,\,p} \left(\frac{\mu}{2m+2} + \frac{6m+2}{\mu} \right) \binom{2p+1}{p-m} A_{n-m,\,2m+2} \right\} \right] \qquad (1.26\text{k})$$

$$\begin{Bmatrix} S_{2p+1}^{2n+1} \\ W_{2p+1}^{2n+1} \end{Bmatrix} = \frac{2p+2}{2^{2p+2}} \left[\mp \frac{2n+1}{2n+2} \binom{2p+1}{p} A_{n,1} \right.$$

$$+ \frac{1}{\mu} \sum_{m=0}^{n,\,p} \frac{4m+2}{2m+2} \binom{2p+1}{p-m} A_{n-m,\,2m+2} \qquad (1.26\text{l})$$

$$\left. + \frac{2}{2n+2} \sum_{m=0}^{n+1,\,p} \left(\frac{2m}{\mu} \pm 1 \right) \binom{2p+1}{p-m} A_{n-m+1,\,2m+1} \right]$$

Two values for the lower limits of the summations correspond to the two quantities in the lefthand sides, and for the upper limits the smaller value should be taken. The quantities A in equations (1.26) are defined as the coefficients of the following expansion

$$[Z - (Z^2 - 1)^{\frac{1}{2}}]^m = \sum_{n=0}^{\infty} A_{n,m} Z^{-(2n+m)} \quad (|Z| > 1) \qquad (1.27)$$

and their closed form expressions are [7]

$$A_{n-m,\,2m} = \frac{m}{2^{2n}n} \binom{2n}{n-m} \qquad (1.28\text{a})$$

$$A_{n-m,\,2m+1} = \frac{2m+1}{2^{2n+1}(2n+1)} \binom{2n+1}{n-m} \qquad (1.28\text{b})$$

The general forms of the complex potentials in equations (1.24) together with the results in equations (1.25) and (1.26) are convenient for solving boundary value problems in general. In those expressions, the crack tip singularities do not appear explicitly, but in the final stage of the analysis they can be obtained in closed forms.

The method will be illustrated by a simple example consisting of a wide plate with an isolated crack subjected to $M_y^\infty = M$. The doubly symmetric nature of the problem suggests that

$$\phi(z) = \sum_{n=0}^{\infty} (F_{2n}^{\cdot} z^{-(2n+1)} + L_{2n}^{\cdot} z^{2n+1}) \tag{1.29a}$$

$$\psi(z) = - D_0^{\cdot} \log z + \sum_{n=1}^{\infty} D_{2n}^{\cdot} z^{-2n} + \sum_{n=0}^{\infty} K_{2n}^{\cdot} z^{2n+2} \tag{1.29b}$$

The boundary conditions at infinity together with equations (1.10) give

$$L_0^{\cdot} = - \frac{M}{4(1+v)}, \; K_0^{\cdot} = \frac{M}{4(1-v)} \tag{1.30a}$$

$$L_{2n}^{\cdot} = K_{2n}^{\cdot} = 0 \quad (n \geq 1) \tag{1.30b}$$

and the traction free relations reduce equations (1.25) to

$$D_{2n}^{\cdot} = a^{2n+2} (P_0^{2n} K_0^{\cdot} + R_0^{2n} L_0^{\cdot}) \tag{1.30c}$$

$$F_{2n}^{\cdot} = a^{2n+2} (Q_0^{2n} K_0^{\cdot} + S_0^{2n} L_0^{\cdot}) \tag{1.30d}$$

in which the coefficients P_0^{2n} to S_0^{2n} are given by equations (1.26). The complex function $\phi(z)$, required for calculating the stress intensity factor, is transformed as follows:

$$\phi(z) = L_0^{\cdot} z + \sum_{n=0}^{\infty} a^{2n+2} (Q_0^{2n} K_0^{\cdot} + S_0^{2n} L_0^{\cdot}) z^{-(2n+1)}$$

$$= L_0^{\cdot} z + \sum_{n=0}^{\infty} a^{2n+2} \left[K_0^{\cdot} \frac{1}{\mu} + L_0^{\cdot} \frac{1}{2} \left(\frac{1}{\mu} - 1 \right) \right] A_{n,1} \, z^{-(2n+1)}$$

$$= - \frac{Mz}{4(1+v)} + \frac{Ma}{2(3+v)} \sum_{n=0}^{\infty} A_{n,1} \left(\frac{a}{z} \right)^{2n+1}$$

$$= - \frac{Mz}{4(1+v)} + \frac{M}{2(3+v)} [z - (z^2 - a^2)^{\frac{1}{2}}] \tag{1.31}$$

Substituting equation (1.31) into (1.15), the following stress intensity factor is obtained:

$$k_1 = \sigma_B \sqrt{a} \tag{1.32a}$$

where

$$\sigma_B = \frac{6M}{h^2} \tag{1.32b}$$

1.4   Arbitrary array of cracks in wide plate

Let a wide plate be subjected to bending at infinity. An arbitrary array of cracks are located with their centers at points O_j ($j = 1, 2, \ldots, N$). The axes X, Y are placed in the middle plane of the plate while the corresponding polar coordinates are denoted by (ρ, β). The location of the center of the *j-th* crack is (ρ_j, β_j), its length is a_j and inclination angle is α_j. Introduce the N coordinate systems (X_j, Y_j) ($j = 1, 2, \ldots, N$) with their origins at O_j and with X_j-axes along the cracks. For convenience the following dimensionless coordinates, complex variables and parameters are defined:

$$x = \frac{X}{d},\ y = \frac{Y}{d},\ z = x + iy \tag{1.33a}$$

$$x_j = \frac{X_j}{d},\ y_j = \frac{Y_j}{d},\ z_j = x_j + iy_j \tag{1.33b}$$

$$\lambda_j = \frac{a_j}{d},\ r_j = \frac{\rho_j}{d},\ r_{jk} = \frac{\rho_{jk}}{d}\quad (j, k = 1, 2, \ldots, N) \tag{1.33c}$$

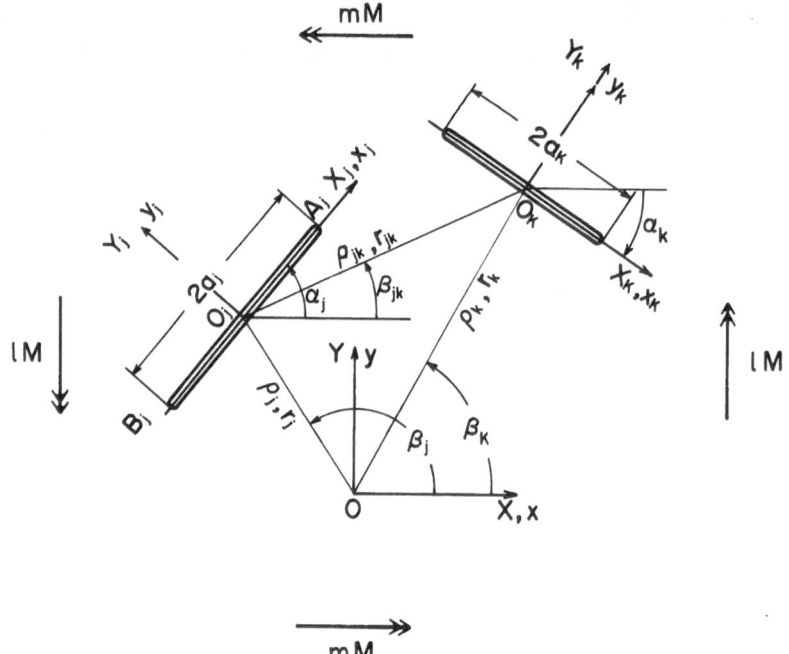

Figure 1.3. Arbitrary array of crack in wide plate

in which d is a reference length which may be arbitrarily specified. In the numerical examples of section 1.5, d will be taken as the crack spacing.

Attention is now turned to the new definition of z which is different from that in the preceding section. The radius vector between O_j and O_k is given by

$$r_{jk}e^{i\beta_{jk}} = r_k e^{i\beta_k} - r_j e^{i\beta_j} \tag{1.34}$$

The plate is assumed to be of infinite extent and subjected to uniform bending moment at infinity:

$$M_x^\infty = lM, \, M_y^\infty = mM \tag{1.35}$$

Assume a displacement function w of the form:

$$w = w_0 + \sum_{k=1}^{N} w_k \tag{1.36}$$

where w_0 is the displacement for the uncracked plate under the same moments at infinity, and hence as

$$Dw_0 = Md^2[\bar{z}\phi_0(z) + \psi_0(z)] \tag{1.37a}$$

in which

$$\phi_0(z) = L_0 z, \, \psi_0(z) = K_0 z^2 \tag{1.37b}$$

and

$$L_0 = -\frac{m+l}{4(1+v)}, \, K_0 = \frac{m-l}{4(1-v)} \tag{1.37c}$$

The displacement functions w_k ($k=1, 2, \ldots, N$) in equation (1,36) contain the crack tip singularities and can be written as:

$$Dw_k = Md^2 \, \text{Re}[\bar{z}_k\phi_k(z_k) + \psi_k(z_k)] \tag{1.38a}$$

such that

$$\phi_k(z_k) = \sum_{n=0}^{\infty} (\dot{F}_{n,k} + iF'_{n,k})z_k^{-(n+1)} \tag{1.38b}$$

$$\psi_k(z_k) = -\dot{D}_{0,k} \log z_k + \sum_{n=1}^{\infty} (\dot{D}_{n,k} + iD'_{n,k})z_k^{-n} \tag{1.38c}$$

where the coefficients with dots and primes denote real and imaginary parts of the quantities involved. They should be determined from the boundary conditions on the crack. The boundary conditions at infinity are satisfied by w_0 giving rise to the applied bending moments as w_k gives rise to no stresses there. Therefore, there remains only the consideration of traction free conditions along the edges of all cracks.

In order to consider the j-th crack, it would be convenient to express w in term of the only complex variable z_j. From simple geometry, the following relations hold:

$$z = z_j e^{i\alpha_j} + r_j e^{i\beta_j} \tag{1.39a}$$

$$z_k = z_j e^{i(\alpha_j - \alpha_k)} - r_{jk} e^{i(\beta_{jk} - \alpha_k)} \tag{1.39b}$$

Substituting equations (1.39) into (1.37) and (1.38), w_0 and w_k $(k \neq j)$ can be written in terms of z_j as follows:

$$Dw_0 = Md^2 \operatorname{Re}\left[\bar{z}_j \Phi_0^*(z_j) + \Psi_0^*(z_j)\right] \tag{1.40a}$$

with

$$\Phi_0^*(z_j) = L_0 z_j, \quad \Psi_0^*(z_j) = K_0 z_j^2 \tag{1.40b}$$

and

$$Dw_k = Md^2 \operatorname{Re}\left[\bar{z}_j \Phi_k^*(z_j) + \Psi_k^*(z_j)\right] \tag{1.41a}$$

with

$$\Phi_k^*(z_j) = e^{i(\alpha_k - \alpha_j)} \phi_k\left[z_j e^{i(\alpha_j - \alpha_k)} - r_{jk} e^{i(\beta_{jk} - \alpha_k)}\right] \tag{1.41b}$$

$$\Psi_k^*(z_j) = \psi_k\left[z_j e^{i(\alpha_j - \alpha_k)} - r_{jk} e^{i(\beta_{jk} - \alpha_k)}\right] - r_{jk} e^{i(\alpha_j - \beta_{jk})} \Phi_k^*(z_j) \tag{1.41c}$$

The complex potentials in equations (1.38) are reexpanded in power series of z_j assuming that $|z_j| < r_{jk}$. Hence w_0 and w_k are expressible in positive power series of z_j. In this way, the total displacement function w becomes:

$$Dw = Md^2\left[\bar{z}_j \Phi_j(z_j) + \Psi_j(z_j)\right] \tag{1.42a}$$

in which

$$\Phi_j(z_j) = \sum_{n=0}^{\infty} \left[(F_{n,j}^{\cdot} + iF_{n,j}')z_j^{-(n+1)} + (L_{n,j}^{\cdot} + iL_{n,j}')z_j^{n+1}\right] \tag{1.42b}$$

$$\Psi_j(z_j) = -D_{0,j}^{\cdot} \log z_j + \sum_{n=1}^{\infty} (D_{n,j}^{\cdot} + iD_{n,j}')z_j^{-n} + \sum_{n=0}^{\infty} (K_{n,j}^{\cdot} + iK_{n,j}')z_j^{n+2} \tag{1.42c}$$

The coefficients in the complex potentials are given by

$$L_{n,j}^{\bullet} = L_0 \Delta_n^0 + \sum_{p=0}^{\infty} \sum_{k \neq j}^{N} (e_{n,j}^{p,k} F_{p,k}^{\bullet} + f_{n,j}^{p,k} F_{p,k}')$$ (1.43a)

$$L_{n,j}' = \sum_{p=0}^{\infty} \sum_{k \neq j}^{N} (-f_{n,j}^{p,k} F_{p,k}^{\bullet} + e_{n,j}^{p,k} F_{p,k}')$$ (1.43b)

$$K_{n,j}^{\bullet} = K_0 \Delta_n^0 + \sum_{p=0}^{\infty} \sum_{k \neq j}^{N} (a_{n,j}^{p,k} D_{p,k}^{\bullet} + b_{n,j}^{p,k} D_{p,k}' + c_{n,j}^{p,k} F_{p,k}^{\bullet} + d_{n,j}^{p,k} F_{p,k}')$$ (1.43c)

$$K_{n,j}' = \sum_{p=0}^{\infty} \sum_{k \neq j}^{N} (- b_{n,j}^{p,k} D_{p,k}^{\bullet} + a_{n,j}^{p,k} D_{p,k}' - d_{n,j}^{p,k} F_{p,k}^{\bullet} + c_{n,j}^{p,k} F_{p,k}')$$ (1.43d)

Δ_n^0 are the Kronecker deltas and the coefficients in the series are geometric constants which are defined by

$$a_{n,j}^{o,k} = \frac{\cos\left[(n+2)(\beta_{jk} - \alpha_j)\right]}{(n+2)(r_{jk})^{n+2}}, \quad b_{n,j}^{o,k} = \frac{\sin\left[(n+2)(\beta_{jk} - \alpha_j)\right]}{(n+2)(r_{jk})^{n+2}}$$ (1.44a)

$$\begin{Bmatrix} a_{n,j}^{p,k} \\ b_{n,j}^{p,k} \end{Bmatrix} =$$

$$(-1)^p \binom{n+p+1}{n+2} \begin{Bmatrix} \cos \\ \sin \end{Bmatrix} \frac{[p(\alpha_j - \alpha_k) + (n+p+2)(\beta_{jk} - \alpha_j)]}{(r_{jk})^{n+p+2}} (p \geq 1)$$ (1.44b)

$$\begin{Bmatrix} c_{n,j}^{p,k} \\ d_{n,j}^{p,k} \end{Bmatrix} =$$

$$(-1)^p \binom{n+p+2}{n+2} \begin{Bmatrix} \cos \\ \sin \end{Bmatrix} \frac{[(p+2)(\alpha_j - \alpha_k) + (n+p+4)(\beta_{jk} - \alpha_j)]}{(r_{jk})^{n+p+2}}$$ (1.44c)

$$\begin{Bmatrix} e_{n,j}^{p,k} \\ f_{n,j}^{p,k} \end{Bmatrix} =$$

$$(-1)^{p+1} \binom{n+p+1}{n+1} \begin{Bmatrix} \cos \\ \sin \end{Bmatrix} \frac{[(p+2)(\alpha_j - \alpha_k) + (n+p+2)(\beta_{jk} - \alpha_j)]}{(r_{jk})^{n+p+2}}$$ (1.44d)

The complex potentials have been reduced to the same forms as those in equations (1.24), and the relations for traction free cracks are written in forms given by equations (1.25). The above procedure holds for each of the cracks, and the relations to be used on crack edges can be summarized:

$$D^{\bullet}_{2n,\,j} = \sum_{p=0}^{\infty} \lambda_j^{2n+2p+2}(P^{2n}_{2p} K^{\bullet}_{2p,\,j} + R^{2n}_{2p} L^{\bullet}_{2p,\,j}) \tag{1.45a}$$

$$F^{\bullet}_{2n,\,j} = \sum_{p=0}^{\infty} \lambda_j^{2n+2p+2}(Q^{2n}_{2p} K^{\bullet}_{2p,\,j} + S^{2n}_{2p} L^{\bullet}_{2p,\,j}) \tag{1.45b}$$

$$D^{\bullet}_{2n+1,\,j} = \sum_{p=0}^{\infty} \lambda_j^{2n+2p+4}(P^{2n+1}_{2p+1} K^{\bullet}_{2p+1,\,j} + R^{2n+1}_{2p+1} L^{\bullet}_{2p+1,\,j}) \tag{1.45c}$$

$$F^{\bullet}_{2n+1,\,j} = \sum_{p=0}^{\infty} \lambda_j^{2n+2p+4}(Q^{2n+1}_{2p+1} K^{\bullet}_{2p+1,\,j} + S^{2n+1}_{2p+1} L^{\bullet}_{2p+1,\,j}) \tag{1.45d}$$

$$D'_{2n,\,j} = - \sum_{p=0}^{\infty} \lambda_j^{2n+2p+2}(T^{2n}_{2p} K'_{2p,\,j} + V^{2n}_{2p} L'_{2p,\,j}) \tag{1.45e}$$

$$F'_{2n,\,j} = - \sum_{p=0}^{\infty} \lambda_j^{2n+2p+2}(U^{2n}_{2p} K'_{2p,\,j} + W^{2n}_{2p} L'_{2p,\,j}) \tag{1.45f}$$

$$D'_{2n+1,\,j} = - \sum_{p=0}^{\infty} \lambda_j^{2n+2p+4}(T^{2n+1}_{2p+1} K'_{2p+1,\,j} + V^{2n+1}_{2p+1} L'_{2p+1,\,j}) \tag{1.45g}$$

$$F'_{2n+1,\,j} = - \sum_{p=0}^{\infty} \lambda_j^{2n+2p+4}(U^{2n+1}_{2p+1} K'_{2p+1,\,j} + W^{2n+1}_{2p+1} L'_{2p+1,\,j}) \tag{1.45h}$$

where the coefficients P^{2n}_{2p} to W^{2n+1}_{2p+1} are given by equations (1.26).

The next step in the analysis is to determine the unknown coefficients $F^{\bullet}_{n,j}, F'_{n,j}, L^{\bullet}_{n,j}, L'_{n,j}, D^{\bullet}_{n,j}, D'_{n,j}, K^{\bullet}_{n,j}, K'_{n,j}$ $(n = 0, 1, \ldots; j = 1, 2, \ldots, N)$ in equations (1.43) and (1.45). To do this, a perturbation technique is applied. For convenience, the ratio of the crack lengths is fixed by the constants s_j $(j = 1, 2, \ldots, N)$. Let $\lambda_j = s_j \lambda$ and all the unknowns are expanded in power series of the only parameter λ:

$$D^{\bullet}_{2n,\,j} = \sum_{p=n+1}^{\infty} D^{\bullet\,(2p)}_{2n,\,j} \lambda^{2p}, \qquad D^{\bullet}_{2n+1,\,j} = \sum_{p=n+2}^{\infty} D^{\bullet\,(2p)}_{2n+1,\,j} \lambda^{2p}, \tag{1.46a}$$

$$D'_{2n,\,j} = \sum_{p=n+1}^{\infty} D'^{(2p)}_{2n,\,j} \lambda^{2p}, \qquad D'_{2n+1,\,j} = \sum_{p=n+2}^{\infty} D'^{(2p)}_{2n+1,\,j} \lambda^{2p}, \tag{1.46b}$$

$$F^{\bullet}_{2n,\,j} = \sum_{p=n+1}^{\infty} F^{\bullet\,(2p)}_{2n,\,j} \lambda^{2p} \qquad F^{\bullet}_{2n+1,\,j} = \sum_{p=n+2}^{\infty} F^{\bullet\,(2p)}_{2n+1,\,j} \lambda^{2p} \tag{1.46c}$$

$$F'_{2n,j} = \sum_{p=n+1}^{\infty} F'^{(2p)}_{2n,j} \lambda^{2p} \qquad F'_{2n+1,j} = \sum_{p=n+2}^{\infty} F'^{(2p)}_{2n+1,j} \lambda^{2p} \tag{1.46d}$$

$$K^{\bullet}_{n,j} = K_0 \Delta^0_n + \sum_{p=1}^{\infty} K^{\bullet(2p)}_{n,j} \lambda^{2p}, \qquad K'_{n,j} = \sum_{p=1}^{\infty} K'^{(2p)}_{n,j} \lambda^{2p} \tag{1.46e}$$

$$L^{\bullet}_{n,j} = L_0 \Delta^0_n + \sum_{p=1}^{\infty} L^{\bullet(2p)}_{n,j} \lambda^{2p} \qquad L'_{n,j} = \sum_{p=1}^{\infty} L'^{(2p)}_{n,j} \lambda^{2p} \tag{1.46f}$$

Substitute equations (1.46) into (1.43) and (1.45), rearrange both sides of the resulting equations in ascending power series of λ and equate terms of the same powers. The final results are:

$$L^{\bullet(0)}_{0,j} = L_0, \ K^{\bullet(0)}_{0,j} = K_0 \tag{1.47a}$$

$$L^{\bullet(0)}_{n,j} = L'^{(0)}_{n,j} = 0, \ K^{\bullet(0)}_{n,j} = K'^{(0)}_{n,j} = 0 \quad (n \geq 1) \tag{1.47b}$$

$$D^{\bullet(2n+2)}_{2n,j} = s^{2n+2}_j (P^{2n}_0 K^{\bullet(0)}_{0,j} + R^{2n}_0 L^{\bullet(0)}_{0,j}) \tag{1.47c}$$

$$F^{\bullet(2n+2)}_{2n,j} = s^{2n+2}_j (Q^{2n}_0 K^{\bullet(0)}_{0,j} + S^{2n}_0 L^{\bullet(0)}_{0,j}) \tag{1.47d}$$

$$D'^{(2n+2)}_{2n,j} = F'^{(2n+2)}_{2n,j} = 0 \tag{1.47e}$$

$$L^{\bullet(2)}_{n,j} = \sum_{\substack{k=1 \\ k \neq j}}^{N} e^{0,k}_{n,j} F^{\bullet(2)}_{0,k} \tag{1.47f}$$

$$K^{\bullet(2)}_{n,j} = \sum_{\substack{k=1 \\ k \neq j}}^{N} (a^{0,k}_{n,j} D^{\bullet(2)}_{0,k} + c^{0,k}_{n,j} F^{\bullet(2)}_{0,k}) \tag{1.47g}$$

$$L'^{(2)}_{n,j} = -\sum_{\substack{k=1 \\ k \neq j}}^{N} f^{0,k}_{n,j} F^{\bullet(2)}_{0,k} \tag{1.47h}$$

$$K'^{(2)}_{n,j} = -\sum_{\substack{k=1 \\ k \neq j}}^{N} (b^{0,k}_{n,j} D^{\bullet(2)}_{0,k} + d^{0,k}_{n,j} F^{\bullet(2)}_{0,k}) \tag{1.47i}$$

$$D^{\bullet(2n+2q)}_{2n,j} = \sum_{p=0}^{q-1} s^{2n+2p+2}_j [P^{2n}_{2p} K^{\bullet(2q-2p-2)}_{2p,j} + R^{2n}_{2p} L^{\bullet(2q-2p-2)}_{2p,j}] \tag{1.47j}$$

$$F^{\bullet(2n+2q)}_{2n,j} = \sum_{p=0}^{q-1} s^{2n+2p+2}_j [Q^{2n}_{2p} K^{\bullet(2q-2p-2)}_{2p,j} + S^{2n}_{2p} L^{\bullet(2q-2p-2)}_{2p,j}] \tag{1.47k}$$

$$D'^{(2n+2q)}_{2n,j} = -\sum_{p=0}^{q-1} s^{2n+2p+2}_j [T^{2n}_{2p} K'^{(2q-2p-2)}_{2p,j} + V^{2n}_{2p} L'^{(2q-2p-2)}_{2p,j}] \tag{1.47l}$$

$$F'^{(2n+2q)}_{2n,j} = -\sum_{p=0}^{q-1} s^{2n+2p+2}_j [U^{2n}_{2p} K'^{(2q-2p-2)}_{2p,j} + W^{2n}_{2p} L'^{(2q-2p-2)}_{2p,j}] \tag{1.47m}$$

$$D_{2n+1, j}^{* (2n+2q)} = \sum_{p=0}^{q-2} s_j^{2n+2p+4} [P_{2p+1}^{2n+1} K_{2p+1, j}^{* (2q-2p-4)} + R_{2p+1}^{2n+1} L_{2p+1, j}^{* (2q-2p-4)}] \quad (1.47\text{n})$$

$$F_{2n+1, j}^{* (2n+2q)} = \sum_{p=0}^{q-2} s_j^{2n+2p+4} [Q_{2p+1}^{2n+1} K_{2p+1, j}^{* (2q-2p-4)} + S_{2p+1}^{2n+1} L_{2p+1, j}^{* (2q-2p-4)}] \quad (1.47\text{p})$$

$$D_{2n+1, j}^{\prime (2n+2q)} = - \sum_{p=0}^{q-2} s_j^{2n+2p+4} [T_{2p+1}^{2n+1} K_{2p+1, j}^{\prime (2q-2p-4)} + V_{2p+1}^{2n+1} L_{2p+1, j}^{\prime (2q-2p-4)}] \quad (1.47\text{q})$$

$$F_{2n+1, j}^{\prime (2n+2q)} = - \sum_{p=0}^{q-2} s_j^{2n+2p+4} [U_{2p+1}^{2n+1} K_{2p+1, j}^{\prime (2q-2p-4)} + W_{2p+1}^{2n+1} L_{2p+1, j}^{\prime (2q-2p-4)}] \quad (1.47\text{r})$$

$$K_{n, j}^{* (2q)}$$

$$= \sum_{k \neq j}^{N} \sum_{p=0}^{2q-2} [a_{n, j}^{p, k} D_{p, k}^{* (2q)} + b_{n, j}^{p, k} D_{p, k}^{\prime (2q)} + c_{n, j}^{p, k} F_{p, k}^{* (2q)} + d_{n, j}^{p, k} F_{p, k}^{\prime (2q)}] \quad (1.47\text{s})$$

$$L_{n, j}^{* (2q)} = \sum_{k \neq j}^{N} \sum_{p=0}^{2q-2} [e_{n, j}^{p, k} F_{p, k}^{* (2q)} + f_{n, j}^{p, k} F_{p, k}^{\prime (2q)}] \quad (1.47\text{t})$$

$$K_{n, j}^{\prime (2q)}$$

$$= \sum_{k \neq j}^{N} \sum_{p=0}^{2q-2} [-b_{n, j}^{p, k} D_{p, k}^{* (2q)} + a_{n, j}^{p, k} D_{p, k}^{\prime (2q)} - d_{n, j}^{p, k} F_{p, k}^{* (2q)} + c_{n, j}^{p, k} F_{p, k}^{\prime (2q)}] \quad (1.47\text{u})$$

$$L_{n, j}^{\prime (2q)} = \sum_{k \neq j}^{N} \sum_{p=0}^{2q-2} [-f_{n, j}^{p, k} F_{p, k}^{* (2q)} + e_{n, j}^{p, k} F_{p, k}^{\prime (2q)}] \quad (1.47\text{v})$$

in which $j = 1, 2, \ldots, N$ and $q = 2, 3, \ldots,$. Taking $j = 1, 2, \ldots, N$, the expansion coefficients in equations (1.46) can be calculated successively to any desired accuracy.

Having expressed the complex potentials $\Phi_j(z_j)$ $(j = 1, 2, \ldots, N)$ in series form, the stress intensity factor at the crack tip A_j $(z_j = \lambda_j)$ can be found from equation (1.15) as

$$[k_1 - ik_2]_{Aj} = - \frac{12(2d)^{\frac{1}{2}}(3 + \nu)M}{h^2} \lim_{z_j \to \lambda_j} [(z_j - \lambda_j)^{\frac{1}{2}} \Phi'_j(z_j)] \quad (1.48)$$

Putting equations (1.46) into (1.42) and using the calculated results of $F_{2 n, j}^{*}$ etc., $\Phi_j'(z)$ in equation (1.48) may be written as a power series in λ:

$$\Phi_j'(z_j) = \sum_{n=0}^{\infty} (n + 1) [-(F_{n, j}^{*} + iF_{n, j}^{\prime})z_j^{-(n+2)} + (L_{n, j}^{*} + iL_{n, j}^{\prime})z_j^{n}]$$

$$= \sum_{q=0}^{\infty} \lambda^{2q} \sum_{p=0}^{q} s_j^{2p} [K_{2p,j}^{*(2q-2p)} \sum_{n=0}^{\infty} (2n+1)Q_{2p}^{2n}Z_j^{-(2n+2)}$$

$$+ L_{2p,j}^{*(2q-2p)} \left\{ (2p+1)Z_j^{2p} + \sum_{n=0}^{\infty} (2n+1)S_{2p}^{2n}Z_j^{-(2n+2)} \right\}$$

$$- iK_{2p,j}^{'(2q-2p)} \sum_{n=0}^{\infty} (2n+1)U_{2p}^{2n}Z_j^{-(2n+2)}$$

$$- iL_{2p,j}^{'(2q-2p)} \left\{ (2p+1)Z_j^{2p} + \sum_{n=0}^{\infty} (2n+1)W_{2p}^{2n}Z_j^{-(2n+2)} \right\}$$

$$+ \sum_{q=0}^{\infty} \lambda^{2q+1} \sum_{p=0}^{q} s_j^{2p-1} [K_{2p+1,j}^{*(2q-2p)} \sum_{n=0}^{\infty} (2n+2)Q_{2p+1}^{2n+1}Z_j^{-(2n+3)}$$

$$+ L_{2p+1,j}^{*(2q-2p)} \left\{ (2p+2)Z_j^{2p+1} + \sum_{n=0}^{\infty} (2n+2)S_{2p+1}^{2n+1}Z_j^{-(2n+3)} \right\}$$

$$- iK_{2p+1,j}^{'(2q-2p)} \sum_{n=0}^{\infty} (2n+2)U_{2p+1}^{2n+1}Z_j^{-(2n+3)}$$

$$- iL_{2p+1,j}^{'(2q-2p)} \left\{ (2p+2)Z_j^{2p+1} + \sum_{n=0}^{\infty} (2n+2)W_{2p+1}^{2n+1}Z_j^{-(2n+3)} \right\} \Bigg] \tag{1.49}$$

where z_j stands for

$$Z_j = \frac{z_j}{\lambda_j} \tag{1.50}$$

The infinite series containing Q_{2p}^{2n} to W_{2p+1}^{2n+1} can be summed up in closed forms with the help of equations (1.26) and the relations

$$\sum_{n=m}^{\infty} (2n+1)A_{n-m,\,2m+1}Z^{-(2n+2)} = \frac{2m+1}{(Z^2-1)^{\frac{1}{2}}}[Z-(Z^2-1)^{\frac{1}{2}}]^{2m+1} \tag{1.51a}$$

$$\sum_{n=m}^{\infty} (2n+1)A_{n-m,\,2m+2}Z^{-(2n+2)} = \left[\frac{(2m+2)Z}{(Z^2-1)^{\frac{1}{2}}}-1\right][Z-(Z^2-1)^{\frac{1}{2}}]^{2m+2} \tag{1.51b}$$

together with

$$\sum_{n=m}^{\infty} (2n+2)A_{n-m,\,2m+2}Z^{-(2n+3)} = \frac{2m+2}{(Z^2-1)^{\frac{1}{2}}}[Z-(Z^2-1)^{\frac{1}{2}}]^{2m+2} \tag{1.52a}$$

$$\sum_{n=m}^{\infty} (2n+2)A_{n-m+1,\,2m+1}Z^{-(2n+3)} = \left[\frac{(2m+1)Z}{(Z^2-1)^{\frac{1}{2}}}-1\right][Z-(Z^2-1)^{\frac{1}{2}}]^{2m+1}$$

$$(1.52b)$$

which are easily derived from equation (1.27).

Substituting the obtained expressions of $\Phi'_j(z_j)$ into equation (1.48), the stress intensity factor at the crack tip A_j can be given as a power series whose coefficients are evaluated exactly. Computer program has been prepared on the basis of the above formulations. It automatically generates the stress intensity factors at all crack tips for any given geometric and mechanical parameters.

The stress intensity factors for the jth crack in power series form as follows:

$$k_{1,\,Aj} = \sigma_B \sqrt{a_j}\, F_{Aj}, \qquad F_{Aj} = \sum_{n=0}^{s} P_n \lambda^n \qquad (1.53a)$$

$$k_{1,\,Bj} = \sigma_B \sqrt{a_j}\, F_{Bj}, \qquad F_{Bj} = \sum_{n=0}^{s} P_n(-\lambda)^n \qquad (1.53b)$$

$$k_{2,\,Aj} = \sigma_B \sqrt{a_j}\, F'_{Aj}, \qquad F'_{Aj} = \sum_{n=0}^{s} Q_n \lambda^n \qquad (1.53c)$$

$$k_{2,\,Bj} = \sigma_B \sqrt{a_j}\, F'_{Bj}, \qquad F'_{Bj} = \sum_{n=0}^{s} Q_n(-\lambda)^n \qquad (1.53d)$$

where

$$\sigma_B = \frac{6M}{h^2}$$

The factor $\sigma_B \sqrt{a_j}$ is k_1 for an isolated crack under bending at infinity. Therefore, F_{Aj}, F_{Bj}, F'_{Aj} and F'_{Bj} may be regarded as the magnification factors, owing to crack interaction effects.

The analysis is simplified for the special case of an infinite row of periodic cracks. Consider an infinite row of equal parallel cracks distributed with a stagger angle α as shown in Figure 1.4. For simplicity, consider only three adjacent cracks. Let the distance between two adjacent crack centers be the reference length d. The coordinates (X_j, Y_j), dimensionless coordinates (x_j, y_j), complex coordinates and parameters as shown in Figure 1.4 are defined as

$$x_j = \frac{X_j}{d}, \quad y_j = \frac{Y_j}{d}, \quad z_j = x_j + iy_j \qquad (1.54a)$$

$$\lambda = \frac{2a}{d} \qquad\qquad\qquad (1.54b)$$

in which $j = -\infty, \ldots, -1, 0, 1, \ldots, \infty$.

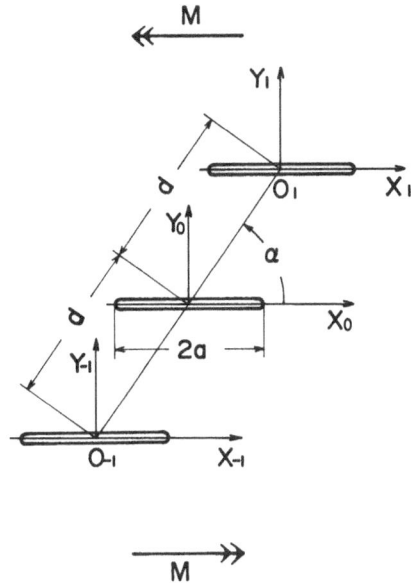

Figure 1.4. Periodic parallel cracks

Due to the periodic nature of the problem, the complex potentials $\phi_k(z_k)$ and $\psi_k(z_k)$ must have a common form for each crack. Equations (1.36) and (1.37) are replaced by

$$w = w_{(0)} + \sum_{k=-\infty}^{\infty} w_k \qquad\qquad\qquad (1.55)$$

where

$$Dw_{(0)} = Md^2 \mathrm{Re}[\bar{z}_0 \phi_{(0)}(z_0) + \psi_{(0)}(z_0)] \qquad\qquad (1.56a)$$

and

$$\phi_{(0)}(z_0) = -\frac{z}{4(1+v)}, \quad \psi_{(0)}(z_0) = \frac{z^2}{4(1-v)} \qquad (1.56b)$$

Moreover, w_k in equation (1.55) becomes

$$Dw_k = Md^2 \, \mathrm{Re}[\bar{z}_k \phi_k(z_k) + \psi_k(z_k)] \tag{1.57}$$

and

$$\phi_k(z_k) = \sum_{n=0}^{\infty} F_{2n} z_k^{-(2n+1)} \tag{1.58a}$$

$$\psi_k(z_k) = -D_0^{\cdot} \log z_k + \sum_{n=0}^{\infty} D_{2n} z_k^{-2n} \tag{1.58b}$$

where F_{2n} and D_{2n} are generally complex constants. Equation (1.55) permits consideration of the boundary conditions of any one of the cracks. The zeroth crack will be taken for convenience. Expansion of the displacement function w around the center of this crack yields the following expression in terms of z_0:

$$Dw = Md^2 \, \mathrm{Re} \left[\bar{z}_0 \Phi_0(z_0) + \Psi_0(z_0) \right] \tag{1.59}$$

in which

$$\Phi_0(z_0) = \sum_{n=0}^{\infty} (F_{2n} z_0^{-(2n+1)} + L_{2n} z_0^{2n+1}) \tag{1.60a}$$

$$\Psi_0(z_0) = -D_0^{\cdot} \log z_0 + \sum_{n=1}^{\infty} D_{2n} z_0^{-2n} + \sum_{n=0}^{\infty} K_{2n} z_0^{2n+2} \tag{1.60b}$$

and the coefficients L_{2n} and K_{2n} are

$$L_{2n} = -\frac{1}{4(1+\nu)} \Delta_n^0 - \sum_{p=0}^{\infty} \gamma_{2p}^{2n} F_{2p} \tag{1.61a}$$

$$K_{2n} = \frac{1}{4(1-\nu)} \Delta_n^0 + \sum_{p=0}^{\infty} (\alpha_{2p}^{2n} D_{2p} + \beta_{2p}^{2n} F_{2p}) \tag{1.61b}$$

In equations (1.61), the following contractions have been made:

$$\alpha_0^{2n} = \frac{1}{2n+2} S_{2n+2} \, e^{-i(2n+2)\alpha} \tag{1.62a}$$

$$\alpha_{2p}^{2n} = \binom{2n+2p+1}{2n+2} S_{2n+2p+2} e^{-i(2n+2p+2)\alpha} \quad (p \geq 1) \tag{1.62b}$$

$$\beta_{2p}^{2n} = \binom{2n+2p+2}{2n+2} S_{2n+2p+2} \, e^{-i(2n+2p+4)\alpha} \tag{1.62c}$$

$$\gamma_{2p}^{2n} = \binom{2n + 2p + 1}{2n + 1} S_{2n+2p+2} \, e^{-i(2n+2p+2)\alpha} \tag{1.62d}$$

provided that

$$S_{2q} = 2 \sum_{n=1}^{q} \frac{1}{n^{2q}} = \frac{(-1)^{q+1} B_{2q}(2\pi)^{2q}}{(2q)!} \tag{1.63}$$

The Bernoulli's numbers are B_{2q} with the following first values and recurrent formula

$$B_2 = \tfrac{1}{6}, \quad B_4 = -\tfrac{1}{30}, \cdots \tag{1.64}$$

$$B_{2q} = \frac{1}{2q+1} \left[q - \tfrac{1}{2} - \sum_{m=1}^{q-1} \binom{2q+1}{2m} B_{2m} \right]$$

For this problem, equations (1.25) pertaining to the traction free conditions become

$$D_{2n}^{\bullet} = \sum_{p=0}^{\infty} \lambda^{2n+2p+2} (P_{2p}^{2n} K_{2p}^{\bullet} + R_{2p}^{2n} L_{2p}^{\bullet}) \tag{1.65a}$$

$$F_{2n}^{\bullet} = \sum_{p=0}^{\infty} \lambda^{2n+2p+2} (Q_{2p}^{2n} K_{2p}^{\bullet} + S_{2p}^{2n} L_{2p}^{\bullet}) \tag{1.65b}$$

$$D_{2n}' = -\sum_{p=0}^{\infty} \lambda^{2n+2p+2} (T_{2p}^{2n} K_{2p}' + V_{2p}^{2n} L_{2p}') \tag{1.65c}$$

$$F_{2n}' = -\sum_{p=0}^{\infty} \lambda^{2n+2p+2} (U_{2p}^{2n} K_{2p}' + W_{2p}^{2n} L_{2p}') \tag{1.65d}$$

The perturbation technique can now be applied to solve equations (1.61) and (1.65) and the stress intensity factors in terms of power series in λ can then be evaluated from equation (1.48).

1.5 Numerical results

Fortran computer program has been prepared on the basis of the above formulation. The values of F_{Aj}, F_{Bj}, F_{Aj}' and F_{Bj}' can be obtained automatically for loading parameters l, m, dimensionless coordinates of crack centers (r_j, β_j), angles of inclination α_j and crack length ratios s_j ($j = 1, 2, \ldots, N$). The proposed method involves neither solutions of simultaneous equations nor numerical integrations which causes deterioration of accuracy

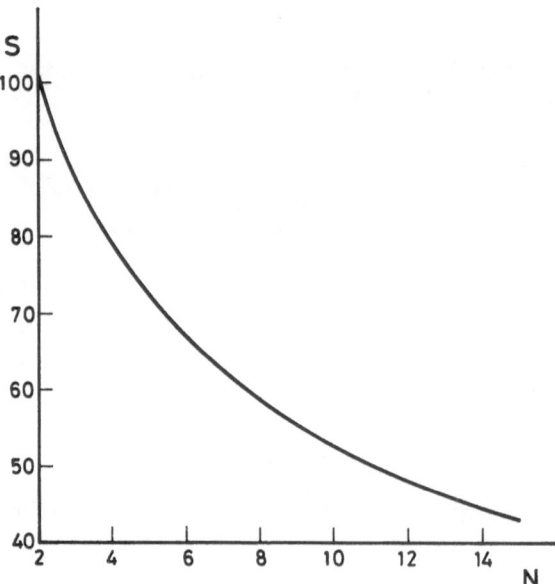

Figure 1.5. Relation between N and the upper limit of S

with increasing N and S, number of cracks and degree of the power series
of equations (1.53). Therefore N and S can be enlarged up to the limit of
the core capacity of the computer without loss of accuracy. Figure 1.5
shows the relation between N and the upper limit of S for the current computer system at the Kyushu University. In the present scheme of numerical
calculations, however, S is chosen to be 74 for periodic cracks and 43 for
other cases regardless of N such that reasonable computing cost and numerical accuracy are achieved. Discussions on the accuracy of the results are
given in section 1.6. Numerical results for some typical interaction problems
will now be treated. They have been obtained by assuming the Poisson's
ratio v to be 0.3 except for the results of collinear cracks which are independent of v.

Some inclined cracks. Figure 1.6 gives the results for a pair of inclined
cracks where the moment vector is perpendicular to the axis of symmetry.
Solid and dashed curves correspond to F_A and F_B for the inner and outer
crack tips. The magnification factors of k_1 are defined in equations (1.53).
The general trend of the curves is similar to that in the corresponding plane
problem [1, 2] except that less interaction effect is observed for the bending

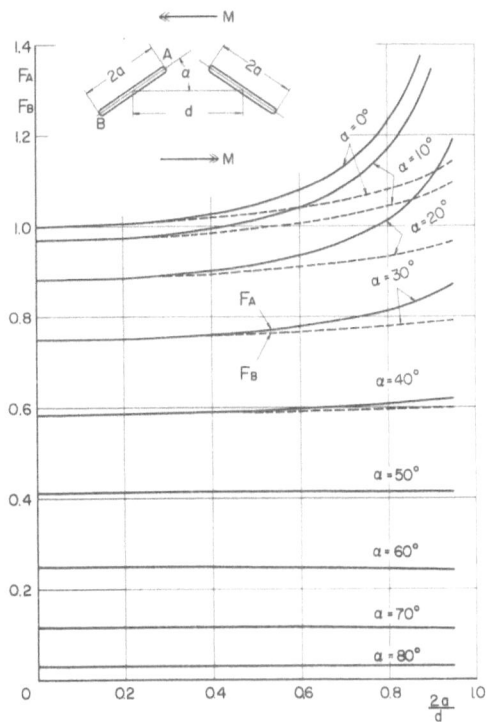

Figure 1.6. A pair of inclined cracks under bending parallel to axis of symmetry

problem for large angles of inclination. Figure 1.7 is for the case where the bending vector is parallel to the symmetric axis, showing a small interaction effect for a wide range of α. The results of uniform bending for equal length cracks radially distributed at equal angles are shown in Figure 1.8. For each configuration, F_A for the inner crack tip is higher than F_B for the outer crack tip. As the number of cracks increases, both F_A and F_B tend to decrease for the same values of a/d used in the calculation.

Collinear cracks. The magnification factors F for collinear cracks under bending at infinity are the same as those for tension, inplane shear and antiplane shear loads. They are defined as the corresponding stress intensity factors divided by those of an isolated crack subjected to loadings at infinity. Before discussing the present results, available closed form solutions on

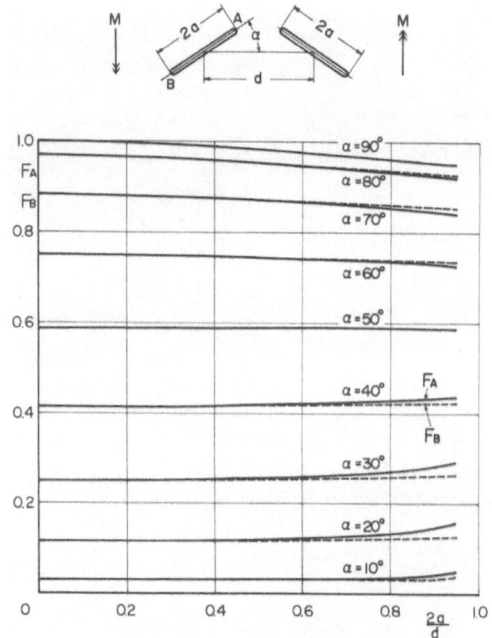

Figure 1.7. A pair of inclined cracks under bending vertical to axis of symmetry

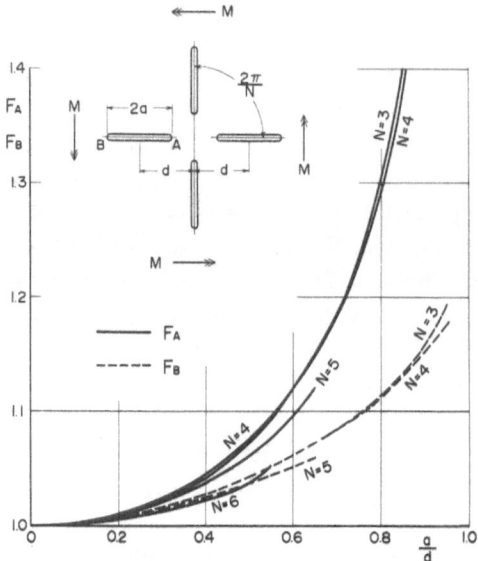

Figure 1.8. Radially distributed cracks under uniform bending

collinear cracks will be listed. The notations in the original papers are changed as that the results will be consistent with those in the present paper.

(1) Two cracks [8, 9] (Figure 1.9)

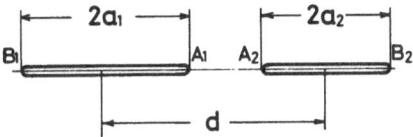

Figure 1.9. Two collinear cracks

$$F_{A1} = \left(\frac{1 - \lambda_1 + \lambda_2}{1 - \lambda_1 - \lambda_2}\right)^{\frac{1}{2}} \left[1 + \frac{1 + \lambda_1 - \lambda_2}{2\lambda_1} \left\{\frac{E(m_1)}{K(m_1)} - 1\right\}\right] \tag{1.66a}$$

$$F_{A2} = \left(\frac{1 + \lambda_1 - \lambda_2}{1 - \lambda_1 - \lambda_2}\right)^{\frac{1}{2}} \left[1 + \frac{1 - \lambda_1 + \lambda_2}{2\lambda_2} \left\{\frac{E(m_1)}{K(m_1)} - 1\right\}\right] \tag{1.66b}$$

in which

$$m_1 = 2\left(\frac{\lambda_1 \lambda_2}{(1 + \lambda_1 - \lambda_2)(1 - \lambda_1 + \lambda_2)}\right)^{\frac{1}{2}}, \quad \lambda_1 = \frac{a_1}{d}, \quad \lambda_2 = \frac{a_2}{d}$$

In the special case, when $a_1 = a_2 = a$, equations (1.66) simplify to

$$F_A = \frac{1}{2\lambda(1 - \lambda)^{\frac{1}{2}}} \left[(1 + \lambda)^2 \frac{E(m_2)}{K(m_2)} - (1 - \lambda)^2\right] \tag{1.67a}$$

$$F_B = \frac{(1 + \lambda)^{\frac{1}{2}}}{2\lambda} \left[1 - \frac{E(m_2)}{K(m_2)}\right] \tag{1.67b}$$

and

$$m_2 = \frac{2\sqrt{\lambda}}{1 + \lambda}, \quad \lambda = \frac{2a}{d}$$

(2) Three symmetric cracks [10] (Figure 1.10)

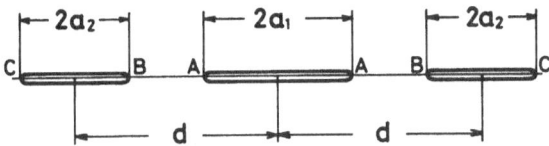

Figure 1.10 Three symmetric cracks

$$F_A = \left(\frac{(1+\lambda_2)^2 - \lambda_1^2}{(1-\lambda_2)^2 - \lambda_1^2}\right)^{\frac{1}{2}} \cdot \frac{E(m_3)}{K(m_3)} \tag{1.68a}$$

$$F_B = \frac{[(1-\lambda_2)\{(1-\lambda_2)^2 - \lambda_1^2\}]^{\frac{1}{2}}}{2\lambda_2}\left[1 - \frac{(1+\lambda_2)^2 - \lambda_1^2}{(1-\lambda_2)^2 - \lambda_1^2} \cdot \frac{E(m_3)}{K(m_3)}\right] \tag{1.68b}$$

$$F_C = \frac{[(1+\lambda_2)\{(1+\lambda_2)^2 - \lambda_1^2\}]^{\frac{1}{2}}}{2\lambda_2}\left[1 - \frac{E(m_3)}{K(m_3)}\right] \tag{1.68c}$$

where

$$m_3 = 2\left(\frac{\lambda_2}{(1+\lambda_2)^2 - \lambda_1^2}\right)^{\frac{1}{2}}, \quad \lambda_1 = \frac{a_1}{d}, \quad \lambda_2 = \frac{a_2}{d}$$

(3) Periodic cracks [11] (Figure 1.11)

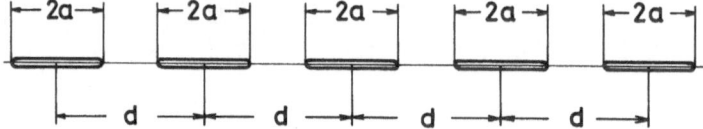

Figure 1.11. Periodic collinear cracks

$$F = \left(\frac{2}{\pi\lambda}\tan\frac{\pi\lambda}{2}\right)^{\frac{1}{2}}, \quad \lambda = \frac{2a}{d} \tag{1.69}$$

In the foregoing expressions, $K(m)$ and $E(m)$ are complete elliptic integrals of the first and second kinds.

No other closed form solutions are available for collinear cracks subjected to loadings at infinity. The present method applies to any array of cracks, and gives the results in power series whose coefficients can be evaluated exactly. As an example, the F-series for periodic collinear cracks shown by Figure 1.11 is given as follows:

$$\begin{aligned}
F(\lambda) = 1 &+ 0.4112\lambda^2 + 0.3213\lambda^4 + 0.2732\lambda^6 + 0.2413\lambda^8 + 0.2183\lambda^{10}\\
&+ 0.2007\lambda^{12} + 0.1867\lambda^{14} + 0.1753\lambda^{16} + 0.1657\lambda^{18} + 0.1576\lambda^{20}\\
&+ 0.1505\lambda^{22} + 0.1443\lambda^{24} + 0.1388\lambda^{26} + 0.1339\lambda^{28} + 0.1295\lambda^{30}\\
&+ 0.1255\lambda^{32} + 0.1218\lambda^{34} + 0.1185\lambda^{36} + 0.1154\lambda^{38} + 0.1125\lambda^{40}\\
&+ 0.1098\lambda^{42} + 0.1074\lambda^{44} + 0.1050\lambda^{46} + 0.1029\lambda^{48} + 0.1008\lambda^{50}\\
&+ 0.0989\lambda^{52} + 0.0971\lambda^{54} + 0.0953\lambda^{56} + 0.0937\lambda^{58} + 0.0922\lambda^{60}\\
&+ 0.0907\lambda^{62} + 0.0893\lambda^{64} + 0.0879\lambda^{66} + 0.0866\lambda^{68} + 0.0854\lambda^{70}\\
&+ 0.0842\lambda^{72}
\end{aligned} \tag{1.70}$$

This series coincides with the expansion of the closed form solution in equation (1.69).

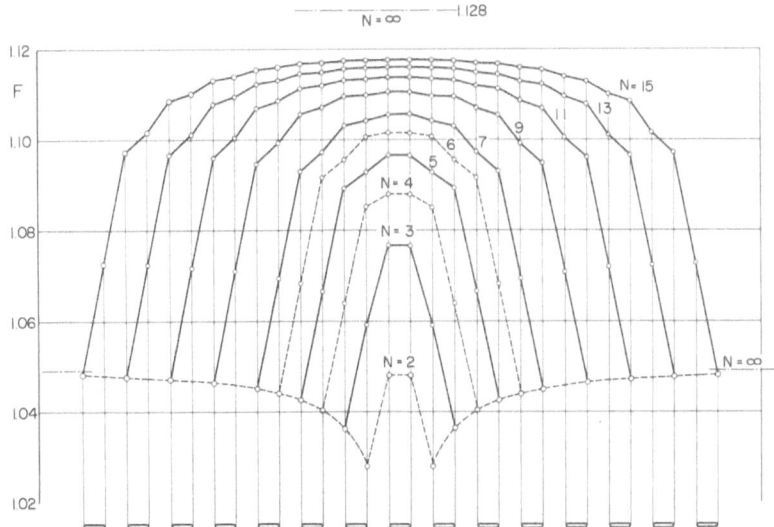

Figure 1.12 Variation of F with crack tip stations for equal collinear cracks ($2a/d = 0.5$)

In what follows, the results of equally spaced collinear cracks of equal size will be examined in some detail. The values of F tend to decrease with increasing distance from the center of the crack array. Figure 1.12 gives the results for $N \leq 15$ and $2a/d = 0.5$, where the solid and dashed lines correspond to odd and even numbers of cracks respectively.

TABLE 1.1

F_A of extreme inner tips of equal collinear cracks

λ \ N	2 [9]	3 [10]	4	5	7	9	11	13	15	$13, 15 \to \infty$	∞ [11]
0.1	1.0013	1.0025	1.0028	1.0031	1.0034	1.0036	1.0037	1.0038	1.0038	1.00414	1.00414
0.2	1.0057	1.0103	1.0117	1.0129	1.0140	1.0147	1.0151	1.0154	1.0156	1.01698	1.01698
0.3	1.0138	1.0241	1.0274	1.0302	1.0329	1.0344	1.0354	1.0361	1.0366	1.03983	1.03983
0.4	1.0272	1.0453	1.0517	1.0569	1.0621	1.0650	1.0669	1.0682	1.0691	1.07532	1.07533
0.5	1.0480	1.0766	1.0880	1.0966	1.1056	1.1106	1.1138	1.1161	1.1177	1.12845	1.12838
0.6	1.0804	1.1232	1.1423	1.1560	1.1707	1.1791	1.1844	1.1881	1.1909	1.20854	1.20846
0.7	1.1333	1.1956	1.2278	1.2491	1.2735	1.2873	1.2962	1.3024	1.3069	1.33626	1.33601
0.8	1.2289	1.3214	1.3782	1.4126	1.4551	1.4793	1.4949	1.5058	1.5137	1.56553	1.56497
0.9	1.4539	1.6068	1.7218	1.7860	1.8767	1.9280	1.9579	1.9811	1.9982	2.10940	2.11331

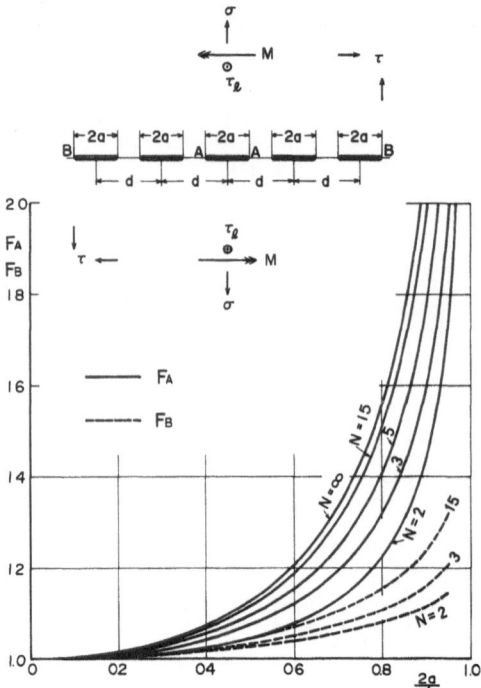

Figure 1.13. *F* of extreme inner and outer tips of equal parallel cracks

Table 1.1. and the solid curves of Figure 1.13 give F_A for various N and $\lambda = 2a/d$, which is the maximum of F at the tip of the middle crack. Values for $N = 2, 3$ and infinity in the table correspond to the exact solutions of equations (1.67), (1.68) and (1.69).

Numerical comparison of the results obtained from the series in equations (1.53) with the corresponding exact solutions will give a measure of the accuracy of the present analysis. This is given in section 1.6.

As shown above, F_A for the same value of $2a/d$ tends to decrease with increasing number of cracks. In order to be more specific, the results in Table 1.1 are plotted in Figure 1.14 by taking $1/N$ as the abscissa. Values for odd and even N are shown by open and solid circles, and those for the same values of $2a/d$ are connected with solid and dashed lines respectively. Note that for $N \geq 6$ both lines tend to converge to almost straight lines representing the limiting values of $N = \infty$ i.e., a periodic array of cracks. To check this limit numerically, values of $N = \infty$ are estimated by assuming

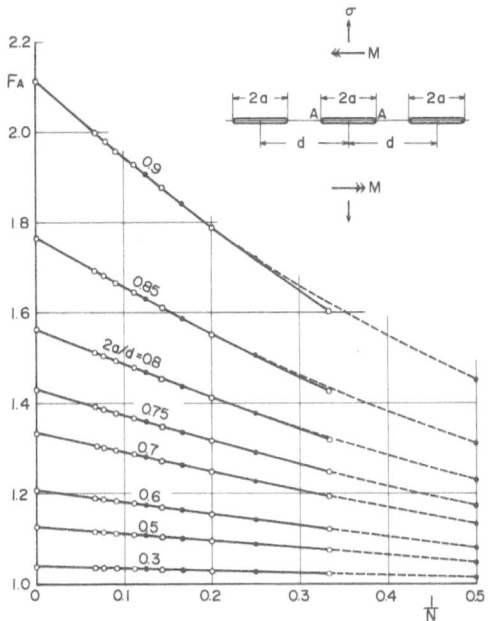

Figure 1.14. Variation of F_A with $1/N$ for constant $2a/d$

straight lines passing through the values for $N = 13$ and 15. The results are given in the column marked '13,15 → ∞' of Table 1.1, and are shown to be remarkably close to the exact ones. Therefore uncalculated values of F_A for N larger than 15 can be estimated by linear interporation of values for $N = 15$ and infinity with excellent accuracy.

TABLE 1.2

F_B of extreme outer tips of equal collinear cracks

λ \ N	2 [9]	3 [10]	4	5	7	9	11	13	15	13, 15 → ∞
0.1	1.0012	1.0015	1.0016	1.0017	1.0018	1.0018	1.0019	1.0019	1.0019	1.0020
0.2	1.0046	1.0058	1.0064	1.0067	1.0071	1.0072	1.0074	1.0074	1.0075	1.0079
0.3	1.0102	1.0130	1.0142	1.0150	1.0158	1.0162	1.0164	1.0166	1.0168	1.0176
0.4	1.0179	1.0230	1.0253	1.0267	1.0282	1.0290	1.0294	1.0298	1.0300	1.0316
0.5	1.0280	1.0363	1.0402	1.0425	1.0449	1.0463	1.0471	1.0477	1.0481	1.0507
0.6	1.0409	1.0538	1.0600	1.0635	1.0674	1.0695	1.0708	1.0717	1.0724	1.0766
0.7	1.0579	1.0772	1.0866	1.0921	1.0982	1.1015	1.1035	1.1049	1.1059	1.1124
0.8	1.0811	1.1103	1.1249	1.1335	1.1432	1.1484	1.1516	1.1538	1.1554	1.1658
0.9	1.1174	1.1644	1.1887	1.2035	1.2206	1.2298	1.2351	1.2390	1.2418	1.2601

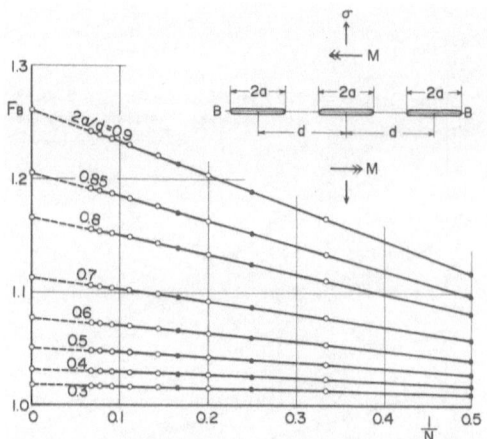

Figure 1.15. Variation of F_B with $1/N$ for constant $2a/d$

Values of F_B, the minimums of F occurring at the extreme outer crack tips are given in Table 1.2 and in Figure 1.13 by dashed lines. Values for $N = 2$ and 3 are the exact solutions of equations (1.67) and (1.68). F_B are shown to increase with increasing N for the same λ. They are replotted in Figure 1.15 by taking $1/N$ as the abscissa, and are connected with almost straight lines. Values for $N = \infty$, where no exact solution of F_B is available,

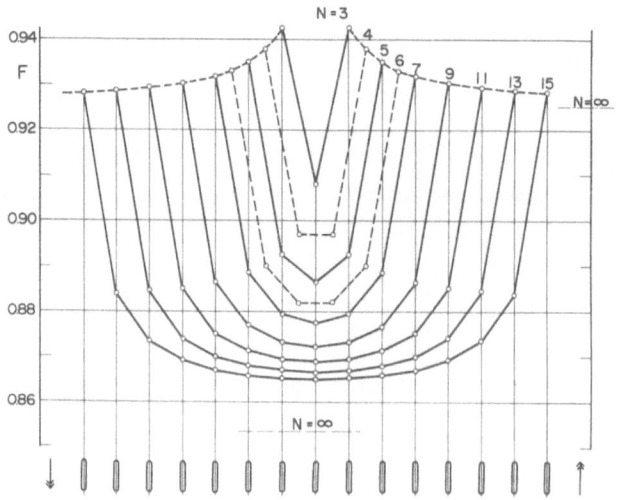

Figure 1.16. Variation of F with crack tip stations for qeual parallel cracks ($2a/d = 0.9$)

are estimated by assuming straight lines passing through the values of $N = 13$ and 15 and are given in the last column of Table 1.2 as well as in Figure 1.15 by dashed lines. The accuracy of those limiting values, however, may not be so good as that in case of F_A, since the linearity of the curves seems to be disturbed in their left parts.

Parallel cracks equally spaced without staggering. Values of F at all the crack tips are smaller than unity, and they decrease monotonically with increasing distance from the center of the crack array. This is contrary to the collinear crack problem. Figure 1.16 gives an example for $N = 3$ to 15 and $\lambda = 0.9$. F_A and F_B, the minimum and maximum of F occurring at the extreme inner and outer crack tips are tabulated in the left columns of Tables 1.3 and 1.4. Table 1.3 also gives the results for periodic parallel cracks, which are in good agreement with Nisitani's previous results [12] shown in the last column. Values in Tables 1.3 and 1.4 are plotted in Figure 1.17 against $2a/d$, and also shown in Figures 1.18 and 1.19 taking $1/N$ as the abscissa. Linearity of the curves towards larger values of N seems excel-

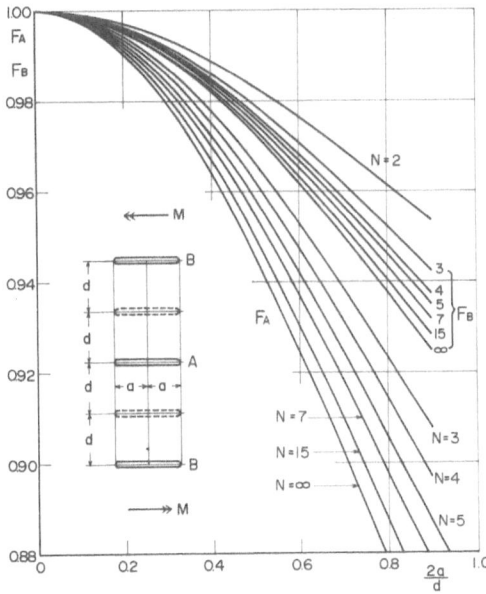

Figure 1.17. F of extreme inner and outer tips of equal parallel cracks

TABLE 1.3

F_A of extreme inner tips of equal parallel cracks

λ \ N	2	3	4	5	7	9	11	13	15	13, 15 → ∞	∞	Ref. [
0.1	0.9993	0.9986	0.9984	0.9982	0.9980	0.9980	0.9979	0.9979	0.9978	0.9976	0.9976	0.997(
0.2	0.9971	0.9943	0.9936	0.9929	0.9923	0.9919	0.9917	0.9915	0.9914	0.9907	0.9907	0.990'
0.3	0.9937	0.9873	0.9858	0.9842	0.9829	0.9821	0.9816	0.9812	0.9810	0.9793	0.9793	0.979·
0.4	0.9889	0.9779	0.9752	0.9726	0.9702	0.9689	0.9680	0.9674	0.9670	0.9641	0.9641	0.964·
0.5	0.9831	0.9664	0.9623	0.9583	0.9547	0.9528	0.9515	0.9506	0.9499	0.9457	0.9457	0.946:
0.6	0.9764	0.9531	0.9475	0.9420	0.9371	0.9344	0.9326	0.9314	0.9305	0.9247	0.9247	–
0.7	0.9690	0.9386	0.9314	0.9242	0.9179	0.9144	0.9121	0.9106	0.9094	0.9019	0.9019	–
0.8	0.9612	0.9232	0.9143	0.9055	0.8977	0.8934	0.8906	0.8887	0.8872	0.8780	0.8779	–
0.9	0.953	0.908	0.897	0.887	0.877	0.872	0.869	0.867	0.865	0.8540	0.8535	–

TABLE 1.4

F_B of extreme outer tips of equal parallel cracks

λ \ N	2	3	4	5	7	9	11	13	15	13, 15 → ∞
0.1	0.9993	0.9991	0.9990	0.9990	0.9989	0.9989	0.9989	0.9989	0.9989	0.9988
0.2	0.9971	0.9964	0.9961	0.9959	0.9958	0.9957	0.9956	0.9955	0.9955	0.9953
0.3	0.9937	0.9921	0.9914	0.9910	0.9906	0.9904	0.9902	0.9901	0.9901	0.9896
0.4	0.9889	0.9862	0.9850	0.9844	0.9836	0.9832	0.9830	0.9828	0.9827	0.9819
0.5	0.9831	0.9790	0.9772	0.9762	0.9751	0.9745	0.9742	0.9739	0.9737	0.9726
0.6	0.9764	0.9708	0.9683	0.9669	0.9654	0.9646	0.9641	0.9637	0.9635	0.9618
0.7	0.9690	0.9617	0.9585	0.9567	0.9547	0.9536	0.9530	0.9525	0.9522	0.9501
0.8	0.9612	0.9521	0.9481	0.9459	0.9434	0.9421	0.9413	0.9407	0.9403	0.9377
0.9	0.953	0.942	0.938	0.935	0.932	0.930	0.929	0.929	0.928	0.925

lent. Values for $N = \infty$ are estimated by linear extrapolation of the results for $N = 13$ and 15 and are given in the columns marked '13,15 → ∞' of Tables 1.3 and 1.4. The estimated values of F_A coincide with the exact ones almost up to four figures, and the linear relationship is verified. Uncalculated values for $N > 15$ can thus be estimated by linear interpolation or extrapolation with excellent accuracy.

Other parallel cracks. Figure 1.20 gives the values of F_A for the inner crack tips of two equal parallel cracks that are staggered. The staggering effects are influenced by the ratios e/f and $2a/d$. Corresponding results for a periodic array of parallel cracks are shown in Figure 1.21. In addition,

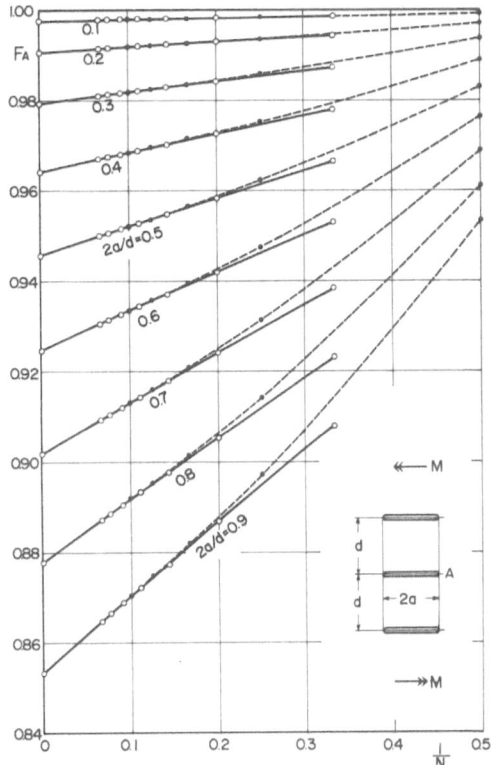

Figure 1.18. Variation of F_A with $1/N$ for constant $2a/d$

TABLE 1.5

Variation of F_A of inner tips of two parallel cracks with λ and stagger angle α

λ \ α	0° [9]	10°	20°	30°	40°	45°	50°	60°	70°	80°	90°
0.1	1.0013	1.0012	1.0008	1.0003	0.9999	0.9997	0.9995	0.9993	0.9993	0.9993	0.9993
0.2	1.0057	1.0050	1.0033	1.0012	0.9993	0.9985	0.9979	0.9973	0.9971	0.9971	0.9971
0.3	1.0138	1.0121	1.0076	1.0023	0.9978	0.9961	0.9949	0.9937	0.9934	0.9935	0.9937
0.4	1.0272	1.0231	1.0134	1.0028	0.9948	0.9921	0.9903	0.9885	0.9882	0.9885	0.9889
0.5	1.0480	1.0394	1.0204	1.0020	0.9898	0.9862	0.9838	0.9817	0.9817	0.9824	0.9831
0.6	1.0804	1.0623	1.0270	0.9986	0.9824	0.9780	0.9754	0.9735	0.9740	0.9751	0.9764
0.7	1.1333	1.0929	1.0307	0.9912	0.9723	0.9677	0.9652	0.9640	0.9652	0.9670	0.9690
0.8	1.2289	1.1288	1.0271	0.9791	0.9595	0.9553	0.9534	0.9533	0.9555	0.9583	0.9612
0.9	1.4539	1.154	1.012	0.962	0.945	0.940	0.940	0.941	0.945	0.950	0.953

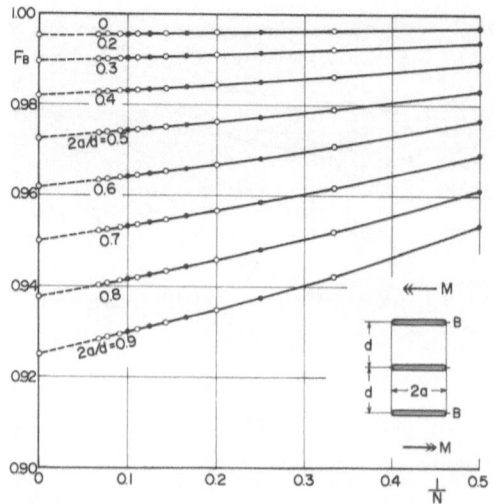

Figure 1.19 .Variation of F_B with $1/N$ for constant $2a/d$

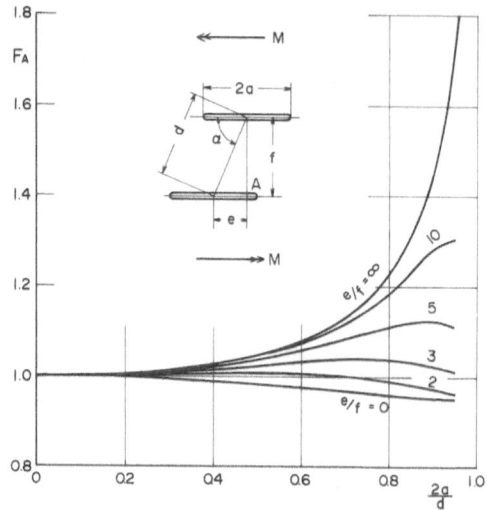

Figure 1.20. F_A of inner crack tips for two parallel cracks staggered

Tables 1.5 and 1.6 give their numerical values calculated for λ and inclination angle $\alpha = \cot^{-1}(e/f)$.

A row of parallel and equal cracks with $\alpha = 45°$ will now be discussed.

TABLE 1.6

Variation of F of periodic parallel cracks with λ and stagger angle α

λ\α	0° [11]	10°	20°	30°	40°	45°	50°	60°	70°	80°	90°
0.1	1.0041	1.0037	1.0026	1.0012	0.9997	0.9991	0.9986	0.9980	0.9977	0.9976	0.9976
0.2	1.0170	1.0152	1.0106	1.0046	0.9988	0.9964	0.9944	0.9919	0.9908	0.9906	0.9907
0.3	1.0398	1.0354	1.0241	1.0098	0.9968	0.9916	0.9874	0.9820	0.9798	0.9793	0.9793
0.4	1.0753	1.0661	1.0431	1.0162	0.9932	0.9843	0.9774	0.9687	0.9651	0.9642	0.9641
0.5	1.1284	1.1101	1.0676	1.0225	0.9872	0.9743	0.9645	0.9523	0.9472	0.9458	0.9457
0.6	1.2085	1.1720	1.0965	1.0271	0.9784	0.9615	0.9488	0.9334	0.9268	0.9249	0.9247
0.7	1.3360	1.2580	1.1269	1.0283	0.9663	0.9459	0.9307	0.9124	0.9046	0.9022	0.9019
0.8	1.5650	1.3735	1.1537	1.0244	0.9510	0.9276	0.9105	0.8899	0.8811	0.8784	0.8779
0.9	2.1133	1.5065	1.1706	1.0148	0.9326	0.9070	0.8885	0.8664	0.8570	0.8541	0.8535

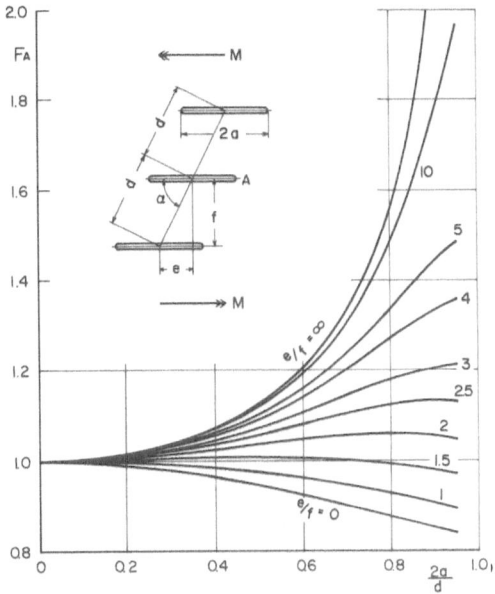

Figure 1.21. F_A of periodic parallel cracks staggered

Note that F_B for the extreme outer tip and F_A for the extreme inner tip are the maximum and minimum of the magnification factors. Numerical results are plotted in Figures 1.22 to 1.24 in the same way as the preceding cases, and the linear relationship between F and $1/N$ is confirmed.

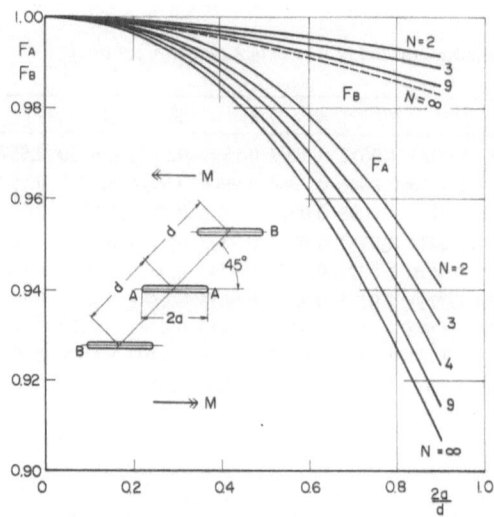

Figure 1.22. *F* of extreme inner and outer crack tips of equal parallel cracks located staggered at an angle of 45°

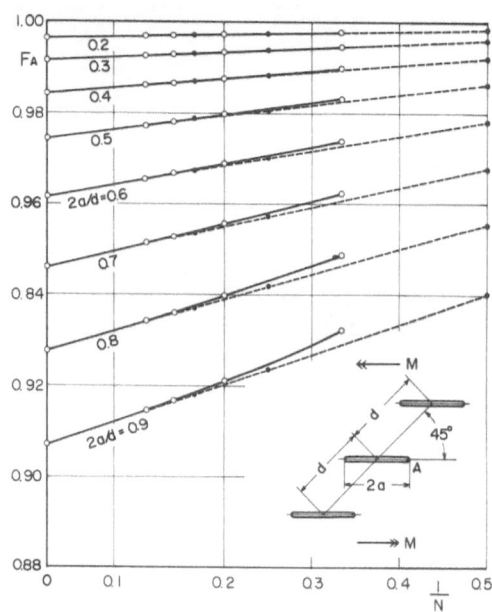

Figure 1.23. Variation of F_A with $1/N$ for constant $2a/d$

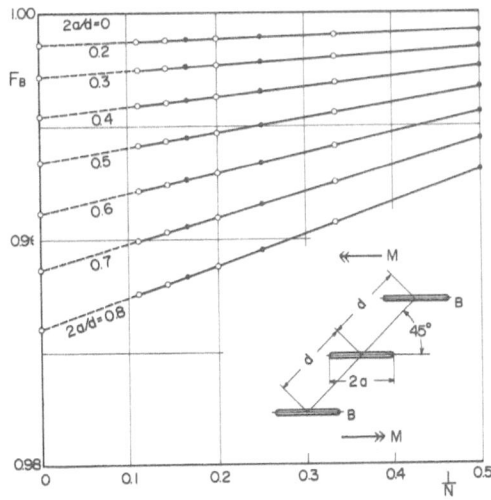

Figure 1.24. Variation of F_A with $1/N$ for constant $2a/d$

The above results pertain to the magnification factors of k_1. Values of k_2 for skew symmetric deformation are small as compared with k_1 in all the examples considered. Comparative large values of k_2 have been observed for a row of parallel cracks with $\alpha = 45°$ as mentioned above. Their maximums occur at the extreme inner crack tips, and the corresponding magnification factors F'_A are given in Table 1.7 for various values of N and λ.

TABLE 1.7

F'_A of extreme inner tips of equal parallel cracks located with stagger angle of 45°

λ \ N	2	3	4	5	6	7	9	∞
0.1	0.0008	0.0015	0.0017	0.0019	0.0020	0.0021	0.0022	0.0025
0.2	0.0003	0.0061	0.0069	0.0076	0.0079	0.0083	0.0086	0.0100
0.3	0.0076	0.0137	0.0155	0.0171	0.0178	0.0186	0.0194	0.0224
0.4	0.0139	0.0243	0.0275	0.0302	0.0316	0.0328	0.0343	0.0395
0.5	0.0220	0.0376	0.0425	0.0467	0.0488	0.0507	0.0529	0.0609
0.6	0.0315	0.0528	0.0599	0.0656	0.0687	0.0713	0.0745	0.0857
0.7	0.0416	0.0689	0.0785	0.0859	0.0900	0.0934	0.0977	0.1126
0.8	0.0511	0.0847	0.0970	0.1062	0.1114	0.1157	0.1210	0.1399
0.9	0.060	0.099	0.114	0.125	0.132	0.137	0.144	0.167

1.6 Discussions

In the present analysis the magnification factors $F(\lambda)$ and $F'(\lambda)$ are given as power series of a relative crack length λ, and their coefficients are evaluated exactly from closed form expressions. Therefore, they correspond to the first terms of the Maxlaurin expansions. An example has been shown by equations (1.69) and (1.70) for the case of a periodic array of collinear cracks.

The accuracy of the present analysis has been established in those cases

TABLE 1.8
A pair of equal collinear cracks

$\dfrac{2a}{d}$	F_A		F_B	
	(1.53) $S = 43$	*(1.67)*	*(1.53)* $S = 43$	*(1.67)*
0.1	1.001322	1.001322	1.001196	1.001196
0.2	1.005660	1.005660	1.004624	1.004624
0.3	1.013831	1.013831	1.010167	1.010167
0.4	1.027171	1.027171	1.017868	1.017868
0.5	1.047960	1.047960	1.027953	1.027953
0.6	1.080404	1.080404	1.040937	1.040937
0.7	1.133262	1.133262	1.057865	1.057865
0.8	1.228931	1.228935	1.081066	1.081067
0.9	1.4521	1.453869	1.11726	1.117412

TABLE 1.9
Three equal collinear cracks

$\dfrac{2a}{d}$	F_A		F_B		F_C	
	(1.53) $S = 62$	Sih (1.68)	(1.53) $S = 62$	Sih (1.68)	(1.53) $S = 62$	Sih (1.68)
0.1	1.002518	1.002518	1.001645	1.001645	1.001503	1.001503
0.2	1.010297	1.010297	1.007017	1.007017	1.005849	1.005849
0.3	1.024070	1.024070	1.017100	1.017100	1.012959	1.012959
0.4	1.045291	1.045291	1.033530	1.033530	1.022973	1.022973
0.5	1.076630	1.076630	1.059132	1.059132	1.036311	1.036311
0.6	1.123160	1.123160	1.099151	1.099151	1.053831	1.053831
0.7	1.195578	1.195578	1.164559	1.164559	1.077235	1.077235
0.8	1.321359	1.321359	1.283484	1.283484	1.110316	1.110316
0.9	1.6066	1.606847	1.5643	1.564536	1.164394	1.164389

TABLE 1.10
F of periodic collinear cracks

$\dfrac{2a}{d}$	(1.53) S = 74	Westergaard (1.69)
0.1	1.004145	1.004145
0.2	1.016982	1.016982
0.3	1.039830	1.039830
0.4	1.075327	1.075327
0.5	1.128379	1.128379
0.6	1.208465	1.208465
0.7	1.336005	1.336005
0.8	1.564974	1.564974
0.9	2.11317	2.113307

where the corresponding closed form solutions are available. Tables 1.8 to 1.10 show the comparison of the present results with their exact counterparts for a pair of equal collinear cracks, three equal collinear cracks and a periodic array of collinear cracks given by equation (1.67), a special case of equation (1.68) and equation (1.69), respectively.

In Table 1.8, A and B denote the inner and outer crack tips, and in Table 1.9 A, B, C represent the inner, middle and outer crack tips respectively. The agreement of the present results with the exact solutions are quite remarkable for $\lambda \leqq 0.8$.

The source of errors in the present analysis is the truncation of the infinite series, but no theoretical means of estimating the upper bounds of the errors is available at the present time. However, according to the author's experience in various problems treated in this paper as well as those in previous works based on the perturbation technique, it appears reasonable to estimate the upper bounds of the numerical errors by assuming some geometic series for the uncalculated terms of higher orders. Another way is to check the partial sums of the obtained series in equations (1.53). An example for the periodic collinear cracks is given in Table 1.11, showing that S may be taken as 10, 20 and 30 in order to get reasonable results for $\lambda = 0.7$, 0.8 and 0.9, respectively. The numerical results in the preceding section have been obtained by assuming $S = 74$ for periodic cracks and $S = 43$ for the other cases, and they are considered to be sufficiently accurate when $\lambda = 0.95$.

For general arrays of cracks, the proposed method appears to give accurate results as long as λ is less than and not too close to unity, where λ is

TABLE 1.11

Partial sums of series (1,53) for F of periodic collinear cracks

λ \ s	10	20	30	40	50	60	70	Eq. (1.69)
0.1	1.00414	1.00414	1.00414	1.00414	1.00414	1.00414	1.00414	1.00414
0.2	1.01698	1.01698	1.01698	1.01698	1.01698	1.01698	1.01698	1.01698
0.3	1.03983	1.03983	1.03983	1.03983	1.03983	1.03983	1.03983	1.03983
0.4	1.07532	1.07533	1.07533	1.07533	1.07533	1.07533	1.07533	1.07533
0.5	1.12832	1.12838	1.12838	1.12838	1.12838	1.12838	1.12838	1.12838
0.6	1.2078	1.20846	1.20846	1.20846	1.20846	1.20846	1.20846	1.20846
0.7	1.331	1.3359	1.33600	1.33601	1.33601	1.33601	1.33601	1.33601
0.8	1.53	1.562	1.5647	1.56495	1.56497	1.56497	1.56497	1.56497
0.9	—	2.04	2.093	2.107	2.1113	2.1127	2.1131	2.11331

the relative crack length defined as the maximum value of $(a_j + a_k)/\rho_{jk}$. This limitation does not prevail in the case of collinear cracks. For parallel cracks, however, λ may take any large values, and the present analysis no longer applies. The difficulty may be resolved by directly solving the simultaneous equations instead of the perturbation precedure.

References

[1] Isida, M., Analysis of Stress Intensity Factors for Plates Containing Random Array of Cracks, *Bulletin of the JSME*, 13, 59, pp. 635–642 (1970).

[2] Isida, M., Method of Laurent Series Expansion for Internal Crack Problems, *Mechanics of Fracture I*, edited by G. C. Sih, Noordhoff International Publishing, Leiden, pp. 56–130 (1973).

[3] Moriguti, S., Plane Problems of Elasticity (in Japanese), Iwanami-koza, B7-a, Iwanami-shoten (1957).

[4] Williams, M. L., The Bending Stress Distribution at the Base of a Stationary Crack, *Journal of Applied Mechanics*, 28, Trans. ASME, E, 82, pp. 78–82 (1961).

[5] Sih, G. C., Paris, P. C. and Erdogan, F., Crack-Tip Stress-Intensity Factors for Plane Extension and Plate Bending Problems, *Journal of Applied Mechanics*, 29, Trans. ASME, E, 83, pp. 306–312 (1962).

[6] Isida, M., On the Stress Function in the Plane Problems of an Elastic Body Containing a Free Elliptic Hole, *Transactions of the Japan Society of Mechanical Engineers*, 21, 107, pp. 502–506 (1955).

[7] Hayashi, T., *Transactions of the Japan Society of Mechanical Engineers*, 25, 159, p. 1133, Ref. (8) (1959).

[8] Yokobori, T., Ohashi, M. and Ichikawa, M., The Interaction of Two collinear Asymmetrical Elastic Cracks, *Reports of the Research Institute for Strength and Fracture of Materials*, Tohoku University, 1, 2, pp. 33–39 (1965).

[9] Erdogan, F., On the stress distribution in plates with collinear cuts under arbitrary loads, *Proceedings of the 4th U.S. National Congress of Applied Mechanics*, Berkeley, 1, pp. 547–553 (1962).

[10] Sih, G. C., Boundary Problems for Longitudinal Shear Cracks, *Proceedings of the Second Conference on Theoretical and Applied Mechanics*, Pergamon Press, pp. 117–130 (1964).

[11] Westergaard, H. M., Bearing Pressure and Cracks, *Journal of Applied Mechanics*, Trans. ASME, pp. A49–53 (1939).

[12] Nisitani, H., *Journal of the Japan Society of Mechanical Engineers*, 71, 589, p. 209 (1968).

R. J. Hartranft

2 | *Improved approximate theories of the bending and extension of flat plates*

2.1 Introduction

The analysis of thin plates for stresses and deformations is a problem whose technical importance has grown for a century. Design refinements and new applications to ships, planes, and spacecraft require improved methods of analysis. As design specifications call for materials of higher strength, the tolerance of structures to flaws is reduced. Thus the design process must include a fracture mechanics analysis. But, although classical plate theories are quite appropriate for ordinary problems without stress singularities, more accurate theories are necessary to adequately model the behavior of a plate near a crack tip.

Linear elasticity formulation.   Although material and kinematical nonlinearities make caution necessary, the linear theory of elasticity would be expected to provide significant results. The particular problem of interest is that of an infinite plate containing a crack and loaded uniformly at infinity. The equations of elasticity are to be solved subject to boundary conditions, which may be written with reference to Figure 2.1 as

$$|y| \to \infty: \tau_{yx} = \tau_{yz} = 0, \quad \sigma_y = \sigma_0 + \frac{12}{h^3} M_0 z \qquad (2.1a)$$

$$|x| \to \infty: \sigma_x = \tau_{xy} = \tau_{xz} = 0 \qquad (2.1b)$$

$$|z| = \frac{h}{2}: \quad \tau_{zx} = \tau_{zy} = \sigma_z = 0 \qquad (2.1c)$$

$$y = 0, |x| < a: \quad \tau_{yx} = \sigma_y = \tau_{yz} = 0 \qquad (2.1d)$$

If the symmetry of the problem is exploited, and superposition is used, the solution of the problem of Figure 2.1 is reduced to the solution of one with the boundary conditions

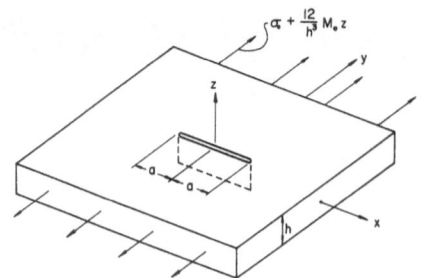

Figure 2.1. Uniformly loaded, cracked, infinite plate

$$y \to \infty: \quad \tau_{yx} = \sigma_y = \tau_{yz} = 0 \tag{2.2a}$$

$$x \to \infty: \quad \sigma_x = \tau_{xy} = \tau_{xz} = 0 \tag{2.2b}$$

$$x = 0: \quad u_x = \tau_{xy} = \tau_{xz} = 0 \tag{2.2c}$$

$$y = 0: \quad \begin{cases} x < a: \tau_{yx} = \tau_{yz} = 0, \sigma_y = -\sigma_0 - \dfrac{12}{h^3} M_0 z \\[2mm] a < x: \tau_{yx} = \tau_{yz} = 0, u_y = 0 \end{cases} \tag{2.2d}$$

The problem of Figure 2.1 is solved by adding the solution above to

$$u_x = -\frac{v}{E} x \left[\sigma_0 + \frac{12}{h^3} M_0 z \right] \tag{2.3a}$$

$$u_y = \frac{1}{E} y \left[\sigma_0 + \frac{12}{h^3} M_0 z \right] \tag{2.3b}$$

$$u_z = -\frac{v}{E} z\sigma_0 + \frac{6M_0}{Eh^3} \left[v(x^2 - z^2) - y^2 \right] \tag{2.3c}$$

$$\sigma_y = \sigma_0 + \frac{12}{h^3} M_0 z \tag{2.3d}$$

$$\sigma_x = \sigma_z = \tau_{xy} = \tau_{xz} = \tau_{yz} = 0 \tag{2.3e}$$

where E and v are Young's modulus and Poisson's ratio for the material.

In equations (2.1) through (2.3), and in subsequant work, σ_0 represents the constant stress at infinity for the extension of the plate and M_0 is the uniform bending couple at infinity in the plate bending problem. Equations (2.3) comprise the linear elastic solution of the problem of Figure 2.1 when no crack is present.

Approximations of linear elasticity. There are several ways of obtaining approximations of the solutions of plate problems. There is the obvious resort to the numerical techniques associated with finite difference or finite element methods. This approach may be characterized as seeking approximations of the solution of the exact equations. This chapter is concerned with the alternative approaches which are characterized as seeking exact solutions of approximating equations. Most of these approaches involve manipulations of the transverse variable, z, to obtain a set of equations in the remaining two in-plane variables.

The theory of generalized plane stress is an example of the technique of averaging the equations of elasticity with respect to the transverse variations. The averages of the stresses, τ_{xz}, τ_{yz}, and σ_z, are then neglected in the resulting equations. Extensions of this method [1–3] use averages weighted by various powers of z. These weighted averages, or moments, are specified by a set of partial differential equations in x and y.

Equations in two-dimensions for the coefficients of series expansions of the solution have also been obtained. In [4], expansions in both power series and trigonometric series in the variable z are used. As indicated in [1], the accuracy of the solution cannot be readily estimated in terms of powers of the plate thickness, h. Expansions in powers of plate thickness have also been used [5, 6]. These lead to boundary layer analyses to compensate for the fact that the boundary conditions of the problems can only be satisfied approximately.

Variational methods have been applied as a numerical technique for obtaining approximate solutions and as a general technique for obtaining approximate equations. Reissner's theory of plate bending has been derived from the theorem of minimum complementary potential energy [7] and from a theorem which permits both stress and displacement to be varied [8]. A modified theory of generalized plane stress has also been obtained by variational methods [9]. The next section contains a more complete discussion of variational methods.

2.2   Approximate theories by variational methods

There are three major variational principles available for obtaining approximate theories. In the first, a class of displacement functions having a suitable form is examined for the one which minimizes the potential energy. If the class is general enough, the exact solution may be found in this way. More often, however, the class of functions examined is restricted, and the minimizing function is an approximation to the exact solution. The fewer the restrictions on the class examined, the better the approximation, and the closer to the actual minimum value of the potential energy one arrives. The techniques used to obtain approximate theories lead to differential equations which the minimizing functions must satisfy. The equations for this first principle may be regarded as approximate equilibrium equations. The stress-displacement equations are satisfied exactly.

The second principle requires a search for the stress state, of all those which satisfy the equilibrium equations, which minimizes the complementary potential energy. Again, restrictions on the class of functions considered leads to a set of differential equations for an approximate stress field. These equations are approximations of the stress-displacement relationships.

In the third principle [8], displacements and stresses, without regard to equilibrium, are examined for those which minimize a certain functional related to the strain energy of the body. The equations obtained approximate the elasticity equations of equilibrium and the stress-displacement equations.

These same three principles form the basis for the stiffness, force, and hybrid finite element methods. Whatever the application, the choice of principle depends on what are considered to be the important features of the problem. If a continuous displacement field is of primary interest, minimum potential energy is appropriate. If local equilibrium is critical, minimum complementary potential energy is indicated. The third principle [8] would be chosen to enable a particular form to be used for both stress and displacement. It is possible, in addition, to incorporate other features if they can be formulated as constraints to be included by the method of Lagrange multipliers [10].

Reissner's varitional principle.   The third principle discussed above is usually named after its originator [8]. In the following statement of the principle, certain combinations of derivatives of displacement could be expressed in terms of the strains, but for clarity of thought, strain is not explicitly used. Consider the expression,

$$R = \int_\tau [T_x \cdot u_{,x} + T_y \cdot u_{,y} + T_z \cdot u_{,z} - 2V - (T_{,x} + T_{,y} + T_{,z} + 2F) \cdot u] d\tau - \int_{\Sigma_T} T_n^* \cdot u d\Sigma + \int_{\Sigma_u} T_n \cdot u^* d\Sigma \qquad (2.4)^*$$

The components of displacement are

$$u = u_x i + u_y i + u_z k \qquad (2.5)$$

while the stress components are the components of the three stress vectors,

$$T_x = \sigma_x i + \tau_{xy} j + \tau_{xz} k \qquad (2.6a)$$

$$T_y = \tau_{yx} i + \sigma_y j + \tau_{yz} k \qquad (2.6b)$$

$$T_z = \tau_{zx} i + \tau_{zy} j + \sigma_z k \qquad (2.6c)$$

If n is the outward normal to the surface,

$$T_n = T_x n_x + T_y n_y + T_z n_z \qquad (2.7)$$

is the force per unit area acting on the surface. The volume occupied by the body is denoted by τ and its surface by Σ. The two parts of the surface, Σ_T and Σ_u, are those on which tractions, T_n^*, and displacements, u^*, are specified, respectively. The strain energy density for a linear elastic, isotropic, homogeneous material is

$$V = \frac{1}{2E} (\sigma_x^2 + \sigma_y^2 + \sigma_z^2) - \frac{v}{E} (\sigma_x\sigma_y + \sigma_x\sigma_z + \sigma_y\sigma_z)$$

$$+ \frac{1 + v}{E}(\tau_{xy}^2 + \tau_{xz}^2 + \tau_{yz}^2) \qquad (2.8)$$

For other materials, the expression for V should be the function of the stress components whose derivatives equal the strains. Finally, F is the body force per unit volume.

The principle states that R (equation (2.4)) is an extremum** for a partic-

---

* Differentiation of a quantity is indicated by a comma followed by the variables with respect to which the derivative is to be taken.
This expression differs from that in Reissner's original paper. This has the advantage that the treatment of boundary conditions on displacement is similar to that on tractions.

** An extremum could be a minimum, but it includes any state in which a small, first-order change of stress and displacement produces only a second-order change in R.

ular stress and displacement field if, and only if, the stresses and displacements satisfy the equations of elasticity,

$$T_{x,x} + T_{y,y} + T_{z,z} + F = 0 \tag{2.9}$$

and

$$Eu_{x,x} = \sigma_x - v(\sigma_y + \sigma_z), \quad E(u_{x,y} + u_{y,x}) = 2(1 + v)\tau_{xy}$$

$$Eu_{y,y} = \sigma_y - v(\sigma_x + \sigma_z), \quad E(u_{x,z} + u_{z,x}) = 2(1 + v)\tau_{xz} \tag{2.10}$$

$$Eu_{z,z} = \sigma_z - v(\sigma_x + \sigma_y), \quad E(u_{y,z} + u_{z,y}) = 2(1 + v)\tau_{yz}$$

and the boundary conditions, $u = u^*$ on Σ_u and $T_n = T^*$ on Σ_T.

This is the principle which will be used in this chapter, for it permits one to discuss both stresses and displacements directly. The others involve averages of one or the other, and, although the results can be equivalent, the physical meaning is not clear. For example, in [9], the principle of minimum complementary energy was used, and weighted averages of the displacements were introduced. The discussion in this chapter results in the same set of equations, but with explicit expressions for the displacements.

Reissner's theory for plate bending. The first of the improved theories was published in the 1940's. It is restricted to plate bending. One considers the class of displacements and stresses given by

$$u_x = \beta_x z \qquad u_y = \beta_y z \qquad u_z = w$$

$$\sigma_x = \frac{12}{h^3} M_x z \qquad \sigma_y = \frac{12}{h^3} M_y z \qquad \tau_{xy} = \frac{12}{h^3} M_{xy} z$$

$$\tag{2.11}$$

$$\tau_{xz} = \frac{3}{2h} V_x[1 - (2z/h)^2] \qquad \tau_{yz} = \frac{3}{2h} V_y[1 - (2z/h)^2]$$

$$\sigma_z = -\frac{p}{4}[2 + 3(2z/h) - (2z/h)^3]$$

where β_x, β_y, w, M_x, M_y, M_{xy}, V_x, V_y and the pressure p on the top ($z = h/2$) of the plate are functions of x and y, only. Since the z dependence is explicit, the integration with respect to z can be performed, and the expression (2.4) for R becomes, for $F = 0$,

$$R = \int_A dA[M_x\beta_{x,x} + M_y\beta_{y,y} + M_{xy}(\beta_{x,y} + \beta_{y,x})$$

$$+ V_x(\beta_x + w_{,x}) + V_y(\beta_y + w_{,y}) + pw$$

$$- (M_{x,x} + M_{xy,y} - V_x)\beta_x - (M_{xy,x} + M_{y,y} - V_y)\beta_y]$$

$$- \frac{1}{(1 - v^2)D} \int_A dA[M_x^2 + M_y^2 - 2vM_xM_y + 2(1 + v)M_{xy}^2 \qquad (2.12)$$

$$+ 2(1 + v)\alpha^2(V_x^2 + V_y^2) + 2v\alpha^2 p(M_x + M_y) + \tfrac{65}{21}\alpha^4 p^2]$$

$$- \int_{S_T} ds[M_n^*\beta_n + M_{ns}^*\beta_s + V_n^*w]$$

$$- \int_{S_u} ds[M_n\beta_n^* + M_{ns}\beta_s^* + V_nw^*]$$

where

$$D = \frac{Eh^3}{12(1 - v^2)}, \ \alpha = h/\sqrt{10} \qquad (2.13)$$

The terms in the line integral over the portion, S_T, of the edge on which stresses are specified are given by

$$M_n^* = \int_{-h/2}^{h/2} \sigma_n^*z\,dz \qquad (2.14a)$$

$$M_{ns}^* = \int_{-h/2}^{h/2} \tau_{ns}^*z\,dz \qquad (2.14b)$$

$$V_n^* = \int_{-h/2}^{h/2} \tau_{nz}^*\,dz \qquad (2.14c)$$

and

$$\beta_n = \beta_x \cos \phi + \beta_y \sin \phi \qquad (2.15a)$$

$$\beta_s = - \beta_x \sin \phi + \beta_y \cos \phi \qquad (2.15b)$$

The stress components, σ_n^*, τ_{ns}^*, and τ_{nz}^*, are shown in Figure 2.2.

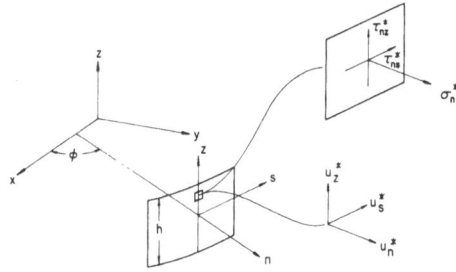

Figure 2.2. Stresses and displacements on the edge of a flat plate

In the line integral over the portion, S_u, where displacements are specified,

$$M_n = \tfrac{1}{2}(M_x + M_y) + \tfrac{1}{2}(M_x - M_y)\cos 2\phi + M_{xy}\sin 2\phi \qquad (2.16a)$$

$$M_{ns} = -\tfrac{1}{2}(M_x - M_y)\sin 2\phi + M_{xy}\cos 2\phi \qquad (2.16b)$$

$$V_n = V_x\cos\phi + V_y\sin\phi \qquad (2.16c)$$

and

$$\beta_n^* = \frac{12}{h^3}\int_{-h/2}^{h/2} u_n^* z\,dz, \qquad (2.17a)$$

$$\beta_s^* = \frac{12}{h^3}\int_{-h/2}^{h/2} u_s^* z\,dz \qquad (2.17b)$$

$$w^* = \frac{3}{2h}\int_{-h/2}^{h/2} u_z^*[1 - (2z/h)^2]\,dz \qquad (2.17c)$$

The displacements, u_n^*, u_s^*, and u_z^*, are shown in Figure 2.2.

A notable feature which recurrs in plate theories is the fact that some details of the variation of the boundary stresses and displacements disappear. In this case, only the integrals (2.14) and (2.17) remain, and any z variation of σ_n^*, τ_{ns}^*, and τ_{nz}^* or u_n^*, u_s^*, and u_z^* is absent from R. As a result, the theory which follows assigns the same solution to two loadings which both have the same weighted averages as given by equations (2.14) and (2.17). The connection between the weighted averages and the unknown functions in equations (2.11) is given by the natural boundary conditions of the theory.

Consider, for example, the variation [10] of the first two terms in the expression (2.12) for R. The variations of the derivatives are eliminated as follows by application of the divergence theorem:

$$\delta\int_A [M_x\beta_{x,x} + M_y\beta_{y,y}]\,dA = \int_A [\beta_{x,x}\delta M_x + \beta_{y,y}\delta M_y + M_x\delta\beta_{x,x}$$

$$+ M_y\delta\beta_{y,y}]\,dA = \int_A [\beta_{x,x}\delta M_x + \beta_{y,y}\delta M_y - M_{x,x}\delta\beta_x - M_{y,y}\delta\beta_y$$

$$+ (M_x\delta\beta_x)_{,x} + (M_y\delta\beta_y)_{,y}]\,dA = \int_A [\beta_{x,x}\delta M_x + \beta_{y,y}\delta M_y - M_{x,x}\delta\beta_x$$

$$- M_{y,y}\delta\beta_y]\,dA + \int_S [M_x\delta\beta_x\cos\phi + M_y\delta\beta_y\sin\phi]\,ds$$

Through similar procedures for the variations of the other derivatives, and by applying equations (2.15) and (2.16), the variation of equation (2.12) may be written in the form,

$$0 = \delta R = 2\int_A dA\{\delta M_x[\beta_{x,x} - (M_x - \nu M_y + \nu\alpha^2 p)/(1 - \nu^2)D]$$

$$+ \delta M_y[\beta_{y,y} - (M_y - \nu M_x + \nu\alpha^2 p)/(1 - \nu^2)D]$$

$$+ \delta M_{xy}[\beta_{x,y} + \beta_{y,x} - 2M_{xy}/(1 - \nu)D]$$

$$+ \delta V_x[\beta_x + w_{,x} - 2\alpha^2 V_x/(1 - \nu)D]$$

$$+ \delta V_y[\beta_y + w_{,y} - 2\alpha^2 V_y/(1 - \nu)D]$$

$$+ \delta w[p - V_{x,x} - V_{y,y}]$$

$$+ \delta\beta_x[V_x - M_{x,x} - M_{xy,y}]$$

$$+ \delta\beta_y[V_y - M_{xy,x} - M_{y,y}]\}$$

$$+ \int_{S_T} ds\{[(M_n - M_n^*)\delta\beta_n + (M_{ns} - M_{ns}^*)\delta\beta_s + (V_n - V_n^*)\delta w]$$

$$- [\beta_n\delta M_n + \beta_s\delta M_{ns} + w\delta V_n]\}$$

$$- \int_{S_u} ds\{[(\beta_n - \beta_n^*)\delta M_n + (\beta_s - \beta_s^*)\delta M_{ns} + (w - w^*)\delta V_n]$$

$$- [M_n\delta\beta_n + M_{ns}\delta\beta_s + V_n\delta w]\} \tag{2.18}$$

Because the eight variations, $\delta M_x, \ldots \delta\beta_y$ are arbitrary in A, their coefficients in equation (2.18) must be separately zero. Thus the basic set of eight first order, linear, partial differential equations are obtained. The boundary integrals in (2.18) will be zero if

$$M_n = M_n^*, \ M_{ns} = M_{ns}^*, \ V_n = V_n^* \text{ on } S_T \tag{2.19a}$$

$$\beta_n = \beta_n^*, \ \beta_s = \beta_s^*, \ w = w^* \text{ on } S_u \tag{2.19b}$$

for then

$$\delta M_n = \delta M_{ns} = \delta V_n = 0 \text{ on } S_T \tag{2.20a}$$

$$\delta\beta_n = \delta\beta_s = \delta w = 0 \text{ on } S_u \tag{2.20b}$$

Generalization of Reissner's theory. If, in the previous discussion, the assumed stresses and displacements of equations (2.11) are replaced by

$$u_x = \beta_x z, \ u_y = \beta_y z, \ u_z = w \tag{2.21}$$

and

$$\sigma_x = M_x f''(z), \ \sigma_y = M_y f''(z), \ \tau_{xy} = M_{xy} f''(z) \tag{2.22a}$$

$$\tau_{xz} = - V_x f'(z), \ \tau_{yz} = - V_y f'(z), \ \sigma_z = p f(z) \tag{2.22b}$$

where

$$f'(\pm h/2) = 0, \quad f(h/2) = -1, \quad f(-h/2) = 0$$

the differential equations and boundary conditions implied by equation (2.18) are reproduced. The meaning of some terms is changed. In equation (2.12), D and α represent

$$D = \frac{Eh^3}{12(1-v^2)} \frac{1}{I_1}, \quad \alpha^2 = \frac{h^2}{10} \frac{I_2}{I_1} \tag{2.23}$$

where

$$I_1 = \frac{h^3}{12} \int_{-h/2}^{h/2} [f''(z)]^2 dz, \quad I_2 = \frac{5h}{6} \int_{-h/2}^{h/2} [f'(z)]^2 dz$$

The quantities involved in the traction boundary conditions are again given by equations (2.14). But the displacement integrals which must be given on the portion, S_u, of the edge are altered from equations (2.17). They are

$$\beta_n^* = \int_{-h/2}^{h/2} u_n^* f''(z) dz, \quad \beta_s^* = \int_{-h/2}^{h/2} u_s^* f''(z) dz,$$

$$w^* = -\int_{-h/2}^{h/2} u_z^* f'(z) dz \tag{2.24}$$

It may be noted that if the specified stresses are of the form

$$\sigma_n^* = \overline{M}_n f''(z), \quad \tau_{ns}^* = \overline{M}_{ns} f''(z), \quad \tau_{nz}^* = -\overline{V}_n f'(z) \tag{2.25}$$

then equations (2.14) verify the expectation that

$$M_n^* = \overline{M}_n, \quad M_{ns}^* = \overline{M}_{ns}, \quad V_n^* = \overline{V}_n \tag{2.26}$$

However, it is not necessary that the boundary stresses have the transverse variation of equations (2.25). As pointed out in the discussion of Reissner's theory, two sets of boundary stresses with the same weighted averages defined by equations (2.14) will lead to the same solution.

Except for notation this theory is the same as that proposed in [11, 12]. In [11], except for some minor differences in notation and some serious misprints, the equations of [11, §2] are equivalent to the above. The symbols A and $B = C$ in [11] are denoted by $12I_1/h^3$ and $1.2I_2/h$, respectively in this article. The equations in [12], after an inversion of the coordinate system, are identical to those of this article if $f_3(z) = f(z)$ and $k_1 = 12I_1/h^3$, $k_2 = 1.2I_2/h$. Reissner's theory is obtained if

$$f(z) = \frac{1}{4}\left(\frac{2z}{h}\right)^3 - \frac{3}{4}\left(\frac{2z}{h}\right) - \frac{1}{2} \tag{2.27}$$

Extended theories of generalized plane stress. The variational principles can be used to obtain two-dimensional, approximate theories for in-plane deformation of plates. The equations of generalized plane stress are usually thought of in terms of averages of equations (2.9) and (2.10). However, as will be seen, the variational method leads to the same equations. The extended theory in [9], as well as a combination discussed in [13], are also presented from a new point of view in this section.

The forms chosen for approximating the displacements and stresses, in terms of an arbitrary function, $f(z)$, are

$$u_x = v_x + \frac{h^3}{12I_3}\, U_x f''(z) \tag{2.28a}$$

$$u_y = v_y + \frac{h^3}{12I_3}\, U_y f''(z) \tag{2.28b}$$

$$u_z = -\frac{h}{2I_2}\, U_z f'(z) \tag{2.28c}$$

and

$$\sigma_x = s_x + S_x f''(z), \quad \tau_{xz} = -Z_x f'(z)$$
$$\sigma_y = s_y + S_y f''(z), \quad \tau_{yz} = -Z_y f'(z) \tag{2.29}$$
$$\tau_{xy} = t_{xy} + T_{xy} f''(z), \quad \sigma_z = Z_z f(z)$$

where $f(\pm h/2) = f'(\pm h/2) = 0$. The two sets of unknowns, v_x, v_y, s_x, s_y, t_{xy} and U_x, U_y, U_z, S_x, S_y, T_{xy}, Z_x, Z_y, Z_z are functions only of the in-plane variables, x and y. If the second set are chosen to be zero, equations (2.28) and (2.29) reduce to the form appropriate for generalized plane stress. If the first set are zero, the form which results in the theory of [9] remains. Except for notation, equations (2.28) and (2.29) are those proposed in [13].

When equations (2.28) and (2.29) are substituted into equation (2.4), and the integrations across the thickness performed, R is found to be

$$R = \int_A dA\{h[s_x v_{x,\,x} + s_y v_{y,\,y} + t_{xy}(v_{x,\,y} + v_{y,\,x})] + S_x U_{x,\,x} + S_y U_{y,\,y}$$
$$+ T_{xy}(U_{x,\,y} + U_{y,\,x}) + Z_x(U_x + U_{z,\,x}) + Z_y(U_y + U_{z,\,y}) + Z_z U_z\}$$

$$- \frac{h}{E} \int_A dA \{s_x^2 + s_y^2 - 2vs_x s_y + 2(1+v)t_{xy}^2 + 2ve^2 Z_z(s_x + s_y)\}$$

$$- \frac{1}{(1-v^2)D} \int_A dA \{S_x^2 + S_y^2 - 2vS_x S_y + 2(1+v)T_{xy}^2$$

$$+ 2v\alpha^2 Z_z(S_x + S_y) + 2(1+v)\alpha^2 (Z_x^2 + Z_y^2) + (\beta^2+1)\alpha^4 Z_z^2\}$$

$$- \int_A dA \{h[(s_{x,x} + t_{xy,y})v_x + (t_{xy,x} + s_{y,y})v_y] + (S_{x,x} + T_{xy,y} - Z_x)U_x$$

$$+ (T_{xy,x} + S_{y,y} - Z_y)U_y + (Z_{x,x} + Z_{y,y} - Z_z)U_z\}$$

$$- \int_{S_T} ds \{h[s_n^* v_n + t_{ns}^* v_s] + S_n^* U_n + T_{ns}^* U_s + Z_n^* U_z\}$$

$$+ \int_{S_u} ds \{h[s_n v_n^* + t_{ns} v_s^*] + S_n U_n^* + T_{ns} U_s^* + Z_n U_z^*\} \qquad (2.30)$$

where

$$D = \frac{Eh^3}{12(1-v^2)} \frac{1}{I_3}, \ \alpha^2 = \frac{h^2}{6} \frac{I_2}{I_3}$$

$$\qquad (2.31)$$

$$\beta^2 = \frac{3I_1 I_3}{2I_2^2} - 1, \ e^2 = -I_4$$

where

$$I_1 = \frac{2}{h} \int_{-h/2}^{h/2} [f(z)]^2 dz, \ I_2 = \frac{h}{2} \int_{-h/2}^{h/2} [f'(z)]^2 dz$$

$$I_3 = \frac{h^3}{12} \int_{-h/2}^{h/2} [f''(z)]^2 dz, \ I_4 = \frac{1}{h} \int_{-h/2}^{h/2} f(z) dz$$

The specified quantities in the boundary integrals of equation (2.30) are

$$s_n^* = \frac{1}{h} \int_{-h/2}^{h/2} \sigma_n^* dz, \ t_{ns}^* = \frac{1}{h} \int_{-h/2}^{h/2} \tau_{ns}^* dz$$

$$S_n^* = \frac{h^3}{12I_3} \int_{-h/2}^{h/2} \sigma_n^* f''(z) dz \qquad (2.32a)$$

$$T_{ns}^* = \frac{h^3}{12I_3} \int_{-h/2}^{h/2} \tau_{ns}^* f''(z) dz \qquad (2.32b)$$

$$Z_n^* = -\frac{h}{2I_2} \int_{-h/2}^{h/2} \tau_{nz}^* f'(z) dz \qquad (2.32c)$$

on S_T and

$$v_n^* = \frac{1}{h}\int_{-h/2}^{h/2} u_n^* dz, \quad v_s^* = \frac{1}{h}\int_{-h/2}^{h/2} u_s^* dz$$

$$U_n^* = \int_{-h/2}^{h/2} u_n^* f''(z)dz, \quad U_s^* = \int_{-h/2}^{h/2} u_s^* f''(z)dz \tag{2.33}$$

$$U_z^* = -\int_{-h/2}^{h/2} u_z^* f'(z)dz,$$

on S_u. See Figure 2.2 for the stresses and displacements on the boundary. The remaining terms in the boundary integrals are related to the unknowns in equations (2.28) and (2.29) by

$$v_n = v_x \cos \phi + v_y \sin \phi, \; v_s = -v_x \sin \phi + v_y \cos \phi$$

$$\tag{2.34}$$

$$U_n = U_x \cos \phi + U_y \sin \phi, \; U_s = -U_x \sin \phi + U_y \cos \phi$$

and

$$s_n = \tfrac{1}{2}(s_x + s_y) + \tfrac{1}{2}(s_x - s_y)\cos 2\phi + t_{xy}\sin 2\phi \tag{2.35a}$$

$$t_{ns} = -\tfrac{1}{2}(s_x - s_y)\sin 2\phi + t_{xy}\cos 2\phi \tag{2.35b}$$

$$S_n = \tfrac{1}{2}(S_x + S_y) + \tfrac{1}{2}(S_x - S_y)\cos 2\phi + T_{xy}\sin 2\phi \tag{2.35c}$$

$$T_{ns} = -\tfrac{1}{2}(S_x - S_y)\sin 2\phi + T_{xy}\cos 2\phi \tag{2.35d}$$

$$Z_n = Z_x \cos \phi + Z_y \sin \phi \tag{2.35e}$$

Application of the usual variational methods to equation (2.30) gives

$$\delta R = 2 \int_A dA\{h\delta s_x[v_{x,x} - (s_x - vs_y + ve^2 Z_z)/E]$$

$$+ h\delta s_y[v_{y,y} - (s_y - vs_x + ve^2 Z_z)/E]$$

$$+ h\delta t_{xy}[v_{x,y} + v_{y,x} - 2(1 + v)t_{xy}/E]$$

$$- h\delta v_x[s_{x,x} + t_{xy,y}] - h\delta v_y[t_{xy,x} + s_{y,y}]$$

$$+ \delta S_x[U_{x,x} - (S_x - vS_y + v\alpha^2 Z_z)/(1 - v^2)D]$$

$$+ \delta S_y[U_{y,y} - (S_y - vS_x + v\alpha^2 Z_z)/(1 - v^2)D]$$

$$+ \delta T_{xy}[U_{x,y} + U_{y,x} - 2T_{xy}/(1 - v)D]$$

$$+ \delta Z_x[U_x + U_{z,x} - 2\alpha^2 Z_x/(1 - v)D] + \delta Z_y[U_y + U_{z,y} - 2\alpha^2 Z_y/(1 - v)D]$$

$$+ \delta Z_z[U_z - ve^2 h(s_x + s_y)/E - \alpha^2\{(\beta^2 + 1)\alpha^2 Z_z + v(S_x + S_y)\}/(1 - v^2)D]$$

$$+ \delta U_x[Z_x - S_{x,x} - T_{xy,y}] + \delta U_y[Z_y - T_{xy,x} - S_{y,y}]$$

$$+ \delta U_z[Z_z - Z_{x,x} - Z_{y,y}]\}$$

$$+ \int_{S_T} ds\{h[(s_n - s_n^*)\delta v_n + (t_{ns} - t_{ns}^*)\delta v_s - (v_n \delta s_n + v_s \delta t_{ns})]$$

$$+ (S_n - S_n^*)\delta U_n + (T_{ns} - T_{ns}^*)\delta U_s + (Z_n - Z_n^*)\delta U_z - (U_n \delta S_n + U_s \delta T_{ns}$$

$$+ U_z \delta Z_n)\} - \int_{S_U} ds\{h[(v_n - v_n^*)\delta s_n + (v_s - v_s^*)\delta t_{ns} - (s_n \delta v_n$$

$$+ t_{ns}\delta v_s)] + (U_n - U_n^*)\delta S_n + (U_s - U_s^*)\delta T_{ns} + (U_z - U_z^*)\delta Z_n$$

$$- (S_n \delta U_n + T_{ns}\delta U_s + Z_n \delta U_z)\} \tag{2.36}$$

The variation δR in equation (2.36) must be zero for arbitrary variations δs_x, etc. Therefore in the area integral, each term in square brackets must be zero everywhere in A. Thus a set of fourteen equations for the unknown variables are obtained. One is an algebraic relationship, and the remaining thirteen are linear, first-order, partial differential equations. The equations of [13] are obtained if the displacement terms are eliminated and a stress function is introduced for s_x, s_y, and t_{xy}.

As was noted in [9] for the case in which $s_x = s_y = t_{xy} = v_x = v_y = 0$, replacing $f(z)$ by a multiple of $f(z)$ does not lead to new stresses and displacements in equations (2.28) and (2.29). Therefore, $f(z)$ can be multiplied by a constant chosen so that $I_3 = 1$ and $I_4 < 0$. For example, the function suggested in [13] would be normalized to the form

$$f(z) = -\frac{1}{16}\sqrt{15}\left[1 - \left(\frac{2z}{h}\right)^2\right]^2 \tag{2.37}$$

and give

$$I_1 = \frac{1}{21}, I_2 = \frac{1}{7}, I_3 = 1, I_4 = -\frac{\sqrt{15}}{30}$$

$$D = \frac{Eh^3}{12(1 - v^2)}, \alpha^2 = \frac{h^2}{42}, \beta^2 = \frac{5}{2}, e^2 = \frac{\sqrt{15}}{30} \tag{2.38}$$

In this connection, note that there has been a normalization in equations (2.22) making $f(h/2) = -1$. No additional requirement can be imposed.

The first five equations resulting from equating δR (equation (2.36)) to zero are nearly identical to those of generalized plane stress.

$$Ev_{x,x} = s_x - vs_y + ve^2 Z_z \tag{2.39a}$$

$$Ev_{y,y} = s_y - vs_x + ve^2 Z_z \tag{2.39b}$$

$$E(v_{x,y} + v_{y,x}) = 2(1 + v)t_{xy} \tag{2.39c}$$

$$s_{x,x} + t_{xy,y} = 0 \tag{2.39d}$$

$$t_{xy,x} + s_{y,y} = 0$$

The terms involving Z_z are the only ones coupling these equations to the remainder. The associated boundary conditions will require

$$s_n = s_n^*, \ t_{ns} = t_{ns}^* \text{ on } S_T \tag{2.40a}$$

$$v_n = v_n^*, \ v_s = v_s^* \text{ on } S_u \tag{2.40b}$$

Whatever the actual transverse variation of σ_n^*, τ_{ns}^*, u_n^*, and u_s^*, only the weighted averages defined by equations (2.32) and (2.33) affect the solution of the problem.

An algebraic equation also connects the two sets of unknowns.

$$(1 - v^2)DU_z = \alpha^2[(\beta^2 + 1)\alpha^2 Z_z + v(S_x + S_y)] + \frac{h^4}{12I_3} ve^2(s_x + s_y) \tag{2.41}$$

If the initial assumptions of equations (2.29) had used $Z_z = 0$, it would not be necessary to satisfy equation (2.41). In such a case the two sets of unknowns are completely decoupled. There would also be decoupling if Poisson's ratio were zero ($v=0$), and the remains of equation (2.41) would be included with the equations for the second set of inknowns.

If the initial assumptions included $v_x = v_y = s_x = s_y = t_{xy} = 0$, equations (2.39) would not be necessary. If $U_x = U_y = U_z = S_x = S_y = T_{xy} = Z_x = = Z_y = Z_z = 0$, then equation (2.41) and the equations (2.42) in the next paragraph would not have to be satisfied. In the general case both sets are required.

The remaining differential equations are

$$(1 - v^2)DU_{x,x} = S_x - vS_y + v\alpha^2 Z_z \tag{2.42a}$$

$$(1 - v^2)DU_{y,y} = S_y - vS_x + v\alpha^2 Z_z \tag{2.42b}$$

$$(1 - v^2)D(U_{x,y} + U_{y,x}) = 2(1 + v)T_{xy} \tag{2.42c}$$

$$(1 - v)D(U_x + U_{z,x}) = 2\alpha^2 Z_x \tag{2.42d}$$

$$(1 - v)D(U_y + U_{z,y}) = 2\alpha^2 Z_y \tag{2.42e}$$

$$S_{x,x} + T_{xy,y} = Z_x \tag{2.42f}$$

$$T_{xy,x} + S_{y,y} = Z_y \tag{2.42g}$$

$$Z_{x,x} + Z_{y,y} = Z_z \tag{2.42h}$$

The boundary conditions for these variables involve the stress and displacement averages defined in equations (2.32) and (2.33).

$$S_n = S_n^*, \; T_{ns} = T_{ns}^*, \; Z_n = Z_n^* \text{ on } S_T \tag{2.43a}$$

$$U_n = U_n^*, \; U_s = U_s^*, \; U_z = U_z^* \text{ on } S_u \tag{2.43b}$$

For those cases in which it is convenient to use a function of a dimensionless variable to describe the transverse variations, the following formulas are listed:

$$f(z) = f\left(\frac{h}{2}\zeta\right) = g(\zeta) = g\left(\frac{2}{h}z\right)$$

$$f'(z) = \frac{2}{h} g'(\zeta), f''(z) = \frac{4}{h^2} g''(\zeta)$$

$$I_1 = \int_{-1}^{1} [g(\zeta)]^2 d\zeta, \; I_2 = \int_{-1}^{1} [g'(\zeta)]^2 d\zeta \tag{2.44}$$

$$I_3 = \tfrac{2}{3} \int_{-1}^{1} [g''(\zeta)]^2 d\zeta, \; I_4 = \tfrac{1}{2} \int_{-1}^{1} g(\zeta) d\zeta$$

The Hartranft-Sih theory of [9] consists of equations (2.42) and (2.41) with $v_x = v_y = s_x = s_y = t_{xy} = 0$ in equations (2.28) and (2.29).

2.3 Applications to crack problems

The quantities of most importance in linear elastic fracture mechanics are the stress intensity factors, which appear in the crack front stress field singularity. Most work to date [14, 15] indicates that, except for the variation of the stress intensity factors along the crack front, the form of the stress field near the front is always the same. The only exceptions occur when classical plate or shell theories involving a Kirchoff type boundary condition are used [16, 17].

The improved theories considered in this chapter all result in some version of the singular behavior,

$$4Gu_x = k_1^D(2r)^{\frac{1}{2}} \cos\frac{\theta}{2}\left[\kappa - 1 + 2\sin^2\frac{\theta}{2}\right] \tag{2.45a}$$

$$+ k_2^D(2r)^{\frac{1}{2}} \sin\frac{\theta}{2}\left[\kappa + 1 + 2\cos^2\frac{\theta}{2}\right] + O(r)$$

$$4Gu_y = k_1^D(2r)^{\frac{1}{2}} \sin\frac{\theta}{2}\left[\kappa + 1 - 2\cos^2\frac{\theta}{2}\right]$$

$$- k_2^D(2r)^{\frac{1}{2}}\cos\frac{\theta}{2}\left[\kappa - 1 - 2\sin^2\frac{\theta}{2}\right] + O(r) \tag{2.45b}$$

$$Gu_z = k_3^D(2r)^{\frac{1}{2}}\sin\frac{\theta}{2} + O(r) \tag{2.45c}$$

for the displacements, and

$$\sigma_x = \frac{k_1}{(2r)^{\frac{1}{2}}}\cos\frac{\theta}{2}\left[1 - \sin\frac{\theta}{2}\sin\frac{3\theta}{2}\right] - \frac{k_2}{(2r)^{\frac{1}{2}}}\sin\frac{\theta}{2}\left[2 + \cos\frac{\theta}{2}\cos\frac{3\theta}{2}\right] + O(1) \tag{2.46a}$$

$$\sigma_y = \frac{k_1}{(2r)^{\frac{1}{2}}}\cos\frac{\theta}{2}\left[1 + \sin\frac{\theta}{2}\sin\frac{3\theta}{2}\right] + \frac{k_2}{(2r)^{\frac{1}{2}}}\sin\frac{\theta}{2}\cos\frac{\theta}{2}\cos\frac{3\theta}{2} + O(1) \tag{2.46b}$$

$$\sigma_z = 2\nu\left[\frac{k_1^T}{(2r)^{\frac{1}{2}}}\cos\frac{\theta}{2} - \frac{k_2^T}{(2r)^{\frac{1}{2}}}\sin\frac{\theta}{2}\right] + O(1) \tag{2.46c}$$

$$\tau_{xy} = \frac{k_1}{(2r)^{\frac{1}{2}}}\sin\frac{\theta}{2}\cos\frac{\theta}{2}\cos\frac{3\theta}{2} + \frac{k_2}{(2r)^{\frac{1}{2}}}\cos\frac{\theta}{2}\left[1 - \sin\frac{\theta}{2}\sin\frac{3\theta}{2}\right] + O(1) \tag{2.46d}$$

$$\tau_{xz} = -\frac{k_3}{(2r)^{\frac{1}{2}}}\sin\frac{\theta}{2} + O(1) \tag{2.46e}$$

$$\tau_{yz} = \frac{k_3}{(2r)^{\frac{1}{2}}}\cos\frac{\theta}{2} + O(1) \tag{2.46f}$$

for the stresses. Figure 2.3 shows the coordinate system and polar coordinates, r and θ, at the crack front. The xz-plane is the plane of the crack,

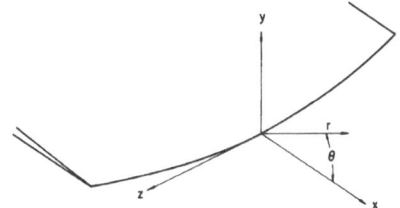

Figure 2.3. Coordinate system near the crack front

and the xy-plane is normal to the crack front. For the problems of this chapter, the z-axis coincides with the straight crack front.

In equations (2.45) for the displacements, $2G = E/(1+v)$, and $\kappa = 3-4v$ is the parameter which applies when the state near the crack front is one of plane strain. For generalized plane stress, $\kappa = (3-v)/(1+v)$. According to [14], an exact solution incorporates $\kappa = 3-4v$ and $k_i^D = k_i$ ($i = 1, 2, 3$). In addition, $k_i^T = k_i$ ($i = 1, 2$), which implies that $\sigma_z = v(\sigma_x+\sigma_y)$ as $r \to 0$. The various approximate theories fail to possess one or more of these characteristics.

Reissner's theory for plate bending. The sine or cosine transforms, as appropriate for a problem symmetric about the yz plane, are used to convert the equations implied by expression (2.18) into ordinary differential equations in the variable y [18]. These are satisfied by solutions of exponential form, and when the inverse transforms are applied, an integral form of the solution is obtained. Further application of the boundary conditions,

$$M_{xy}(x, 0) = V_y(x, 0) = 0 \tag{2.47}$$

leads to an integral representation in terms of one unknown function, $A(s)$. Thus, as in [19, 20], for $y \geq 0$,

$$(1 - v)Dw = \int_0^\infty \frac{1}{s^2} A(s) \{(1 - v)sy - (1 + v)\}e^{-sy} \cos(sx)ds \tag{2.48a}$$

$$(1 - v)D\beta_x = \int_0^\infty \frac{1}{s} A(s) \{[(1 - v)sy - (1 + v) - 4\alpha^2 s^2]e^{-sy}$$
$$+ 4\alpha^2 s^2 m e^{-msy}\} \sin(sx)ds \tag{2.48b}$$

$$(1 - v)D\beta_y = \int_0^\infty \frac{1}{s} A(s) \{[(1 - v)sy - 2 - 4\alpha^2 s^2]e^{-sy}$$
$$+ 4\alpha^2 s^2 e^{-msy}\} \cos(sx)ds \tag{2.48c}$$

and

$$M_x = -\int_0^\infty A(s) \{[(1 - v)(1 - sy) + 4\alpha^2 s^2]e^{-sy}$$
$$- 4\alpha^2 s^2 m e^{-msy}\} \cos(sx)ds \tag{2.49a}$$

$$M_y = \int_0^\infty A(s) \{[3 + v - (1 - v)sy + 4\alpha^2 s^2]e^{-sy}$$
$$- 4\alpha^2 s^2 m e^{-msy}\} \cos(sx)ds \tag{2.49b}$$

$$M_{xy} = \int_0^\infty A(s)\, \{[2 - (1 - v)sy + 4\alpha^2 s^2]e^{-sy}$$
$$+ (2 + 4\alpha^2 s^2)e^{-msy}\}\sin(sx)\mathrm{d}s \tag{2.49c}$$

$$V_x = 2\int_0^\infty sA(s)\,[me^{-msy} - e^{-sy}]\sin(sx)\mathrm{d}s \tag{2.49d}$$

$$V_y = 2\int_0^\infty sA(s)\,[e^{-msy} - e^{-sy}]\cos(sx)\mathrm{d}s \tag{2.49e}$$

where

$$m = m(\alpha s) = \left(1 + \frac{1}{\alpha^2 s^2}\right)^{\frac{1}{4}} \tag{2.50}$$

and α and D are defined by equations (2.13).

The remaining boundary conditions

$$\beta_y(x, 0) = 0,\ |x| > a;\ M_y(x, 0) = -M(x),\ |x| < a \tag{2.51}$$

express symmetry and continuity across the uncracked portion of the *xz*-plane and loading on the cracked portion. When the stress applied to the crack is denoted by $\sigma_y^*(x, 0, z)$, equations (2.14) and (2.51) involve

$$-M(x) = \int_{-h/2}^{h/2} \sigma_y^*(x, 0, z)z\,\mathrm{d}z \tag{2.52}$$

From the discussion of section 2.1 and the bending portion of equations (2.2), one finds, for

$$\sigma_y^*(x, 0, z) = -\frac{12}{h^3} M_0 z \tag{2.53}$$

the special form,

$$M(x) = M_0 \tag{2.54}$$

a constant.

The first of equations (2.51) gives the condition

$$(1 - v)D\beta_y(x, 0) = -2\int_0^\infty \frac{1}{s} A(s)\cos(sx)\mathrm{d}s = 0,\ |x| > a \tag{2.55}$$

Let

$$u(x) = \tfrac{1}{2}(1 - v)D\beta_y(x, 0) \tag{2.56}$$

Then; from inverting equation (2.55,)

$$\frac{1}{s} A(s) = -\frac{2}{\pi}\int_0^a u(x)\cos(sx)\mathrm{d}x \tag{2.57}$$

The second of equations (2.51),

$$M_y(x, 0) = \int_0^\infty t(\alpha s)\, A(s) \cos (sx)ds = -M(x),\ |x| < a \qquad (2.58)$$

where

$$t(\alpha s) = 3 + v + 4\alpha^2 s^2[1 - m(\alpha s)] \qquad (2.59)$$

can be integrated between zero and an arbitrary point, $|x| < a$, to become

$$\int_0^\infty t(\alpha s)\, \frac{1}{s}\, A(s) \sin (sx)ds = -\int_0^x M(\xi)d\xi,\ |x| < a \qquad (2.60)$$

In order to solve equations (2.57) and (2.60), the form

$$u(x) = \int_x^a \frac{\phi(t)t\,dt}{(t^2 - x^2)^{\frac{1}{2}}},\ |x| < a \qquad (2.61)$$

is chosen for the rotation in equation (2.56). This representation has the form appropriate to a crack problem near $x = a$. If equation (2.61) is substituted into equation (2.57) and then (2.60), one additional step results in a Fredholm integral equation of the second kind,

$$(1 + v)\left[\phi(t) + \int_0^a F_1(t, \tau)\phi\,(\tau)\tau\, d\tau\right] = \frac{2}{\pi} \int_0^t \frac{M(x)dx}{(t^2 - x^2)^{\frac{1}{2}}} \qquad (2.62)$$

where

$$F_1(t, \tau) = \int_0^\infty sg(\alpha s)J_0\,(ts)J_0\,(\tau s)ds \qquad (2.63)$$

and

$$g(\alpha s) = \frac{1}{1 + v}t(\alpha s) - 1 \qquad (2.64)$$

For

$$\Phi(\xi) = (1 + v)\sqrt{\xi}\,\phi(a\xi)/M_0 \qquad (2.65)$$

equation (2.62) can be written in the dimensionless form,

$$\phi(\xi) + \int_0^1 F(\xi, \eta)\,\Phi(\eta)d\eta = \frac{2}{\pi}\, \frac{\sqrt{\xi}}{M_0} \int_0^\xi \frac{M(a\rho)d\rho}{(\xi^2 - \rho^2)^{\frac{1}{2}}} \qquad (2.66)$$

where

$$F(\xi, \eta) = (\xi\eta)^{\frac{1}{2}} \int_0^\infty sg\left(\frac{\alpha}{a}\, s\right)J_0\,(\xi s)J_0\,(\eta s)ds \qquad (2.67)$$

The non-zero stress intensity factors for equations (2.45) (in which $\kappa = (3-v)/(1+v)$) and (2.46) can be written as

$$k_1 = k_1^D = k_1^* \frac{2z}{h} \tag{2.68}$$

where

$$k_1^* = \frac{6}{h^2} \Phi(1) M_0 \sqrt{a} \tag{2.69}$$

is the value of k_1 at $z = h/2$. In terms of the maximum bending stress at infinity in Figure 2.1,

$$\sigma_b = \frac{6}{h^2} M_0 \tag{2.70}$$

$$k_1^* = \Phi(1) \sigma_b \sqrt{a} \tag{2.71}$$

The coefficient, $\Phi(1)$, in equations (2.69) and (2.71) is the solution of the integral equation (2.66) evaluated at $\xi = 1$.

It is shown in [19] that the limiting value of the solution of equation (2.66) for large α/a is

$$\Phi(\xi) \to \frac{2}{\pi} \frac{\sqrt{\xi}}{M_0} \int_0^\xi \frac{M(a\rho)d\rho}{(\xi^2 - \rho^2)^{\frac{1}{2}}} \quad \text{as} \quad \frac{\alpha}{a} \to \infty \tag{2.72}$$

and

$$\Phi(\xi) = \frac{1+v}{3+v} \frac{2}{\pi} \frac{\sqrt{\xi}}{M_0} \int_0^\xi \frac{M(a\rho)d\rho}{(\xi^2 - \rho^2)^{\frac{1}{2}}} \quad \text{for} \quad \frac{\alpha}{a} = 0 \tag{2.73}$$

Although the theory is not expected to give good results for thick plates, equation (2.72) verifies the asymptotic behavior which is found numerically in Figure 2.4. It is further shown that the value of $\Phi(1)$ given by equation (2.73) is approached with an infinite slope as $\alpha/a \to 0$.

In general, the solution depends on the parameters of the loading, $M(x)$. For the particular case of the bending problem of Figure 2.1, $M(x)$ is given by equation (2.54). Then

$$\int_0^\xi \frac{M(a\rho)d\rho}{(\xi^2 - \rho^2)^{\frac{1}{2}}} = \frac{\pi}{2} M_0 \tag{2.74}$$

The solution of equation (2.66) depends only on v and α/a. The values of

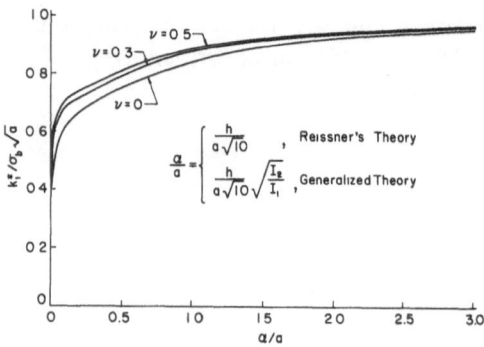

Figure 2.4. Stress intensity factor for plate bending

$\Phi(1)$ for this case are shown on Figure 2.4. Note the infinite slope which occurs for $\alpha/a = 0$.

The value for vanishingly thin plates

$$\Phi(1) = \frac{1 + v}{3 + v} \text{ for } \frac{\alpha}{a} = 0 \tag{2.75}$$

first reported in [21] should not be used. Because of the infinite slope, the correct value of $\Phi(1)$ for a small, but finite, thickness will be significantly different from that of equation (2.75).

Further results in symmetric loading can be found in [19]. Twisting of a cracked plate has also been considered [22]. All results have the proper singularity of equations (2.45) and (2.46). The improved results, as compared with classical theory, should have been expected. The Kirchoff boundary conditions of classical theory are said to affect the solution only near the edges. But it is just these edges which give the crack problem its character, and the solution near the edges is the source of the stress intensity factors.

Generalization of Reissner's theory. The discussion of the generalized Reissner plate bending theory leads to the conclusion that equations (2.48) and (2.49) apply provided that D and α are defined by equations (2.23). In the boundary condition of equation (2.51), the quantity

$$- M(x) = \int_{-h/2}^{h/2} \sigma_n^*(x, 0, z)z \, dz \tag{2.76}$$

as required by equations (2.14), should be used.

The procedure outlined by equations (2.55) through (2.67) applies with the new meanings of α, D, and with $M(x)$ defined by equation (2.76). But since equations (2.52) and (2.76) are identical for the same given $\sigma_y^*(x, 0, z)$ (or different applied stresses with the same moment), the resulting integral equation will be identical to equation (2.66). Therefore, the resulting non-zero stress intensity factors for equations (2.46) can be expressed as

$$k_1 = k_1^* \frac{h^2}{6} f''(z) \tag{2.77}$$

where k_1^* is defined in equations (2.69) and (2.71) and $\Phi(1)$ is the same function of α/a shown in Figure 2.4.

There are several reasons for not taking the z-dependence in equation (2.77) seriously. In the first place, it can be no more accurate than the assumptions involved in equations (2.21) and (2.22). Furthermore, if equations (2.45) (in which $\kappa = (3-v)/(1+v)$) are used, a completely different transverse variation.

$$k_1^D = k_1^* \frac{2z}{h}, \tag{2.78}$$

is obtained. The boundary condition (2.76) is identical for applied stresses with different z-variations if the moments are the same. Thus, two stress distributions are equivalent if they have equal moments. In this sense, equations (2.77) and (2.78) are equivalent, i.e.,

$$\int_{-h/2}^{h/2} k_1(z) z \, \mathrm{d}z = \int_{-h/2}^{h/2} k_1^D(z) z \, \mathrm{d}z \tag{2.79}$$

To obtain $k_1^*/\sigma_b \sqrt{a}$ (see equations (2.70) and (2.71)), it is necessary to choose a particular form for $f(z)$. Any choice, when substituted into equations (2.23), leads to a value of α/a. The value of the stress intensity factor is then determined from Figure 2.4.

A previous solution to this problem [23] used Fourier sine and cosine series to represent the transverse variations of equations (2.21) and (2.22). These are equivalent to the closed form expression for the variation of the transverse normal stress,

$$f(z) = -\frac{1}{2} - \frac{1}{2} \frac{\sinh (2nz/h) - (2nz/h) \cosh n}{\sinh n - n \cosh n} \tag{2.80}$$

where n is an arbitrary parameter chosen equal to

$$n_0 = \frac{\pi h}{2a} \tag{2.81}$$

in [23]. As required, $f(z)$ is -1 or zero for $z = h/2$ or $-h/2$. The variation of the transverse shear stresses is given by

$$f'(z) = -\frac{n}{h} \frac{\cosh(2nz/h) - \cosh n}{\sinh n - n \cosh n} \tag{2.82}$$

which is zero at $z = \pm h/2$. The in-plane stresses vary as

$$f''(z) = -\frac{2n^2}{h^2} \frac{\sinh(2nz/h)}{\sinh n - n \cosh n} \tag{2.83}$$

For small values of n, these reduce to the variations of equations (2.11) of Reissner's theory.

The maximum values of $f'(z)$ and $f''(z)$ are shown on Figure 2.5 as functions of n. The limiting values for large and small n are also given. Figure 2.6 shows the normalized values of the functions of equations (2.80), (2.82), and (2.83) for various values of n. Note the rapid decrease of transverse shear and the rapid increase of in-plane stresses near the surface for large values of n. This behavior can be used to model a boundary layer effect at the surface. The integrals of equations (2.23) can be evaluated to obtain

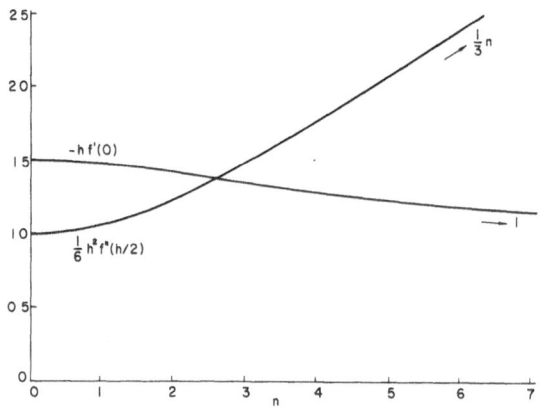

Figure 2.5. Maximum values of $f'(z)$ and $f''(z)$ in generalized plate bending

$$I_1 = \frac{n^3}{12} \frac{\sinh 2n - 2n}{(\sinh n - n \cosh n)^2}$$

$$\alpha^2 = \frac{h^2}{10} \frac{5}{2n^2} \frac{4n + 2n \cosh 2n - 3 \sinh 2n}{\sinh 2n - 2n}$$
(2.84)

which are shown on Figure 2.7.

For a particular choice of the parameter n, equations (2.84) can be used in conjunction with Figure 2.4 to obtain $k_1^*/\sigma_b\sqrt{a}$ as a function of h/a. For

$$n = Pn_0 \tag{2.85}$$

where P is a constant and n_0 is given by equation (2.81), Figure 2.8 results. When $P=0$, the curves correspond to Reissner's theory. The curves for $P=1$ are shown in [23].

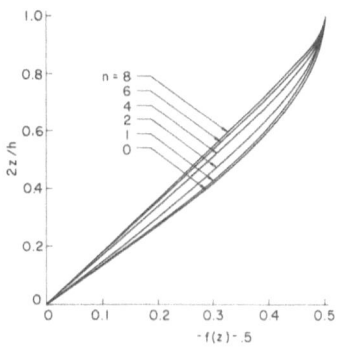

Figure 2.6. Transverse varations in generalized plate bending
(a) Transverse normal stress

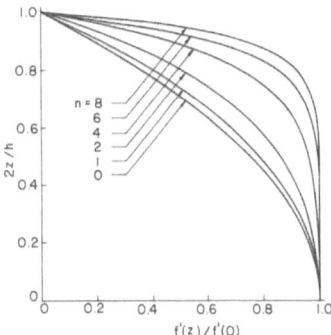

Figure 2.6. Transverse variations in generalized plate bending
(b) Transverse shear stresses

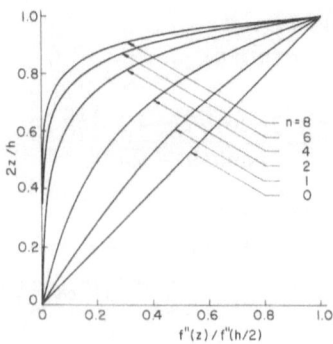

Figure 2.6. Transverse variations in generalized plate bending
(c) In-plane stresses

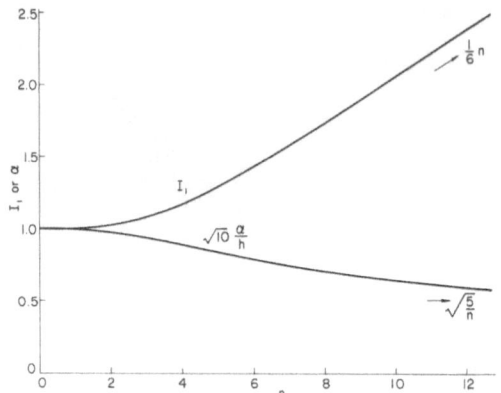

Figure 2.7. Parameters of the generalized bending theory

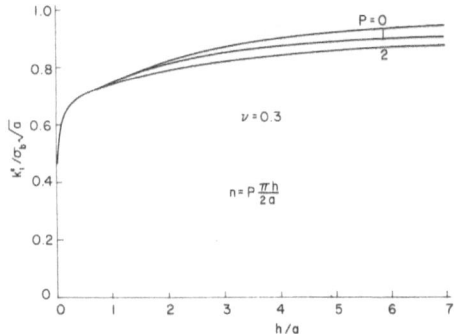

Figure 2.8. Stress intensity factor for generalized plate bending

Hartranft-Sih theory for plate extension. This special version [9] of the general theory of extended plane stress consists of a field of stress and displacement given by equations (2.28) and (2.29), but with $v_x = v_y = s_x = s_y = t_{xy} = 0$. The governing equations then become (2.41) and (2.42). Standard methods of integral transforms lead to the solution [9]

$$(1 - v)DU_x = \alpha^2 \int_0^\infty \left\{ 2(1 - v)\beta m \alpha^2 s^2 e^{-msy} \right.$$
$$\left. - \text{Im}\left[(1 + i\varepsilon\beta)\, \frac{q}{p}\, e^{-psy} \right] \right\} sA(s) \sin sx\, ds \quad (2.86a)$$

$$(1 - v)DU_y = \alpha^2 \int_0^\infty \left\{ 2(1 - v)\beta \alpha^2 s^2 e^{-msy} \right.$$
$$\left. - \text{Im}[(1 + i\varepsilon\beta)q e^{-psy}] \right\} sA(s) \cos sx\, ds \quad (2.86b)$$

$$(1 - v)DU_z = \alpha^2 \int_0^\infty \text{Im}\left[(1 - i\varepsilon\beta)\, \frac{q}{p}\, e^{-psy} \right] A(s) \cos sx\, ds \quad (2.86c)$$

for the displacements. The parameter m is given by equation (2.50) while the complex variable

$$p = \left(1 + \frac{1 - v - i\varepsilon\beta}{1 - v^2 + \beta^2}\, \frac{1 + v}{\alpha^2 s^2} \right)^{\frac{1}{2}} \quad (2.87)$$

In addition,

$$\varepsilon = \left(\frac{1 - v}{1 + v} \right)^{\frac{1}{2}}, q = (1 - v^2)^{\frac{1}{2}} \left[1 + (1 + i\varepsilon\beta)\alpha^2 s^2 \right] \quad (2.88)$$

The unknown function, $A(s)$, is real-valued and must be determined so that

$$S_y(x, 0) = - S(x), |x| < a$$
$$U_y(x, 0) = 0, |x| > a \quad (2.89)$$

The other boundary conditions,

$$T_{xy}(x, 0) = Z_y(x, 0) = 0 \quad (2.90)$$

are satisfied by the solution expressed by equations (2.86). This may be verified by putting $y = 0$ into

$$S_x = \int_0^\infty \left\{ 2(1 - v)\beta m a^4 s^4\, e^{-msy} \right.$$
$$\left. - \text{Im}\left[\left\{ -\frac{v\beta}{\beta - i\,(1 - v^2)^{\frac{1}{2}}} + (1 + i\varepsilon\beta)\alpha^2 s^2 \right\} \frac{q}{p}\, e^{-psy} \right] \right\} A(s) \cos sx\, ds$$

$(2.91a)$

$$S_y = \int_0^\infty \left\{ - 2(1 - v)\beta m\alpha^4 s^4 e^{-msy} \right.$$

$$\left. + \frac{1}{(1 - v^2)^{\frac{1}{2}}} \text{Im} \left[\frac{q^2}{p} e^{-psy} \right] \right\} A(s) \cos sx \, ds \qquad (2.91b)$$

$$T_{xy} = \int_0^\infty \left\{ -(1 - v)\beta(1 + 2\alpha^2 s^2)\alpha^2 s^2 e^{-msy} \right.$$

$$\left. + \text{Im}[(1 + i\varepsilon\beta)qe^{-psy}]\alpha^2 s^2 \right\} A(s) \sin sx \, ds \qquad (2.91c)$$

$$Z_x = \int_0^\infty \left\{ (1 - v)\beta m\alpha^2 s^2 e^{-msy} \right.$$

$$\left. - \text{Im} \left[\frac{q}{p} e^{-psy} \right] \right\} sA(s) \sin sx \, ds \qquad (2.91d)$$

$$Z_y = \int_0^\infty \left\{ (1 - v)\beta\alpha^2 s^2 e^{-msy} - \text{Im}[qe^{-psy}] \right\} sA(s) \cos sx \, ds \qquad (2.91e)$$

$$Z_z = \frac{(1 - v^2)^{\frac{1}{2}}}{\alpha^2(1 - v^2 + \beta^2)} \int_0^\infty \text{Im} \left[((1 - v^2)^{\frac{1}{2}} - i\beta) \frac{q}{p} e^{-psy} \right] A(s) \cos sx \, ds \qquad (2.91f)$$

In terms of the function,

$$u(x) = \tfrac{1}{2} DU_y(x, 0) \qquad (2.92)$$

which, by the second of equations (2.89), is zero for $|x| > a$,

$$sA(s) = - \frac{2}{\pi} \frac{2}{\beta\alpha^2} \int_0^a u(x) \cos sx \, dx \qquad (2.93)$$

In view of

$$S_y(x, 0) = \tfrac{1}{2}\beta\alpha^2 \int_0^\infty s^2 t(\alpha s)A(s) \cos sx \, ds \qquad (2.94)$$

where

$$t(\alpha s) = - 4(1 - v)m\alpha^2 s^2 + \frac{2}{\beta(1 - v^2)^{\frac{1}{2}}} \frac{1}{\alpha^2 s^2} \text{Im} \left[\frac{q^2}{p} \right] \qquad (2.95)$$

and the first of equations (2.89),

$$\int_0^\infty st(\alpha s)A(s) \sin sx \, ds = - \frac{2}{\beta\alpha^2} \int_0^x S(\xi)d\xi, \ |x| < a \qquad (2.96)$$

When the function of equation (2.92) is represented by

$$u(x) = \int_x^a \frac{\phi(t)t dt}{(t^2 - x^2)^{\frac{1}{2}}}, \ |x| < a \qquad (2.97)$$

equation (2.93) becomes

$$sA(s) = -\frac{2}{\beta\alpha^2} \int_0^a \phi(t)J_0(st)t\,dt \tag{2.98}$$

Further use of equations (2.96) and (2.98) lead to an integral equation of the Abel type which is readily converted to the Fredholm integral equation,

$$\Phi(\xi) + \int_0^1 F(\xi, \eta)\Phi(\eta)d\eta = \frac{2}{\pi}\frac{\sqrt{\xi}}{S_0} \int_0^\xi \frac{S(a\rho)d\rho}{(\xi^2 - \rho^2)^{\frac{1}{2}}} \tag{2.99}$$

where

$$F(\xi, \eta) = (\xi\eta)^{\frac{1}{2}} \int_0^\infty sg\left(\frac{\alpha}{a}s\right)J_0(\xi s)J_0(\eta s)ds \tag{2.100}$$

$$g(\alpha s) = \frac{1 - v^2 + \beta^2}{(1 - v^2)(\beta^2 + 1)}t(\alpha s) - 1 \tag{2.101}$$

$$\Phi(\xi) = \frac{(1 - v^2)(\beta^2 + 1)}{1 - v^2 + \beta^2}\sqrt{\xi}\phi(a\xi)/S_0 \tag{2.102}$$

If the displacements and stresses near the crack tip are evaluated and compared with equations (2.45) and (2.46), the non-zero stress intensity factors are found to be

$$k_1 = k_1^D = k_1^* f''(z) \tag{2.103a}$$

$$k_1^T = -\frac{k_1^*}{(\beta^2 + 1)\alpha^2}f(z) = -\frac{4}{h^2}P^2 k_1^* f(z) \tag{2.103b}$$

where

$$k_1^* = \Phi(1)S_0\sqrt{a} \tag{2.104}$$

and

$$P^2 = \frac{h^2}{4\alpha^2(\beta^2 + 1)} = \frac{I_2}{I_1} \tag{2.105}$$

and the parameter in the displacement field of equations (2.45) is

$$\kappa = \left[\frac{3 - v}{1 + v}\beta^2 + 3 - 4v\right]/(\beta^2 + 1) \tag{2.106}$$

The limiting values of equation (2.106) for zero or infinite values of β correspond, respectively, to the cases of plane strain and generalized plane stress.

The parameter, $\Phi(1)$, in equation (2.104) depends on Poisson's ratio, v, and α/a and β. It is computed from equation (2.99) for the particular loading $S(x) = S_0$ and shown in Figure 2.9 for $v = 0.30$. A particular choice of the function $f(z)$ and application of equations (2.31) gives β and the ratio α/h.

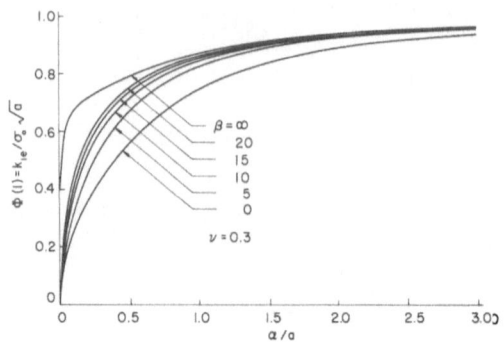

Figure 2.9. Equivalent constant stress intensity factor in plate extension

Thus the applicable curve in Figure 2.9 can be used with an appropriate change of scale of the abscissa to show $\Phi(1)$ as a function of h/a.

The function $f(z)$ was chosen in [9] to satisfy the plane strain condition as closely as possible. In order to have $k_1 = k_1^T$ in equations (2.103), the function must be of the form

$$f(z) = -B \cos\left(P \; \frac{2z}{h}\right) \tag{2.107}$$

This expression is incapable of satisfying the conditions $f(\pm h/2) = f'(\pm h/2) = 0$. Therefore, a boundary layer of thickness $\varepsilon h/2$ is introduced in which the condition of plane strain is violated. Within the boundary layers, a form is desired which matches equation (2.107) and its first and second derivatives at the edges, $z = \pm(1-\varepsilon)h/2$, of the boundary layer. It was considered appropriate to relax the requirement that $f'(z)$ and $f''(z)$ represent the derivatives of $f(z)$ within the boundary layers. Thus, the three transverse variations finally chosen were

$$f(z) = \begin{cases} -\,B\cos(2Pz/h),\ 0 \le z \le (1-\varepsilon)h/2 \\[6pt] -\,B(1-\zeta)^2\,[(1+2\zeta)\cos(1-\varepsilon)P - \zeta P\varepsilon \sin(1-\varepsilon)P], \\[6pt] (1-\varepsilon)\dfrac{h}{2} \le z \le \dfrac{h}{2} \end{cases} \tag{2.108}$$

(for σ_z) where $\zeta = (\varepsilon - 1 + 2z/h)\,/\,\varepsilon$,

$$f'(z) = \begin{cases} \dfrac{2}{h}\,PB\sin(2Pz/h) \\[10pt] \dfrac{2}{h}\,PB(1-\zeta)\,[(1+\zeta)\sin(1-\varepsilon)P + \zeta P\varepsilon \cos(1-\varepsilon)P + C\zeta^2] \end{cases} \tag{2.109}$$

(for τ_{xz} and τ_{yz}) where C is chosen to satisfy equation (2.105), and

$$f''(z) = \frac{4}{h^2}\,P^2 B\cos(2Pz/h),\ 0 < z < h/2 \tag{2.110}$$

(for σ_x, σ_y, and τ_{xy}). When the coefficient B is chosen to give $I_3 = 1$, equations (2.108), (2.109), and (2.110) represent the transverse variations shown in Figure 2.10. Parts (a), (b), and (c) illustrate the effect of changing the parameter, P, for a fixed boundary layer thickness, $\varepsilon = 0.1$. The relation

$$\varepsilon = 1/(2 + 8h/a) \tag{2.111}$$

was introduced in [9] to simulate a boundary layer thickness which becomes constant for large h. Parts (d), (e), and (f) of Figure 2.10 show the effect of changing thickness for $P = 0.3$.

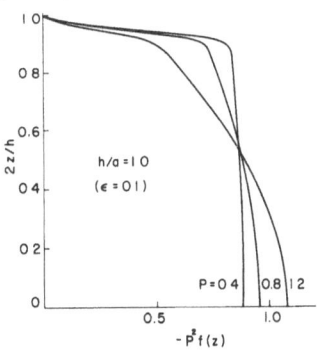

Figure 2.10. Transverse variations in plate extension
(a) Transverse normal stress

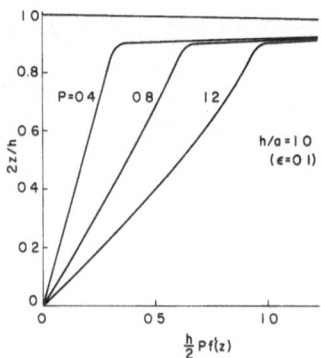

Figure 2.10. Transverse variations in plate extension
(b) Transverse shear stresses

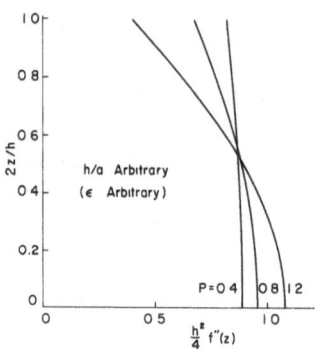

Figure 2.10. Transverse variations in plate extension
(c) In-plane stresses

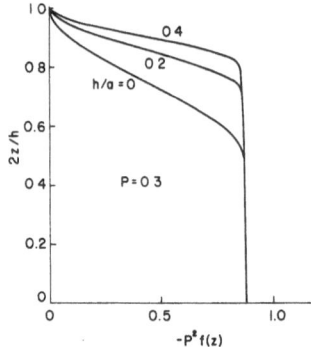

Figure 2.10. Transverse variations in plate extension
(d) Transverse normal stress

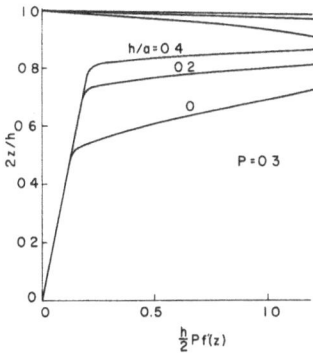

Figure 2.10. Transverse variations in plate extension
(e) Transverse shear stresses

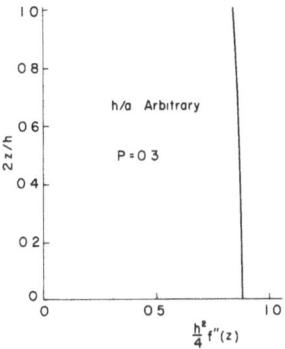

Figure 2.10. Transverse variations in plate extension
(f) In-plane stresses

Additional details of the computation of the parameters α and β are given in [9]. Equations (2.108) to (2.111) result in β and α/h as functions of h/a with P as a parameter as shown in Figure 2.11. The larger values of P give smaller values of β so that equation (2.106) is close to the appropriate form for plane strain. In addition, the large values of P involve very little variation of $f(z)$ and $f''(z)$ and a nearly linear variation of $f'(z)$ except within the boundary layers.

Figures 2.9 and 2.11 are combined in Figure 2.12 to show $\Phi(1)$ as a function of h/a for $v = 0.30$ and several values of P. The small values of $\Phi(1)$ for thin plates indicate that the crack does not degrade their strength as much

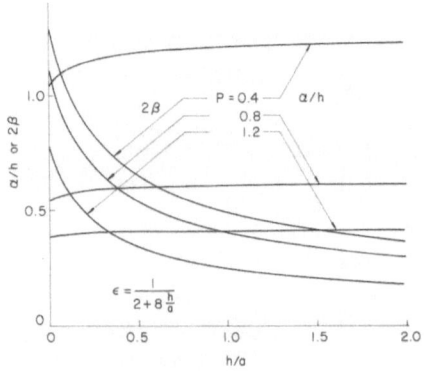

Figure 2.11. Parameters of the modified generalized plane stress
theory for plate extension

as thicker ones. The quantitative details depend on the physical interpretation
of the constant S_0. According to equations (2.32),

$$S_0 = \frac{h^3}{12I_3} \sigma_0 \int_{-h/2}^{h/2} f''(z)dz \tag{2.112}$$

If $f(z)$ satisfied all of the conditions imposed on it, equation (2.112) would
result in $S_0 = 0$.*

However, equations (2.110) and (2.112) lead to

$$S_0 = \sigma_0 \frac{h^4}{12I_3} PB \sin P \tag{2.113}$$

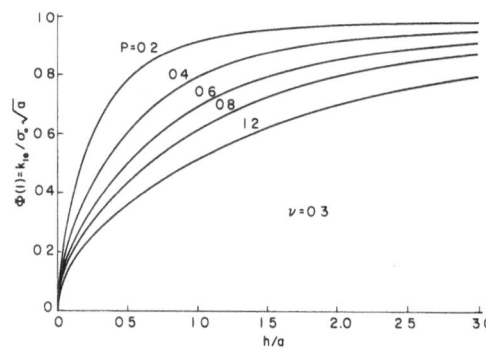

Figure 2.12. Equivalent constant stress intensity factor in plate extension

* It must be noted that this seems to indicate that the full two-term representation of
equations (2.28) and (2.29) and the complete set of equations (2.39), (2.41) and (2.42)
should be used.

One aid in interpreting the results is the idea that applied stresses with identical weighted averages as defined by equations (2.32) are equivalent. Extend the idea to stress intensity factors, and consider the constant, k_{1e}, which is equivalent in the same sense to equation (2.103). It is found that the equivalent constant stress intensity factor is

$$k_{1e} = \Phi(1)\sigma_0 \sqrt{a} \qquad (2.114)$$

Thus, Figure 2.12 is a plot of $k_{1e}/\sigma_0\sqrt{a}$ as a function of h/a.

Alternatively, the usual average,

$$k_{1a} = \frac{1}{h} \int_{-h/2}^{h/2} k_1(z)dz \qquad (2.115)$$

can be discussed. It is found that

$$k_1(z) = \frac{Pk_{1a}}{\sin P} \cos(2Pz/h) \qquad (2.116)$$

where

$$k_{1a} = \frac{4}{P} \frac{\sin^2 P}{2P + \sin 2P} k_{1e} \qquad (2.117)$$

The curves of Figure 2.13 showing the dependence of the average stress intensity factor on plate thickness are obtained by combining Figure 2.12 and equation (2.117).

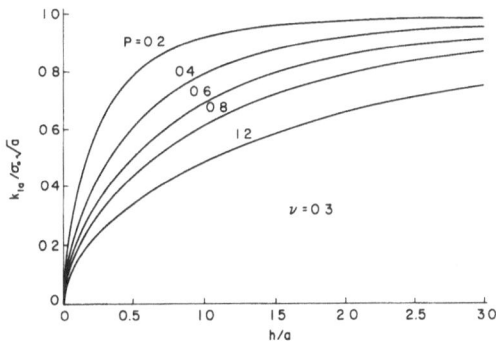

Figure 2.13. Average stress intensity factor in plate extension

Either measure, equation (2.114) or (2.117), of the stress intensity factor or indeed some other such as the maximum

$$k_1(0) = Pk_{1a}/\sin P \tag{2.118}$$

may be used to discuss the amount of reduction in load carrying capacity produced by the crack. For definiteness, consider the average value k_{1a} to be appropriate for a failure theory. That is, failure by crack growth is presumed to occur when

$$k_{1a} = k_c \tag{2.119}$$

where k_c is a material constant. Then Figure 2.13 implies that as plate thickness is decreased, the stress σ_{of} required to cause failure increases as shown in Figure 2.14. This effect is limited by the general yielding which would occur for highly stressed plates.

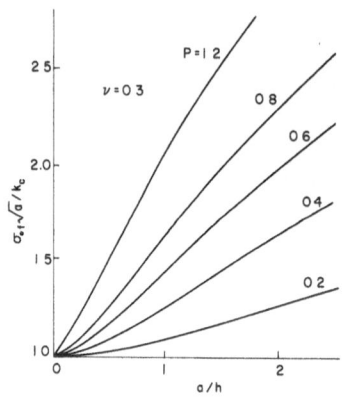

Figure 2.14. Stress at failure using the average stress intensity factor

Photoelastic analysis of thick plates. The stretching of thick cracked plates has been subjected recently [24] to a photoelastic investigation. The results for the transverse variation of the stress intensity factor were directly compared with equation (2.116). Except near the surface, equation (2.116) fits the experimental data well if the value of the parameter P is chosen properly.

A similar study for the bending problem has also been conducted [25]. For the range of plate thicknesses studied, there is very little deviation from the linear form of equation (2.68). In fact, the observed deviations seem to be opposite in direction from those of equation (2.77).

2.4 Guidelines for practical applications

The various plate theories discussed in this chapter and others which may be developed in a similar way call for the use of engineering judgement. This is best developed by actual experience working with the theories, but one of the aims of this chapter has been to show explicitly the approximations which are involved and the reliability of particular results. In this concluding section, general considerations are discussed from the viewpoint adopted for deriving the theories.

Basic to the theories is the initial form assumed for the solution. A particular transverse variation is used and a set of differential equations is obtained. For the generalized theories, several transverse variations can lead to the same set of equations, for it is the integrals of the variations which dominate. This feature makes it possible to model certain characteristics, either observed experimentally or obtained from more exact analyses, with judicious choice of the transverse variation. For example, the form chosen for extension of a cracked plate was the one which permits the plane strain crack tip characteristic to be included.

In a similar vein, the fact that only certain weighted averages of boundary stresses and displacements affect the boundary value problem is an aid in interpreting results. Much more confidence can be placed in the averages predicted by the theories than in the detailed transverse variations. The transverse variations must be judged by comparison with experiment or by conformity with known characteristics.

In general, the entire body of results is applicable to thin plates provided that the initial assumptions are reasonable. However, there is the danger for thin plates that yielding will begin at relatively low loads. On the other hand, only the averages can be reliably used in the analysis of moderately thick plates. But because there is less danger of yielding, the theory may actually agree better with experiment than for thin plates.

For the crack problem, in particular, the discrepancies between the theoretical form of the crack tip solution (equations (2.45) and (2.46)) and that expected [14] are notable. Whereas the plane strain value of κ is expected, the generalized plane stress value is obtained for the plate bending theories. In the case of extension, according to the Hartranft-Sih theory, the plane strain value is closely approximated for some sets of transverse variations. For one of the bending theories, different stress intensity factors were obtained for the displacements than for the stresses. Finally, the transverse normal stress is zero in the case of bending and is given by the appropriate plane strain value for the interior portion of the crack front in the case of extension.

References

[1] Tiffen, R., An Investigation of the Transverse Displacement Equation of Elastic Plate Theory, *Quarterly Journal of Mechanics and Applied Mathematics*, 14, pp. 59–74 (1961).

[2] Tiffen, R. and Lowe, P. G., An Exact Theory of Generally Loaded Elastic Plates in Terms of Moments ofthe Fundamental Equations, *Prodeecings of the London Mathematical Society*, 13, pp. 653–671 (1963).

[3] Sayer, F. P. and Calder, C. C., Plane Stress Theories, *Proceedings of the Cambridge Philosophical Society*, 63, pp. 1369–1378 (1967).

[4] Green, A. E., The Elastic Equilibrium of Isotropic Plates and Cylinders, *Proceedings of the Royal Society of London*, Series A, 195, pp. 533–552 (1949).

[5] Friedrichs, K. O., Kirchoff's Boundary Condition and the Edge Effect for Elastic Plates, *Proceedings of the Symposium on Applied Mathematics*, 3, pp. 117–124 (1950).

[6] Friedrichs, K. O. and Dressler, R. F., A Boundary-Layer Theory for Elastic Plates, *Communications on Pure and Applied Mathematics*, 14, pp. 1–33 (1961).

[7] Reissner, E., On Bending of Elastic Plates, *Quarterly of Applied Mathematics*, 5, pp. 55–68 (1947).

[8] Reissner, E., On a Variational Theorem in Elasticity, *Journal of Mathematics and Physics*, 29, pp. 90–95 (1950).

[9] Hartranft, R. J. and Sih, G. C., An Approximate Three-Dimensional Theory of Plates with Application to Crack Problems, *International Journal of Engineering Science*, 8, pp. 711–729 (1970).

[10] Courant, R. and Hilbert, D., *Methods of Mathematical Physics*, 1, Ch. 4, pp. 164–274, Interscience (1953).

[11] Goldenweiser, A. L., On Reissner's Plate Theory, *Izvestia Akademii Nauk SSSR*, O.T.N. 4, pp. 102–109 (1958).

[12] Fersht, S., An Extended Reissner Thin Plate Theory, *Israel Journal of Technology*, 2, pp. 312–317 (1964).

[13] Reissner, E., On the Calculation of Three-Dimensional Corrections for the Two-Dimensional Theory of Plane Stress, *Proceedings of the Fifteenth Semi-Annual Eastern Photoelasticity Conference*, pp. 23–31, Boston (1942).

[14] Hartranft, R. J. and Sih, G. C., The Use of Eigenfunction Expansions in the General Solution of Three-Dimensional Crack Problems, *Journal of Mathematics and Mechanics*, 19, pp. 123–138 (1969).

[15] Kassir, M. K. and Sih, G. C., Three-Dimensional Stress Distribution Around an Elliptical Crack Under Arbitrary Loadings, *Journal of Applied Mechanics*, 33, pp. 601–611 (1966).

[16] Williams, M. L., The Bending Stress Distribution at the Base of a Stationary Crack, *Journal of Applied Mechanics*, 28, pp. 78-82 (1961).

[17] Sih, G. C., Paris, P. C. and Erdogan, F., Crack-Tip Stress-Intensity Factors for Plane Extension and Plate Bending Problems, *Journal of Applied Mechanics*, 29, pp. 306-312 (1962).

[18] Sih, G. C., editor, *Methods of Analysis and Solutions of Crack Problems*, Noordhoff International Publishing, Leyden (1973).

[19] Hartranft, R. J. and Sih, G. C., Effect of Plate Thickness on the Bending Stress Distribution Around Through Cracks, *Journal of Mathematics* and *Physics*, 47, pp. 276-291 (1968).

[20] Wang, N. M., Effects of Plate Thickness on the Bending of an Elastic Plate Containing a Crack, *Journal of Mathematics and Physics*, 47, pp. 371-390 (1968).

[21] Knowles, J. K. and Wang, N. M., On the Bending of an Elastic Plate Containing a Crack, *Journal of Mathematics and Physics*, 39, pp. 223-236 (1960).

[22] Wang, N. M., Twisting of an Elastic Plate Containing a Crack, *International Journal of Fracture Mechanics*, 6, pp. 367-378 (1970).

[23] Sih, G. C., Bending of a Cracked Plate with an Arbitrary Stress Distribution Across the Thickness, *Journal of Engineering for Industry, Transactions of the ASME*, 92. pp. 350-356 (1970).

[24] Villarreal, G., Sih, G. C. and Hartranft, R. J., Photoelastic Investigation of a Thick Plate with a Transverse Crack, *Journal of Applied Mechanics*, 42, pp. 9-14 (1975).

[25] Mullinix, B. R. and Smith, C. W., Distribution of Local Stresses Across the Thickness of Cracked Plates under Bending Fields, *International Journal of Fracture*, 10, pp. 337-352 (1974).

R. Badaliance, G. C. Sih and E. P. Chen

3 | *Through cracks in multilayered plates*

3.1 Introduction

Analytical modeling of layered plates containing cracks has received very little or no attention in the past. This is mainly because the problem is extremely complicated since it is basically three-dimensional in nature. The presence of cracks or surfaces of discontinuity adds additional difficulties to the development of a tractable theory of layered plates. One of the objectives of this work is to obtain a set of governing equations that can yield effective solutions to the mixed boundary value crack problems. The approach of Reissner [1] and Hartranft and Sih [2] will be followed in that the complementary energy functional of the assumed state of stress will be minimized with respect to the admissible variations of the stress quantities satisfying the equilibrium equations of three-dimensional elasticity and traction boundary conditions. The analysis is approximate as the compatibility or continuity conditions are satisfied only in an average sense. In contrast to the conventional approach adapted in the classical theories of plates, the stress distribution in the thickness direction of the plate will not be assumed arbitrarily but determined from the plane strain condition ahead of the crack as first suggested by Sih [3].

The general formulation is valid for a plate containing any number of layers made of different materials with different thicknesses. A bendingless plate, however, must be constructed with even number of layers and requires the stretching load be applied such that the system possesses material and geometric symmetry with reference to the middle plane of the plate. As an example, the case of a three-layer plate with a through crack subjected to tensile loading is treated. Stress intensity factors are determined for various values of layer thicknesses, material constants, etc. Although the assumptions of isotropy and homogeneity of each layer have been adopted in the development of the theory, the analysis can be easily extended to layered plates whose individual layers are anisotropic and nonhomogeneous.

3.2   Minimum complementary energy applied to a layered plate

The complementary energy principle is developed for a two-material system and the results are easily extended to the corresponding variational principle for a system with any number of layers. Let the complementary energy, Φ, of an isotropic, homogeneous body be defined as the strain energy of that body minus the work done on the portion of the body surface S_u over which the displacements $\bar{u}_i$ are specified, i.e.,

$$\Phi = \int_v \Psi dv - \int_{S_u} T_i \bar{u}_i ds \tag{3.1}$$

where the strain energy density is

$$\Psi = \int_0^{\sigma_{ij}} \varepsilon_{ij} d\sigma_{ij} \tag{3.2}$$

Summation on the indices $i, j, = x, y, z$ are understood.

Consider a two-material body subjected to traction boundary conditions along its external surface as shown in Figure 3.1. For convenience, these

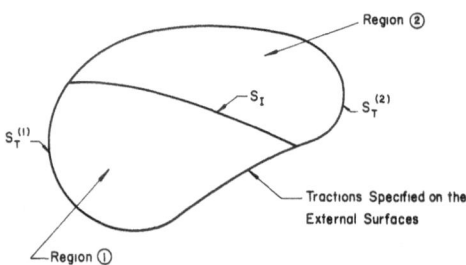

Figure 3.1. A two-phased material

conditions are assumed to prevail over the entire external surface so that the complementary energy reduces to the strain energy and only the interfacial surface S_I must be considered separately. Displacement boundary conditions can be handled in the same way and do not alter the basic development of the analysis. The complementary energy of the two-material system consists of the sum of the strain energies of material (1) and material (2) as follows:

$$\Phi = \int_{v_1} \Psi_1 dv + \int_{v_2} \Psi_2 dv \tag{3.3}$$

This functional Φ is to be minimized with respect to the stress components which are constrained by
(1) the conditions of equilibrium*

$$\sigma_{ij,j}^{(p)} = 0 \tag{3.4}$$

where

$$\sigma_{ij}^{(p)} = \sigma_{ji}^{(p)}$$

(2) the traction boundary conditions

$$\sigma_{ij}^{(p)} n_j^{(p)} = T_i \text{ on } S_T^{(p)} \tag{3.5}$$

(3) and the continuity of tractions across S_I

$$\sigma_{ij}^{(1)} n_j^{(1)} = -\sigma_{ij}^{(2)} n_j^{(2)} \tag{3.6}$$

where $n_j^{(p)}$ are the components of the outward unit normal to the surface of the sub-region (p). For a two-material system, p takes the value of 1 or 2. Thus, on S_I,

$$n_j^{(2)} = -n_j^{(1)} \tag{3.7}$$

The minimization of a functional subject to a set of constraints can be accomplished with the use of Lagrange multipliers for the non-integrable constraint stated in equation (3.4). For this case, the functional Π is introduced as

$$\Pi = \int_{v_1} \Psi_1 dv + \int_{v_2} \Psi_2 dv + \int_{v_1} \lambda_i^{(1)} \sigma_{ij,j}^{(1)} dv + \int_{v_2} \lambda_i^{(2)} \sigma_{ij,j}^{(2)} dv \tag{3.8}$$

where $\lambda_i^{(1)}$ and $\lambda_i^{(2)}$ are the Lagrange multipliers. Minimization of Π with respect to admissible variations of the stress components leads to a set of differential equations and boundary conditions. The variational procedure will be carried out primarily to demonstrate that it leads to continuity conditions for the displacement components across the material interface S_I:

$$\delta\Pi = \int_v \frac{\partial \Psi_1}{\partial \sigma_{ij}^{(1)}} \delta\sigma_{ij}^{(1)} dv + \int_{v_2} \frac{\partial \Psi_2}{\partial \sigma_{ij}^{(2)}} \delta\sigma_{ij}^{(2)} dv$$

* The notation, j denotes derivative with respect to the jth variable x, y, or z.

$$- \int_{v_1} \lambda_i^{(1)} \delta(\sigma_{ij,j}^{(1)}) dv - \int_{v_2} \lambda_i^{(2)} \delta(\sigma_{ij,j}^{(2)}) dv \qquad (3.9)$$

Application of the divergence theorem followed by a regrouping of terms yields

$$\delta \Pi = \int_{v_1} \left(\frac{\partial \Psi_1}{\partial \sigma_{ij}^{(1)}} - \lambda_{i,j}^{(1)} \right) \delta \sigma_{ij}^{(1)} dv + \int_{v_2} \left(\frac{\partial \Psi_2}{\partial \sigma_{ij}^{(2)}} - \lambda_{i,j}^{(2)} \right) \delta \sigma_{ij}^{(2)} dv$$

$$- \int_{S_I} [(\lambda_i^{(1)}) n_j^{(1)} \delta \sigma_{ij}^{(1)} + (\lambda_i^{(2)}) n_j^{(2)} \delta \sigma_{ij}^{(2)}] ds = 0 \qquad (3.10)$$

The stresses must satisfy the symmetry condition which requires that

$$\delta \sigma_{ij}^{(1)} = \delta \sigma_{ji}^{(1)} \text{ in } v_1 \qquad (3.11a)$$

$$\delta \sigma_{ij}^{(2)} = \delta \sigma_{ji}^{(2)} \text{ in } v_2 \qquad (3.11b)$$

In addition, constraint condition in equation (3.6) gives

$$n_j^{(2)} \delta \sigma_{ij}^{(2)} = - n_j^{(1)} \delta \sigma_{ij}^{(1)} \text{ on } S_I \qquad (3.12)$$

Under these considerations, equation (3.9) becomes

$$\delta \Pi = \int_{v_1} \left[2 \frac{\partial \Psi_1}{\partial \sigma_{ij}^{(1)}} - (\lambda_{i,j}^{(1)} + \lambda_{j,i}^{(1)}) \right] dv$$

$$+ \int_{v_2} \left[2 \frac{\partial \Psi_2}{\partial \sigma_{ij}^{(2)}} - (\lambda_{i,j}^{(2)} + \lambda_{j,i}^{(2)}) \right] dv$$

$$+ \int_{S_I} (\lambda_i^{(1)} - \lambda_i^{(2)}) n_j^{(1)} \delta \sigma_{ij}^{(1)} ds = 0 \qquad (3.13)$$

In order that $\delta \Pi = 0$ for all arbitrary combinations of the six stress variations $\delta \sigma_{ij}^{(1)}$ in v_1 and on S_I and of $\delta \sigma_{ij}^{(2)}$ in v_2 and on S_I, it is necessary that

$$\frac{\partial \Psi_1}{\partial \sigma_{ij}^{(1)}} - \tfrac{1}{2}(\lambda_{i,j}^{(1)} + \lambda_{j,i}^{(1)}) = 0, \text{ in } v_1 \qquad (3.14a)$$

$$\frac{\partial \Psi_2}{\partial \sigma_{ij}^{(2)}} - \tfrac{1}{2}(\lambda_{i,j}^{(2)} + \lambda_{j,i}^{(2)}) = 0, \text{ in } v_2 \qquad (3.14b)$$

while

$$(\lambda_i^{(1)} - \lambda_i^{(2)}) = 0 \quad \text{on} \quad S_I \qquad (3.15)$$

is satisfied. From equation (3.2), it is recognized that $\dfrac{\partial \Psi}{\partial \sigma_{ij}} = \varepsilon_{ij}$ and hence equations (3.14) in v_1 and v_2 lead to

$$\lambda_i^{(p)} = + u_i^{(p)}, \ p = 1, 2, \ i = 1, 2, 3 \tag{3.16}$$

Substitution of this result into equation (3.15) renders

$$u_i^{(1)} = u_i^{(2)} \text{ on } S_I \tag{3.17}$$

The complementary energy for a nonhomogeneous body is shown to be the sum of the strain energies of the various constituents minus the work done over the portion of the external boundary over which displacements are prescribed. Minimization of this energy with respect to admissible stress fields leads to the compatibility conditions within each material and continuity of displacements across material interfaces.

3.3 An approximate three-dimensional theory of multi-layered plates

An approximate theory of layered plates will be developed by seeking a stationary value for the complementary energy expression such that the stresses satisfy the equilibrium conditions in the interior of the solid and on that part of the boundary surface where tractions are prescribed. The resulting system of differential equations can be solved to yield effective solutions for the problem of cracks in layered plates.

Consider a layered plate made of n layers with each layer being of different thickness and material properties. The layers are assumed to be perfectly bonded and stacked in even numbers such that the system possesses material

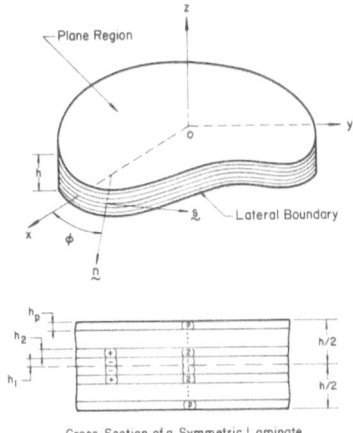

Cross Section of a Symmetric Laminate

Figure 3.2. Geometry of multi-layered plate

symmetry with respect to the middle plane of the laminate as shown in Figure 3.2. In this way, stretching load produces only in-plane deformation.

The way in which the three-dimensional equations of elasticity is relaxed is to assume that the stress components in each layer can be approximated as a product of a function* of the thickness variable z multiplied by a function of the in-plane variables x and y, i.e.,

$$\sigma_{ij}^{(p)} = f_{ij}^{(p)}(z) g_{ij}^{(p)}(x, y), \text{ (no sum on } i, j, p) \tag{3.18}$$

Substitution of the stresses in equation (3.18) into the equilibrium equations lead to the more specific assumption

$$\sigma_z^{(p)} = f_p(z) Z_z^{(p)}(x, y) \tag{3.19a}$$

$$[\tau_{xz}^{(p)}, \tau_{yz}^{(p)}] = -f_p'(z) [Z_x^{(p)}(x, y), Z_y^{(p)}(x, y)] \tag{3.19b}$$

$$[\sigma_x^{(p)}, \sigma_y^{(p)}, \tau_{xy}^{(p)}] = f_p''(z) [S_x^{(p)}(x, y), S_y^{(p)}(x, y), T_{xy}^{(p)}(x, y)] \tag{3.19c}$$

The stresses $\sigma_x^{(p)}, \sigma_y^{(p)}, \ldots, \sigma_z^{(p)}$ are assumed to vary from one layer to the next through the function $f_p(z)$ and its derivatives while $S_x^{(p)}, S_y^{(p)}, \ldots, Z_z^{(p)}$ remain the same for each layer, i.e.,

$$[S_x^{(p)}, S_y^{(p)}, \ldots, Z_z^{(p)}] = [S_x, S_y, \ldots, Z_z] \tag{3.20}$$

Hence, equations (3.19) further reduce to

$$\sigma_z^{(p)} = f_p(z) Z_z(x, y) \tag{3.21a}$$

$$[\tau_{xz}^{(p)}, \tau_{yz}^{(p)}] = -f_p'(z) [Z_x(x, y), Z_y(x, y)] \tag{3.21b}$$

$$[\sigma_x^{(p)}, \sigma_y^{(p)}, \tau_{xy}^{(p)}] = f_p''(z) [S_x(x, y), S_y(x, y), T_{xy}(x, y)] \tag{3.21c}$$

where $f_p(z)$ and its first derivative must match their respective values at the interfaces of the adjoining layers. Making use of equations (3.21), the complementary energy functional becomes

$$\Phi = \sum_{p=1}^{n} \frac{1}{2E_p} \int_{v_p} [\{f_p''^2 [S_x^2 + S_y^2 - 2v_p S_x S_y$$
$$+ 2(1 + v_p) T_{xy}^2] + 2 f_p'^2 (1 + v_p) (Z_x^2 + Z_y^2)$$
$$- 2v_p f_p f_p'' Z_z (S_x + S_y) + f_p^2 Z_z^2\} dv - \int_{S_u^{(p)}} [f_p''(S_n \bar{U}_u^{(p)})$$

---

* This product form of the stress field solution for crack problems was first discussed by Sih et al [4] and later by Hartranft and Sih [5].

$$+ T_{ns}\overline{U}_s^{(p)}) - f_p' Z_n \overline{U}_z^{(p)}]ds] \qquad (3.22)$$

The equilibrium conditions for a layered plate whose surfaces are free of tractions with $f_p = f_p' = 0$ for $z = \pm h/2$ lead to

$$Z_z = Z_{x,x} + Z_{y,y} \qquad (3.23a)$$

$$Z_x = S_{x,x} + T_{xy,y} \qquad (3.23b)$$

$$Z_y = T_{xy,x} + S_{y,y} \qquad (3.23c)$$

When referred to the normal and tangential directions n and s, the following relationships are to be observed:

$$S_n n_x - T_{ns} n_y = S_x n_x + T_{xy} n_y \qquad (3.24a)$$

$$S_n n_y + T_{ns} n_x = T_{xy} n_x + S_y n_y \qquad (3.24b)$$

$$Z_n = Z_x n_x + Z_y n_y \qquad (3.24c)$$

The quantities $\overline{U}_n^{(p)}$, $\overline{U}_s^{(p)}$, and $\overline{U}_z^{(p)}$ in equation (3.22) are the normal, tangential, and transverse components of the displacements prescribed on the edge $S_u^{(p)}$ of layer p as shown in Figure 3.2.

The Lagrange multipliers, λ_i, are now employed to insure satisfaction of the equilibrium equations, and thus the functional

$$\Pi = \Phi - \int_A [\lambda_1(S_{x,x} + T_{xy,y} - Z_x)$$

$$+ \lambda_2(T_{xy,y} + S_{y,y} - Z_y) + \lambda_3(Z_{x,x} + Z_{y,y} - Z_y)]dxdy \qquad (3.25)$$

is formed. The variation of Π with respect to S_x, S_y, T_{xy}, Z_x, Z_y, Z_z, and $f_p(z)$ with the continuity conditions enforced on $f_p(z)$ and $f_p'(z)$ is

$$\delta\Pi = \sum_{p=1}^{n} \left[\frac{1}{2E_p} \int_{v_p} \{f_p''[2S_x\delta S_x + 2S_y\delta S_y - 2v_p(S_x\delta S_y \right.$$

$$+ S_y\delta S_x) + 4(1 + v_p)T_{xy}\delta T_{xy}] + 4f_p'^2(1 + v_p)$$

$$\times (Z_x\delta Z_x + Z_y\delta Z_y) - 2v_p f_p f_p''[Z_z(\delta S_x + \delta S_y)$$

$$+ (S_x + S_y)\delta Z_z] + 2f_p^2 Z_z\delta Z_z\}dv - \int_{S_u^{(p)}} [(f_p''(\delta S_n U_n^{(p)}$$

$$+ \delta T_{ns}\overline{U}_s^{(p)}) - f_p'\delta Z_n \overline{U}_z^{(p)}]ds \right]$$

$$- \int_A [\lambda_1(\delta S_{x,x} + \delta T_{xy,y} - \delta Z_x) \qquad (3.26)$$

$$+ \lambda_2(\delta T_{xy, x} + \delta S_{y, y} - \delta Z_y) + \lambda_3(\delta Z_{x, x} + \delta Z_{y, y} - \delta Z_z)]dxdy$$

$$+ \sum_{p=1}^{n} \left[\frac{1}{2E_p} \int_{v_p} \{2[S_x^2 + S_y^2 - 2v_p S_x S_y \right.$$

$$+ 2(1 + v_p)T_{xy}]f_p''\delta f_p'' + 4(1 + v_p)(Z_x^2 + Z_y^2)f_p'\delta f_p'$$

$$- 2v_p Z_z(S_x + S_y)(f_p\delta f_p'' + f_p''\delta f_p) + 2Z_z^2 f_p\delta f_p\}dv$$

$$\left. - \int_{S_u^{(p)}} [(S_n \overline{U}_n^{(p)} + T_{ns}\overline{U}_s^{(p)})\delta f_p'' - Z_n \overline{U}_z^{(p)}\delta f_p']ds \right]$$

Note that $\delta\Pi$ consists of two parts. The first contains only variations of the in-plane functions S_x, S_y, T_{xy}, Z_x, Z_y, and Z_z; while the second includes only variations of $f_p(z)$ and its derivatives. The integrals containing variations of the in-plane functions can be integrated through the thickness and those containing variations of $f_p(z)$ can be integrated in x and y. Thus, equation (3.26) may be rearranged as

$$\delta\Pi = \int_A \{I_1(S_x\delta S_x + S_y\delta S_y) + 2(I_1 - I_2)T_{xy}\delta T_{xy}$$

$$+ I_2(S_x\delta S_y + S_y\delta S_x) + I_3(Z_x\delta Z_x + Z_y\delta Z_y)$$

$$+ I_4[Z_z(\delta S_x + \delta S_y) + (S_x + S_y)\delta Z_z] + I_5 Z_z\delta Z_z$$

$$- \lambda_1(\delta S_{x, x} + \delta T_{xy, y} - \delta Z_x) - \lambda_2(\delta T_{xy, x} + \delta S_{y, y} - \delta Z_y)$$

$$- \lambda_3(\delta Z_{x, x} + \delta Z_{y, y} - \delta Z_z)\}dxdy$$

$$- \int_{C_u} (\bar{u}_n\delta S_n + \bar{u}_s\delta T_{ns} + \bar{u}_z\delta Z_n)dC$$

$$+ \sum_{p=1}^{n} \int_{t_{p-1}}^{t_p} \{\alpha_1^{(p)}f_p''(z)\delta f_p''(z) + \alpha_2^{(p)}f_p'(z)\delta f_p'(z) \tag{3.27}$$

$$+ \alpha_3^{(p)}[f_p(z)\delta f_p''(z) + f_p''(z)\delta f_p(z)] + \alpha_4^{(p)}f_p(z)\delta f_p(z)$$

$$+ \beta_1^{(p)}\delta f_p''(z) + \beta_2^{(p)}\delta f_p'(z)\}dz$$

where A stands for the plate area and C_u is that part of the contour on which displacement is specified.

In equation (3.27), $\bar{u}_n$, $\bar{u}_s$ and $\bar{u}_z$ are defined by

$$\bar{u}_n = \sum_{p=1}^{n} \int_{t_{p-1}}^{t_p} \overline{U}_n^{(p)}f_p''(z)dz \tag{3.28a}$$

$$\bar{u}_s = \sum_{p=1}^{n} \int_{t_{p-1}}^{t_p} \overline{U}_s^{(p)}f_p''(z)dz \tag{3.28b}$$

$$\bar{u}_z = - \sum_{p=1}^{n} \int_{t_{p-1}}^{t_p} \bar{U}_z^{(p)} f_p''(z) dz \qquad (3.28c)$$

The quantities $I_1, I_2, \ldots, I_5$ stand for

$$I_1 = \sum_{p=1}^{n} \frac{1}{E_p} \int_{t_{p-1}}^{t_p} f_p''^2(z) dz \qquad (3.29a)$$

$$I_2 = - \sum_{p=1}^{n} \frac{v_p}{E_p} \int_{t_{p-1}}^{t_p} f_p''^2(z) dz \qquad (3.29b)$$

$$I_3 = \sum_{p=1}^{n} \frac{2(1+v_p)}{E_p} \int_{t_{p-1}}^{t_p} f_p'^2(z) dz \qquad (3.29c)$$

$$I_4 = - \sum_{p=1}^{n} \frac{v_p}{E_p} \int_{t_{p-1}}^{t_p} f_p(z) f_p''(z) dz \qquad (3.29d)$$

$$I_5 = \sum_{p=1}^{n} \frac{1}{E_p} \int_{t_{p-1}}^{t_p} f_p^2(z) dz \qquad (3.29e)$$

while $\alpha_1^{(p)}, \alpha_2^{(p)}, \ldots, \alpha_4^{(p)}$ are given by

$$\alpha_1^{(p)} = \frac{1}{E_p} \int_A [S_x^2 + S_y^2 - 2v_p S_x S_y + 2(1+v_p) T_{xy}^2] dx dy \qquad (3.30a)$$

$$\alpha_2^{(p)} = \frac{2(1+v_p)}{E_p} \int_A (Z_x^2 + Z_y^2) dx dy \qquad (3.30b)$$

$$\alpha_3^{(p)} = - \frac{v_p}{E_p} \int_A Z_z(S_x + S_y) dx dy \qquad (3.30c)$$

$$\alpha_4^{(p)} = \frac{1}{E_p} \int_A Z_z^2 dx dy \qquad (3.30d)$$

Finally, $\beta_1^{(p)}$ and $\beta_2^{(p)}$ are the integrals

$$\beta_1^{(p)} = \int_{C_u} (S_n \bar{U}_n^{(p)} + T_{ns} \bar{U}_s^{(p)}) dC \qquad (3.31a)$$

$$\beta_2^{(p)} = - \int_{C_u} Z_n \bar{U}_z^{(p)} dC \qquad (3.31b)$$

Integrating equation (3.27) by parts and applying the divergence theorem give

$$\delta\Pi = \int_A [(I_1 S_x + I_2 S_y + I_4 Z_z + \lambda_{1,x}) \delta S_x$$

$$+ (I_1 S_y + I_2 S_x + I_4 Z_z + \lambda_{2,y}) \delta S_y$$

$$+ (2(I_1 - I_2)T_{xy} + \lambda_{1,y} + \lambda_{2,x})\delta T_{xy}$$

$$+ (I_3 Z_x + \lambda_1 + \lambda_{3,x})\delta Z_x$$

$$+ (I_3 Z_y + \lambda_2 + \lambda_{3,y})\delta Z_y$$

$$+ (I_4(S_x + S_y) + I_5 Z_z + \lambda_3)\delta Z_z]dxdy$$

$$- \int_{C_u} [(u_n + \lambda_n)\delta S_n + (u_s + \lambda_s)\delta T_{ns} + (u_z + \lambda_3)\delta Z_n]dc$$

$$+ \sum_{p=1}^{n} \int_{t_{p-1}}^{t_p} [\alpha_1^{(p)} f_p^{iv}(z) + (2\alpha_3^{(p)} - \alpha_2^{(p)})f_p''(z) + \alpha_4^{(p)} f(z)$$

$$+ \beta_{1,zz}^{(p)} - \beta_{2,z}^{(p)}]\tilde{\delta} f_p(z)dz + \sum_{p=1}^{n} (\alpha_1^{(p)} f_p''(z) + \alpha_3^{(p)} f(z)$$

$$+ \beta_1^{(p)})\delta f_p'(z) |_{t_{p-1}}^{t_p} + \sum_{p=1}^{n} (-\alpha_1^{(p)} f_p'''(z) + \alpha_2^{(p)} f_p'(z)$$

$$- \alpha_3^{(p)} f_p'(z) - \beta_{1,z}^{(p)} + \beta_2^{(p)})\delta f_p(z) |_{t_{p-1}}^{t_p} \qquad (3.32)$$

such that

$$\lambda_n = \lambda_1 n_1 + \lambda_2 n_2$$

$$\lambda_s = -\lambda_1 n_2 + \lambda_2 n_2$$

In the above equations, n_1 and n_2 are the components of the unit normal vector n in Figure 3.2. The condition $\delta\Pi = 0$ leads to a system of partial differential equations in the variables x and y:

$$I_1 S_x + I_2 S_y + I_4 Z_z + \lambda_{1,x} = 0 \qquad (3.33a)$$

$$I_1 S_y + I_2 S_x + I_4 Z_z + \lambda_{2,y} = 0 \qquad (3.33b)$$

$$2(I_1 - I_2)T_{xy} + \lambda_{1,y} + \lambda_{2,x} = 0 \qquad (3.33c)$$

$$I_3 Z_x + \lambda_1 + \lambda_{3,x} = 0 \qquad (3.33d)$$

$$I_3 Z_y + \lambda_2 + \lambda_{3,y} = 0 \qquad (3.33e)$$

$$I_4(S_x + S_y) + I_5 Z_z + \lambda_3 = 0 \qquad (3.33f)$$

The functions $S_x, S_y, \ldots, Z_z$ are constrained as

$$Z_z = Z_{\lambda,x} + Z_{y,y} \qquad (3.34a)$$

$$Z_x = S_{x,x} + T_{xy,y} \tag{3.34b}$$

$$Z_y = T_{xy,x} + S_{y,y} \tag{3.34c}$$

with $\lambda_n = -u_n$, $\lambda_s = -u_s$, and $\lambda_3 = -u_z$ on C_u. It is apparent that the boundary conditions associated with the functions S_x, S_y, etc., can be either of the traction or averaged displacement type. Along the contour C of the plate, the prescribed conditions are

$$S_n = \bar{S}_n \quad \text{or} \quad \lambda_n = -\bar{u}_n \tag{3.35a}$$

$$T_{ns} = \bar{T}_{ns} \quad \text{or} \quad \lambda_s = -\bar{u}_s \tag{3.35b}$$

$$Z_n = \bar{Z}_n \quad \text{or} \quad \lambda_3 = -\bar{u}_z \tag{3.35c}$$

The function $f_p(z)$ for the pth layer is governed by

$$\alpha_1^{(p)} f_p^{iv}(z) + (2\alpha_3^{(p)} - \alpha_2^{(p)}) f_p''(z) + \alpha_4^{(p)} f_p(z) = \beta_{2,z}^{(p)} - \beta_{1,zz}^{(p)} \tag{3.36}$$

provided that, $f_p(z)$, $f_p'(z)$, $(\alpha_1^{(p)} f_p''(z) + \alpha_3^{(p)} f_p + \beta_1^{(p)})$ and $[\alpha_1^{(p)} f_p'''(z) + (\alpha_3^{(p)} - \alpha_2^{(p)}) f_p'(z) + \beta_{1,z}^{(p)} - \beta_2^{(p)}]$ are continuous across the interfaces. The traction-free surface conditions are reflected by the requirement that $f_p(z)$ and $f_p'(z)$ must be zero on the plate surfaces.

After a considerable amount of algebra, equations (3.33) can be separated into

$$Z_x - a_6 \nabla^2 Z_x = \frac{\partial}{\partial x} (a_1 \nabla^2 \lambda_3 + a_2 \lambda_3) \tag{3.37a}$$

$$Z_y - a_6 \nabla^2 Z_y = \frac{\partial}{\partial y} (a_1 \nabla^2 \lambda_3 + a_2 \lambda_3) \tag{3.37b}$$

$$a_3 \nabla^4 \lambda_3 + a_4 \nabla^2 \lambda_3 + a_5 \lambda_3 = 0 \tag{3.37c}$$

where extraneous solutions are eliminated by the condition

$$\frac{\partial Z_x}{\partial x} + \frac{\partial Z_y}{\partial y} = -a_7 \nabla^2 \lambda_3 - a_8 \lambda_3 \tag{3.38}$$

The parameters $a_1, a_2, \ldots, a_8$ are defined as

$$
\begin{aligned}
a_1 &= \frac{\{[I_4/(I_1 + I_2)] + [I_2 I_3/(I_1^2 - I_2^2)] - [I_3/(2(I_1 - I_2))]\} I_4}{[I_5(I_1 + I_2) + I_3 I_4 - 2I_4^2]} \\
&\quad + \frac{I_1}{(I_1^2 - I_2^2)}
\end{aligned} \tag{3.39a}
$$

$$a_2 = \frac{\{[I_4/(I_1 + I_2)] + [I_2 I_3/(I_1^2 - I_2^2)] - [I_3/(2(I_1 - I_3))]\} (I_1 + I_2)}{[I_5(I_1 + I_2) + I_3 I_4 - 2I_4^2]}$$

(3.39b)

$$a_3 = I_4^2 - I_1 I_5$$

(3.39c)

$$a_4 = -2I_4(I_1 - I_2) + I_1 I_3$$

(3.39d)

$$a_5 = -(I_1^2 - I_2^2)$$

(3.39e)

$$a_6 = \frac{I_3}{2(I_1 - I_2)}$$

(3.39f)

$$a_7 = \frac{I_4}{[I_5(I_1 + I_2) + I_4(I_3 - 2I_4)]}$$

(3.39g)

$$a_8 = \frac{(I_1 + I_2)}{[I_5(I_1 + I_2) + I_4(I_3 - 2I_4)]}$$

(3.39h)

The remaining unknown functions in terms of the in-plane variables can be expressed in terms of λ_3, Z_x, and Z_y as

$$T_{xy} = \frac{1}{(I_1 - I_2)} \frac{\partial^2 \lambda_3}{\partial x \partial y} + \frac{I_3}{2(I_1 - I_2)} \left(\frac{\partial Z_x}{\partial y} + \frac{\partial Z_y}{\partial x} \right)$$

(3.40a)

$$S_x = \frac{1}{(I_1^2 - I_2^2)} \left\{ I_1 \frac{\partial^2 \lambda_3}{\partial x^2} - I_2 \frac{\partial^2 \lambda_3}{\partial y^2} + [I_1 I_3 - I_4(I_1 - I_2)] \frac{\partial Z_x}{\partial x} \right.$$

$$\left. - [I_2 I_3 + I_4(I_1 - I_2)] \frac{\partial Z_y}{\partial y} \right\}$$

(3.40b)

$$S_y = \frac{1}{(I_1^2 - I_2^2)} \left\{ I_1 \frac{\partial^2 \lambda_3}{\partial y^2} - I_2 \frac{\partial^2 \lambda_3}{\partial x^2} + [I_1 I_3 - I_4(I_1 - I_2)] \frac{\partial Z_y}{\partial y} \right.$$

$$\left. - [I_2 I_3 + I_4(I_1 - I_2)] \frac{\partial Z_x}{\partial x} \right.$$

(3.40c)

The Lagrange multipliers are known through the relations

$$\lambda_1 = -I_3 Z_x - \frac{\partial}{\partial x} (\lambda_3)$$

(3.41a)

$$\lambda_2 = -I_3 Z_y - \frac{\partial}{\partial y} (\lambda_3)$$

(3.41b)

The set of equations (3.37), (3.40) and the condition in equation (3.38) together with the boundary conditions form a complete but approximate theory of layered plates in three dimensions. One of the attractive features of this theory is that the problem may be solved in terms of the variables x and y while the function $f(z)$ governing the stress distribution in the thickness direction may be determined separately.

3.4 Through crack in a layered plate

In the case of a layered plate subjected to loads applied to its edges, the formulation involves the superposition of two separate problems. The first considers the same layered plate and loading with the crack absent. By removing the load on the plate boundary and considering a crack in the plate, the second problem deals with the application of tractions to the crack surfaces. These tractions are equal in magnitude and opposite in direction to those found at the crack site of the first problem. The superposition of these two problems renders the solution to the original case of a cracked plate under remote loading.

In what follows, attention is focused on the nontrivial problem of tractions specified on the crack since the application of integral transforms requires the stresses and displacements to vanish at distances far away from the crack. The crack length is also assumed to be small in comparison with the in-plane dimensions of the plate such that the stress distribution around the crack will not interact with the plate boundary. Hence, the dimensions of the plate boundary may be assumed to extend to infinity. Referring to Figure 3.3, the analysis will deal specifically with a three-layered plate containing a crack of length $2a$ which penetrates through the entire plate thickness

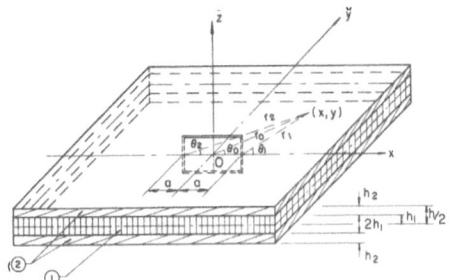

Figure 3.3. Three layer plate containing a through crack

$2(h_1 + h_2)$. A rectangular system of coordinates (x, y, z) are selected such that the crack is centered at the origin and lies in the xz-plane. The problem possesses symmetry about the planes $x = 0$, $y = 0$ and $z = 0$ on which τ_{xy} and τ_{yz} vanish while the displacement components u_x and u_z are zero on the respective planes $x = 0$ and $z = 0$. The conditions to be specified inside and outside the crack at $y = 0$ are

$$\sigma_y(x, 0, z) = p(x, z), 0 \leq x < a \tag{3.42a}$$

$$u_y(x, 0, z) = 0, x > a \tag{3.42b}$$

The equivalent conditions in terms of those functions which depend only on x and y are

$$S_y(x, 0) = p(x), 0 \leq x < a \tag{3.43a}$$

$$\lambda_2(x, 0) = 0, x > a \tag{3.43b}$$

while T_{xy} and Z_y vanish for $y = 0$ and T_{xy}, Z_x and λ_1 vanish for $x = 0$. The symmetry condition about the mid-plane of the layered plate requires $f(z)$ to be an even function of z. The free surface conditions at $z = \pm h/2$, where $h = h_1 + h_2$ are satisfied by requiring $f(z)$ and $f'(z)$ to vanish on these surfaces. Making use of the conditions in equations (3.43), the functions Z_x, Z_y and λ_3 may be determined from equations (3.37) while S_x, S_y and T_{xy} follow from equations (3.40). The remaining unknown Z_z is then given by equation (3.23a).

In-plane variations of the stresses. The Fourier transform will be applied to the space variable x for finding the functions S_x, S_y, etc. Symmetry considerations require S_x, S_y, Z_y, Z_z, λ_2 and λ_3 to be even in x and T_{xy}, Z_x, and λ_1 to be odd in x. Hence, the appropriate Fourier cosine and sine transforms are employed as follows:

$$S_x^c(s, y) = \int_0^\infty S_x(x, y) \cos (sx) dx, \text{ etc.} \tag{3.44a}$$

and

$$T_{xy}^s(s, y) = \int_0^\infty T_{xy}(x, y) \sin (sx) dx, \text{ etc.} \tag{3.44b}$$

Application of the Fourier sine and cosine transforms to the governing equations (3.37) leads to

$$a_3 \frac{d^4 \lambda_3^c}{dy^4} + (a_4 - 2a_3 s^2) \frac{d^2 \lambda_3^c}{dy^2} + (a_3 s^4 - a_4 s^2 + a_5)\lambda_3^c = 0 \tag{3.45a}$$

$$\frac{d^2Z_x^s}{dy^2} - \left(s^2 + \frac{1}{a_6}\right)Z_x^s = s\left[\frac{a_1}{a_6}\frac{d^2\lambda_3^c}{dy^2} + \left(\frac{a_2}{a_6} - \frac{a_1}{a_6}s^2\right)\lambda_3^c\right] \quad (3.45b)$$

$$\frac{d^2Z_y^c}{dy^2} - \left(s^2 + \frac{1}{a_6}\right)Z_y^c = -\frac{a_1}{a_6}\frac{d^3\lambda_3^c}{dy^3} - \left(\frac{a_2}{a_6} - \frac{a_1}{a_6}s^2\right)\frac{d\lambda_3^c}{dy} \quad (3.45c)$$

and equation (3.38) becomes

$$sZ_x^s + \frac{dZ_y^c}{dy} = -a_7\frac{d^2\lambda_3^c}{dy^2} + (a_7s^2 - a_8)\lambda_3^c \quad (3.46)$$

A solution of equations (3.45) that satisfied the regularity conditions at infinity may be written as

$$\lambda_3^c(s, y) = 2\,\text{Re}\left[A_1(s)e^{-q_1y}\right] \quad (3.47a)$$

$$Z_x^s(s, y) = \frac{q_2}{s}B(s)e^{-q_2y} + \text{Re}\left[P_1(s)e^{-q_1y}\right] \quad (3.47b)$$

$$Z_y^c(s, y) = B(s)e^{-q_2y} + \text{Re}\left[P_2(s)e^{-q_1y}\right] \quad (3.47c)$$

in which the parameters q_1 and q_2 stand for

$$q_1 = \pm\left\{s^2 - \frac{a_4}{2a_3} \pm i\left[\frac{a_5}{a_3} - \left(\frac{a_4}{2a_3}\right)^2\right]^{\frac{1}{2}}\right\}^{\frac{1}{2}} \quad (3.48a)$$

$$q_2 = \left(s^2 + \frac{1}{a_6}\right)^{\frac{1}{2}} \quad (3.48b)$$

The quantities P_1 and P_2 in equations (3.47) are given by

$$P_1(s) = -2s\frac{[a_2 - a_1(s^2 - q_1^2)]}{[1 + a_6(s^2 - q_1^2)]}A_1(s) \quad (3.49a)$$

$$P_2(s) = -2q_1\frac{[a_2 - a_1(s^2 - q_1^2)]}{[1 + a_6(s^2 - q_1^2)]}A_1(s) \quad (3.49b)$$

The remaining unknowns may be expressed in terms of λ_3^c, Z_x^s, and Z_y^c. For example, the functions T_{xy}^s, S_y^c, and λ_2^c

$$T_{xy}^s(s, y) = \frac{1}{(I_1 - I_2)}\left[-s\frac{d\lambda_3^c}{dy} + \frac{I_3}{2}\left(\frac{d}{dy}Z_x^s - sZ_y^c\right)\right] \quad (3.50a)$$

$$S_y^c(s, y) = \frac{1}{(I_1^2 - I_2^2)}\left\{I_1\frac{d^2\lambda_3^c}{dy^2} + s^2I_2\lambda_3^c + [I_1I_3 - I_4(I_1 - I_2)]\frac{dZ_y^c}{dy}\right.$$

$$- [I_2 I_3 + I_4 (I_1 - I_2)] s Z_x^s \bigg\} \tag{3.50b}$$

$$\lambda_2^c(s, y) = - I_3 Z_y^c - \frac{d\lambda_3^c}{dy} \tag{3.50c}$$

which are needed for satisfying the required boundary and symmetry conditions. First of all, the condition $Z_y^c(s, 0) = 0$ yields

$$B(s) = \mathrm{Re} \left\{ 2q_1 \left[\frac{a_2 - a_1(s^2 - q_1^2)}{1 + a_6(s^2 - q_1^2)} \right] A_1(s) \right\} \tag{3.51}$$

and $T_{xy}^s(s, 0) = 0$ relates the real and imaginary parts of $A_1(s)$ as

$$\mathrm{Re} \left\{ 2 \left[1 + \frac{I_3}{2} \left(1 - \frac{q_2}{s^2} \right) \left(\frac{a_2 - a_1(s^2 - q_1^2)}{1 + a_6(s^2 - q_1^2)} \right) \right] q_1 A_1(s) \right\} = 0 \tag{3.52}$$

It is convenient to express the above result in the form
$\mathrm{Im}[A_1(s)] = \beta(s)\mathrm{Re}[A_1(s)]$ where

$$\beta(s) = - \left\{ \mathrm{Re}[q_1] \left[s^2 - \frac{I_3}{2a_6} \left(\frac{(a_2 - l_1 a_1)(1 + a_6 l_1) - a_1 a_6 l_2^2}{(1 + a_6 l_1)^2 + (a_6 l_2)^2} \right) \right] \right.$$

$$+ \mathrm{Im}[q_1] \frac{I_3}{2a_6} l_2 \left[\frac{a_1 + a_2 a_6}{(1 + a_6 l_1)^2 + (a_6 l_2)^2} \right] \right\} /$$

$$\left\{ \mathrm{Re}[q_1] \frac{I_3}{2a_6} l_2 \left[\frac{a_1 + a_2 a_6}{(1 + a_6 l_1)^2 + (a_6 l_2)^2} \right] \right.$$

$$- \mathrm{Im}[q_1] \left[s^2 - \frac{I_3}{2a_6} \left(\frac{(a_2 - l_1 a_1)(1 + a_6 l_1) - a_1 a_6 l_2^2}{(1 + a_6 l_1)^2 + (a_6 l_2)^2} \right) \right] \right\} \tag{3.53}$$

In equation (3.53), l_1 and l_2 are defined by

$$l_1 = \frac{a_4}{2a_3} \tag{3.54a}$$

$$- il_2 = \left[\left(\frac{a_4}{2a_3} \right)^2 - \frac{a_5}{a_3} \right]^{\frac{1}{2}} \tag{3.54b}$$

The boundary conditions in equations (3.43) may be expressed in terms of the transformed functions as

$$\frac{2}{\pi} \int_0^\infty S_y^c(s, 0) \cos(sx) ds = - p(x), \quad x < a \tag{3.55a}$$

$$\frac{2}{\pi} \int_0^\infty \lambda_2^c(s, 0) \cos (sx) ds = 0, \, x \geq a \tag{3.55b}$$

The two unknowns $S_y^c(s, 0)$ and $\lambda_2^c(s, 0)$ will be determined in terms of a single function $R(s)$ defined as

$$R(s) = g_1(s) \, \text{Re}[A_1(s)] \tag{3.56}$$

such that

$$\lambda_2^c(s, 0) = R(s) \tag{3.57a}$$

$$S_y^c(s, 0) = \frac{g_2(s)}{g_1(s)} R(s) \tag{3.57b}$$

The functions $g_1(s)$ and $g_2(s)$ stand for

$$g_1(s) = \text{Re}[q_1(1 + i\beta)] \tag{3.58a}$$

$$g_2(s) = \frac{2}{(I_1^2 - I_2^2)} \text{Re}\left[(1 + i\beta)\left\{I_1 q_1 + I_2 s^2 + [-q_1 q_2 (I_1 + I_2) I_3\right.\right.$$

$$+ (q_1^2 I_1 + s^2 I_2) I_3 + (s^2 - q_1^2)(I_1 - I_2) I_4]$$

$$\times \left.\left.\left[\frac{a_2 - a_1 (s^2 - q_1^2)}{1 + a_6 (s^2 - q_1^2)}\right]\right\}\right] \tag{3.58b}$$

Under these considerations, equations (3.55) become

$$\int_0^\infty R(s) \cos (sx) ds = 0, \, x \geq a \tag{3.59a}$$

$$\int_0^\infty sg(s) R(s) \cos (sx) ds = - \frac{\pi p(x)}{2C}, \, x < a \tag{3.59b}$$

in which

$$g(s) = \frac{g_2(s)}{g_1(s)} \frac{1}{sC} \tag{3.60}$$

The constant C is equal to m_6/m_2 where m_2 and m_6 are defined in the Appendix and is obtained from the condition that

$$\lim_{s \to \infty} g(s) \to 1 + O\left(\frac{1}{s^2}\right) \tag{3.61}$$

The system of dual integral equations (3.59) may be solved by reduction to a standard Fredholm integral equation of the second kind. This is accomplished by using Copson's method [6] which involves expressing $R(s)$ in terms of $\Phi(\xi)$:

$$R(s) = -\frac{a^2}{C} \int_0^1 \sqrt{\xi}\, \Phi(\xi)\, J_0(sa\xi) \int_0^\xi \frac{p(a\eta)d\eta}{(\xi^2 - \eta^2)^{\frac{1}{2}}}\, d\xi \tag{3.62}$$

where J_0 is the zero-order Bessel function of the first kind. The Fredholm integral equation is of the form

$$\Phi(\xi) + \int_0^1 K(\xi, \eta)\, \Phi(\eta)d\eta = \sqrt{\xi} \tag{3.63}$$

whose kernel being symmetric in ξ and η is given by

$$K(\xi, \eta) = (\xi\eta)^{\frac{1}{2}} \int_0^\infty s\left[g\left(\frac{s}{a}\right) - 1\right] J_0(s\xi)\, J_0(s\eta)ds \tag{3.64}$$

Equation (3.63) will be approximated by a set of algebraic equations using Simpson's rule and solved numerically for the case of a constant load applied to the crack surfaces, i.e., $p(x) = \sigma_0$.

Stress intensity factor. Those terms of the stresses which are singular at the crack front can be extracted from equation (3.62) and they are associated with $\Phi(1)$ representing the value of $\Phi(\xi)$ evaluated at the crack tip, $\xi = 1$. For convenience, the polar coordinates r_0, r_1, r_2 and $\theta_0, \theta_1, \theta_2$ as shown in Figure 3.3 are introduced so that

$$S_x(x, y) = -\frac{a\sigma_0}{(r_1 r_2)^{\frac{1}{2}}}\, C\Phi(1)\left\{\frac{ay}{r_1 r_2} \sin\left[\tfrac{3}{2}(\theta_1 + \theta_2)\right]\right.$$

$$\left. -\frac{r_0}{a} \cos\left[\theta_0 - \tfrac{1}{2}(\theta_1 + \theta_2)\right]\right\} + O(1) \tag{3.65a}$$

$$S_y(x, y) = +\frac{a\sigma_0}{(r_1 r_2^{\frac{1}{2}})}\, C\Phi(1)\left\{\frac{ay}{r_1 r_2} \sin\left[\tfrac{3}{2}(\theta_1 + \theta_2)\right]\right.$$

$$\left. +\frac{r_0}{a} \cos\left[\theta_0 - \tfrac{1}{2}(\theta_1 + \theta_2)\right]\right\} + O(1) \tag{3.65b}$$

$$T_{xy}(x, y) = -\frac{a\sigma_0}{(r_1 r_2)^{\frac{1}{2}}}\, C\Phi(1)\left\{\frac{ay}{r_1 r_1} \cos\left[\tfrac{3}{2}(\theta_1 + \theta_2)\right]\right\} + O(1) \tag{3.65c}$$

$$Z_z(x, y) = -\frac{a\sigma_0}{(r_1 r_2)^{\frac{1}{2}}}\, C\Phi(1)\left[\frac{e_1(l_1 b_2 + l_2 b_1)}{m_2}\right]$$

$$\times \left\{ \frac{r_0}{a} \cos \left[\theta_0 - \tfrac{1}{2}(\theta_1 + \theta_2) \right] \right\} + O(1) \tag{3.65d}$$

$$Z_x(x, y) = a\sigma_0 \Phi(1) \frac{1}{m_2} \left\{ \left[\frac{e_1}{2} \left(b_1 l_2 + b_2 l_1 - \frac{b_2}{a_6} \right) \right] \right.$$

$$\times \int_0^\infty \frac{1}{s} J_1 (as) \sin (sx) \, e^{-sy} ds$$

$$+ \frac{e_1}{2} \left(b_1 l_2 + b_2 l_1 + \frac{b_2}{a_6} \right) \int_0^\infty y J_1 (as) \sin (sx) \, e^{-sy} ds \right\} \tag{3.65e}$$

$$Z_y(x, y) = a\sigma_0 \Phi(1) \frac{e_1}{2m_2} \left(b_1 l_2 + b_2 l_1 + \frac{b_2}{a_6} \right)$$

$$\times \int_0^\infty y J_1 (as) \cos (sx) \, e^{-sy} \, ds \tag{3.65f}$$

where b_1, b_2, e_1 and m_2 are given in the Appendix. Note that $Z_x(x, y)$ and $Z_y(x, y)$ are nonsingular. Further, approaching the crack tip $(a, 0)$, i.e, taking the limit $r \to a$, $\theta \to 0$, $r_2 \to 2a$, $\theta_2 \to 0$, it follows from equations (3.21) and (3.65) that the near tip stresses may be expressed as

$$\sigma_x = \frac{\alpha \sigma_p(z) \sqrt{a}}{(2r_1)^{\frac{1}{2}}} \left[\cos \left(\frac{\theta_1}{2} \right) - \tfrac{1}{2} \sin (\theta_1) \sin \left(\frac{3\theta_1}{2} \right) \right] + O(1) \tag{3.66a}$$

$$\sigma_y = \frac{\alpha \sigma_p(z) \sqrt{a}}{(2r_1)^{\frac{1}{2}}} \left[\cos \left(\frac{\theta_1}{2} \right) + \tfrac{1}{2} \sin (\theta_1) \sin \left(\frac{3\theta_1}{2} \right) \right] + O(1) \tag{3.66b}$$

$$\tau_{xy} = \frac{\alpha \sigma_p(z) \sqrt{a}}{(2r_1)^{\frac{1}{2}}} \left[\tfrac{1}{2} \sin (\theta_1) \cos \left(\frac{3\theta_1}{2} \right) \right] + O(1) \tag{3.66c}$$

$$\sigma_z = \frac{\alpha \sigma_T(z) \sqrt{a}}{(2r_1)^{\frac{1}{2}}} \left[\cos \left(\frac{\theta_1}{2} \right) \right] + O(1) \tag{3.66d}$$

$$\tau_{xz} = \tau_{yz} = O(1) \tag{3.66e}$$

in which the parameter α stands for

$$\alpha = C\Phi(1) \tag{3.67}$$

The functions $\sigma_p(z)$ and $\sigma_T(z)$ in equations (3.66) depend on the thickness variable z and are

$$\sigma_p(z) = \sigma_0 f''_p(z) \tag{3.68a}$$

$$\sigma_T(z) = \frac{\sigma_0 e_1}{Cm_2} (b_1 l_2 + b_2 l_1) f_p(z) \tag{3.68b}$$

Note that for in-plane loading, the stress components τ_{xz} and τ_{yz} remain finite at the crack edge $r_1 = 0$ while the other stress components are singular of the order of $1/(r_1)^{\frac{1}{2}}$ as $r_1 \to 0$. This behavior and the angular variations of the stresses in θ_1 are in agreement with those found by Sih [3] for the exact three-dimensional solution of a crack in an elastic solid. From equations (3.66), the following stress intensity factors may be defined:

$$k_1^{(p)}(z) = \alpha \sigma_0 f_p''(z) \sqrt{a} \tag{3.69a}$$

$$k_1^{(T)}(z) = \frac{\alpha \sigma_0 e_1}{Cm_2}(b_1 l_2 + b_2 l_1) f_p(z) \sqrt{a} \tag{3.69b}$$

Unlike the two-dimensional solution which is independent of z, these stress intensity factors vary along the crack front through the functions $f(z)$ and $f''(z)$. In view of equation (3.68a), equation (3.69a) may be written as

$$\alpha = \frac{k_1^{(p)}(z)}{\sigma_0(z)\sqrt{a}} \tag{3.70}$$

which may be regarded as a measure of the normalized intensification of the local in-plane stresses. Equation (3.67) shows that α can be computed from the Fredholm integral equation for $\Phi(1)$.

3.5   Stress distribution across the plate thickness

The stress distribution across each layer of the plate as governed by $f_p(z)$ and its derivatives may be determined from equation (3.36). In the immediate vicinity of the crack front, the asymptotic stress solution in equations (3.66) suggests an alternative approach that has also been used successfully elsewhere [3, 7]. More specifically, equations (3.66) indicate that all interior points near the crack front are in a state of plane strain, i.e.,

$$\sigma_z = v(\sigma_x + \sigma_y) \tag{3.71}$$

which can be employed to obtain $f_p(z)$ in each layer. This requirement yields a differential equation for $f_p(z)$:

$$f_p''(z) + \omega_p^2 f_p(z) = 0 \tag{3.72}$$

where ω_p^2 is equal to the asymptotic value of the function $-Z_z/[v_p(S_x + S_y)]$ expanded near the crack tip. The result gives

$$\omega_p^2 = \frac{1}{v_p} \left[\frac{e_1}{2m_6} (b_1 l_2 + b_2 l_1) \right] \tag{3.73}$$

Referring to Figure 3.3, the subscript $p = 1$ will be attached to quantities associated with the middle layer of thickness $2h_1$ while the subscript $p = 2$ refers to the outer layers each of thickness h_2. Solutions of equation (3.72) that satisfy continuity of $f_p(z)$ and $f_p'(z)$ across the material interfaces are

$$f_1(z) = A \cos (\omega_1 z), \ 0 \leqslant z \leqslant h_1 \tag{3.74a}$$

$$f_2(z) = A \left\{ \cos (\omega_1 h_1) \cos [\omega_2(z - h_1)] \right.$$

$$\left. - \frac{\omega_1}{\omega_2} \sin (\omega_1 h_1) \sin [\omega_2(z - h_1)] \right\}, \ h_1 \leqslant z \leqslant (h_1 + h_2) \tag{3.74b}$$

where A is a constant and ω_p ($p = 1,2$) are related to the constants b_1, b_2, etc., as shown in equation (3.73). Refer to the Appendix for the definitions of the constants in equation (3.73).

Equation (3.71) or (3.72) is not expected to hold on the free surfaces $z = \pm h/2$ where $f_2(z)$ and $f_2'(z)$ must vanish since plane strain applies only to the interior region of the plate. For a single layer plate, Hartranft and Sih [2] have suggested to introduce two narrow layers of thickness $\varepsilon h/2$ next to the free surfaces. The quantity $\varepsilon = 1/2[1 + (4h/2)]$ as selected by Hartranft and Sih [2] is referred to as the non-dimensional boundary layer* thickness. For a three-layered plate, ε is chosen as

$$\varepsilon = \frac{1}{2 + 16 \left[(h_1 + h_2)/a \right]} \left(\frac{h_2}{h_1} \right) \tag{3.75}$$

The function

$$f_2(z) = \gamma_1(z - h_1 - h_2)^4 + \gamma_2(z - h_1 - h_2)^3 + \gamma_3(z - h_1 - h_2)^2$$

$$+ \gamma_4(z - h_1 - h_2) + \gamma_5, \ [h_1 + h_2(1 - \varepsilon)] \leqslant z \leqslant (h_1 + h_2) \tag{3.76}$$

is constructed such that the coefficients γ_j ($j = 1, 2, \ldots, 5$) are determined from the continuity of its first and second derivatives across the layer at $z = h_1 + h_2 (1 - \varepsilon)$ and the traction-free conditions are satisfied by requiring $f_2(z) = f_2'(z) = 0$ for $z = \pm(h_1 + h_2)$.

---

* A more detailed discussion on the theoretical and experimental justifications for introducing ε can be found in the work of Villarreal, Sih and Hartranft [7].

The determination of ω_1 in equations (3.74) is accomplished by employing an iterative procedure in which an initial value of ω_1 is assumed for calculating the asymptotic expressions of the functions S_x, S_y and Z_z. The result of imposing equation (3.71) on S_x, S_y and Z_z leads to a new value of ω_1. This procedure is repeated until a set of discrete values of ω_1 is found and equation (3.72) is satisfied. The value of ω_2 follows directly from equation (3.73) since $\omega_2^2 = (v_1/v_2)\omega_1^2$.

Once $f_p(z)$ ($p = 1, 2$) are known, the crack edge stress field as given by equations (3.66) is completely determined. To recapitulate, the in-plane stress components are proportional to $f_p''(z)$, and the normal stress component σ_z is equal to $v_p(\sigma_x + \sigma_y)$ except in the boundary layer while the transverse shear stresses τ_{xz} and τ_{yz} are non-singular. Hence, the intensity of the singular stress field in the vicinity of the crack edge is described by the stress intensity factors in equations (3.69) with the values of ω_1 and ω_2 governing the behavior of $f_p(z)$ and its derivatives.

It should be mentioned that additional boundary layers of the type discussed earlier may also be introduced in the neighborhood of the material interfaces. The effect of such layers on the stress distribution has been discussed by Hilton and Sih [8] and will not be covered here.

3.6   Discussion of numerical results

Recall from equations (3.67) and (3.71) that the stress intensity factor $k_1^{(p)}(z)$ can be computed from the Fredholm integral equation (3.63) for $\Phi(\xi)$ with $\xi = 1$. The results will depend on the geometric and material constants h_2/h_1, h_1/a, E_1/E_2, v_1 and v_2 and presented in a nondimensional form $k_1^{(p)}(z)/\sigma_p(z)\sqrt{a}$, where $\sigma_p(z)$ is the crack face load distribution that varies through the plate thickness.

Figures 3.4 to 3.7 give the values of $k_1^{(p)}(z)$ as a function of h_1/a without the presence of any boundary layers. The parameter ω_1 in $f_p(z)$ may be chosen arbitrarily as the free-surface conditions for $z = \pm(h_1 + h_2)$ are not satisfied. The symbol $\Omega_1 = \omega_1 h_1$ is varied within the range 0.1 and 1.5 with $0 < \omega_1 < \pi/2$. Each of the curves with a specific value of Ω_1 or ω_1 corresponds to a given stress distribution across the layered plate. Note that the normalized stress intensity factor increases as ω_1 is decreased for a fixed ratio h_1/a. In the limit as h_1/a becomes large, $k_1^{(p)}(z)/\sigma_p(z)\sqrt{a}$ approaches unity for all values of ω_1. From equations (3.74), it can be seen that as ω_1 and ω_2 (being directly proportional to ω_1) decrease the stress field that

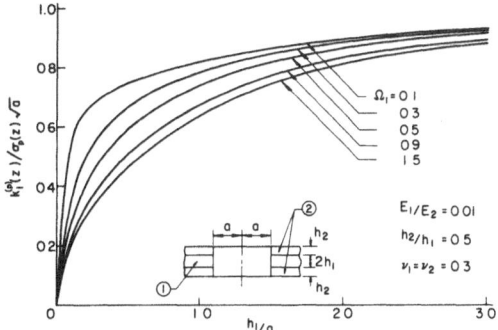

Figure 3.4. Normalized stress intensity factor for $E_1/E_2 = 0.01$ and $h_2/h_1 = 0.5$

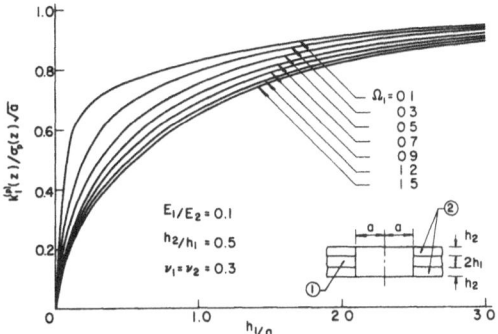

Figure 3.5. Normalized stress intensity factor for $E_1/E_2 = 0.1$ and $h_2/h_1 = 0.5$

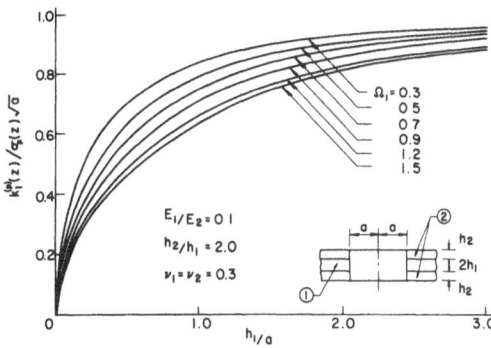

Figure 3.6. Normalized stress intensity factor for $E_1/E_2 = 0.1$ and $h_2/h_1 = 2.0$

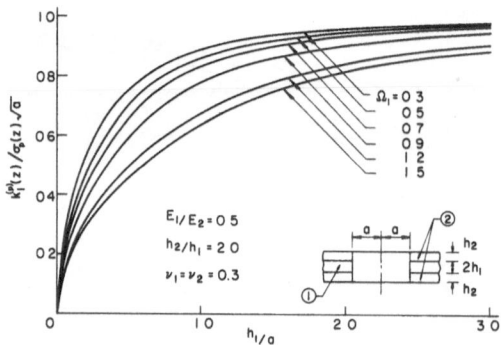

Figure 3.7. Normalized stress intensity factor for $E_1/E_2 = 0.5$ and $h_2/h_1 = 2.0$

is governed by $f_p(z)$ becomes less and less dependent on z. In fact, as ω_1 approaches zero, the crack edge stress field in equations (3.66) becomes independent of z and reduced to the two-dimensional plane strain solution for which $k_1^{(p)} = k_1^{(T)} = \sigma_0\sqrt{a}$.

The dependence of $k_1^{(p)}(z)$ on the change in relative layer thickness h_2/h_1 is also exhibited in Figures 3.4 to 3.7. As h_2/h_1 increases, the stress intensity factor increases for fixed h_1/a until it approaches the limit unity for large h_2/h_1. This is expected since increasing plate thickness results in recovering the plane strain condition that corresponds to large values of h_2/h_1.

In the absence of the boundary layers, the influence of the relative layer stiffness E_1/E_2 is rather weak. No appreciable change in $k_1^{(p)}(z)$ is observed for $E_1/E_2 = 0.01$, 0.1 and 0.5.

Recall that the parameter ω_1 can no longer be chosen arbitrarily if the boundary layers next to the free surfaces are to be included in the analysis. For simplicity, the case $v_1 = v_2$ and hence $\omega_1 = \omega_2$ will be considered. This suggests that $\omega_1(h_1 + h_2)$ is almost constant and approximately equal to $\pi/2$. Numerical results of $k_1^{(p)}(z)$ are presented graphically as function of h_1/a for different values of E_1/E_2, h_2/h_1 with $v_1 = v_2 = 0.3$. Figures 3.8 to 3.12 show that $k_1^{(p)}(z)$ decreases as E_1/E_2 increases while the other parameters are held constant. Increasing E_1/E_2 means that more load is transmitted to the middle layer. In the limit as E_2 goes to zero, the actual plate half thickness is reduced from $h_1 + h_2$ to h_1 and the stress intensity factor is expected to decrease accordingly*. The influence of h_2/h_1 on this effect can

* In the present model, it is not possible to account completely for this effect. The load distribution permitted in this formulation for finite but small E_2 is qualitatively different than for $E_2 = 0$, i.e., for E_2 finite, $p_1 \approx \pi/2\,(1(h_1 + h_2))$ while at $E_2 = 0$, $p_1 \approx \pi/2\,(1/h_1)$.

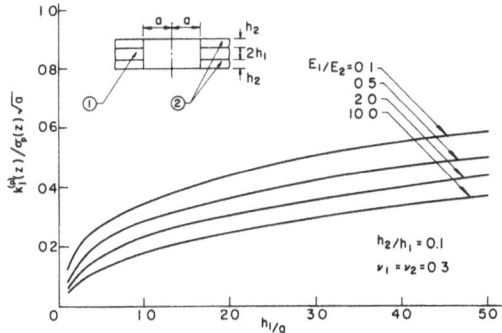

Figure 3.8. Variations of stress intensity factor with h_1/a modulus ratios and $h_2/h_1 = 0.1$ (boundary layers included)

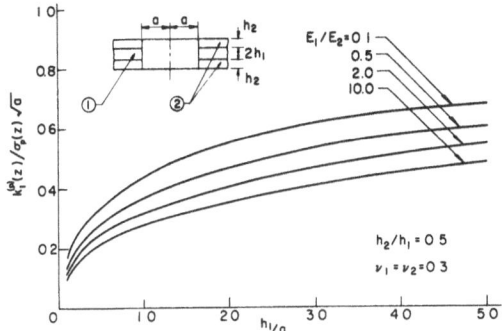

Figure 3.9. Variations of stress intensity factor with h_1/a for different modulus ratios and $h_2/h_1 = 0.5$ (boundary layers included)

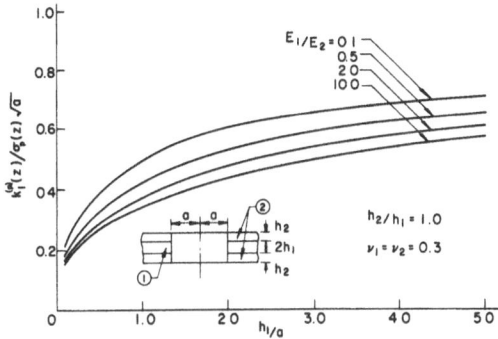

Figure 3.10. Variations of stress intensity factor with h_1/a for different modulus ratios and $h_2/h_1 = 1.0$ (boundary layers included)

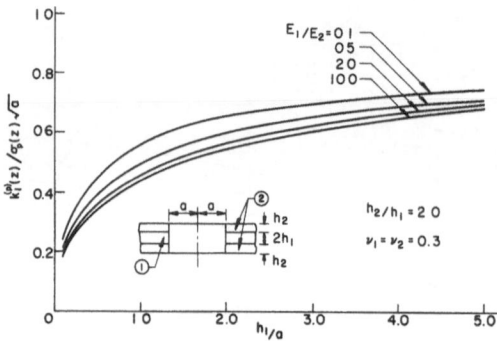

Figure 3.11. Variations of stress intensity factor with h_1/a for different modulus ratios and $h_2/h_1 = 2.0$ (boundary layers included)

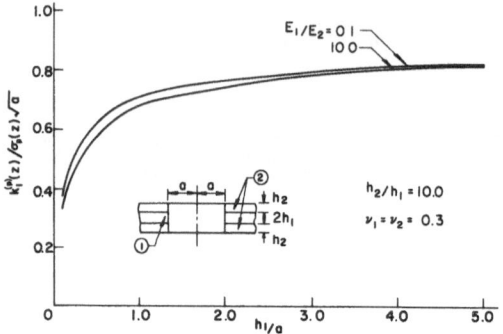

Figure 3.12. Variations of stress intensity factor with h_1/a for different modulus ratios and $h_2/h_1 = 10.0$ (boundary layers included)

be seen by comparing Figures 3.8 to 3.12. As h_2/h_1 increases, the effect of changes of E_1/E_2 on $k_1^{(p)}(z)$ becomes less pronounced. To understand these interactions, recall that at finite E_1/E_2 increasing h_2/h_1 causes the outer layers to carry a larger portion of the applied load and for very large h_2/h_1 the layered plate behavior should be almost independent of the material properties of the inside layer.

Figures 3.13 to 3.16 demonstrate the influence of the parameter h_2/h_1 on $k_1^{(p)}(z)$. If h_1/a is held constant, increasing h_2/h_1 increases the total plate thickness and $k_1^{(p)}(z)$ is observed to increase accordingly.

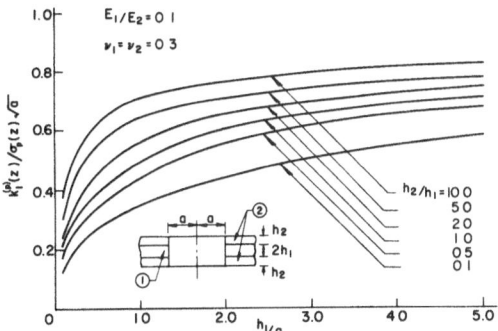

Figure 3.13. Relative layer thickness effect with boundary layers for $E_1/E_2 = 0.1$

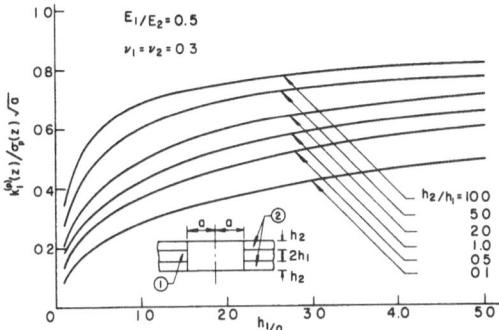

Figure 3.14. Relative layer thickness effect with boundary layers for $E_1/E_2 = 0.5$

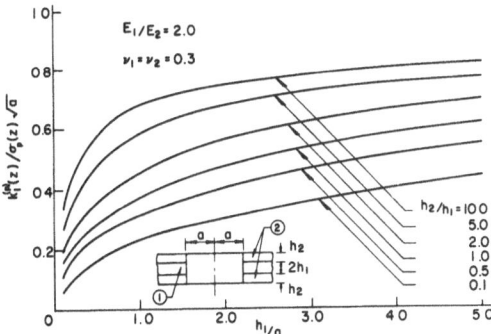

Figure 3.15. Relative layer thickness effect with boundary layers for $E_1/E_2 = 2.0$

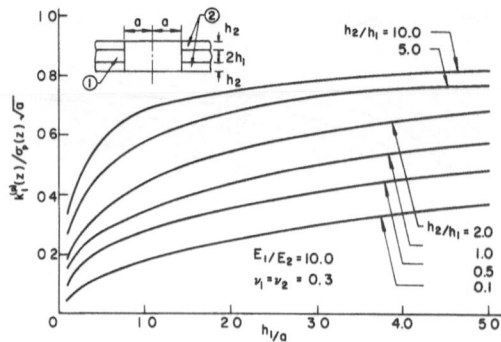

Figure 3.16. Relative layer thickness effect with boundary layers for $E_1/E_2 = 10.0$

Finally, Figure 3.17 gives a plot of $k_1^{(p)}(z)/\sigma_p(z)\sqrt{a}$ against h_1/a for the special case of holding the total plate thickness to crack length constant $(h_1 + h_2)/a) = 3.0$. The dotted curve represents a homogeneous plate where $E_1 = E_2$ and $k_1^{(p)}(z)$ is not expected to vary with h_1/a or h_1/h_2 once $(h_1 + h_2)/a$ is fixed. The slight variation observed is associated with the approximate nature of the constructed boundary layer thickness.

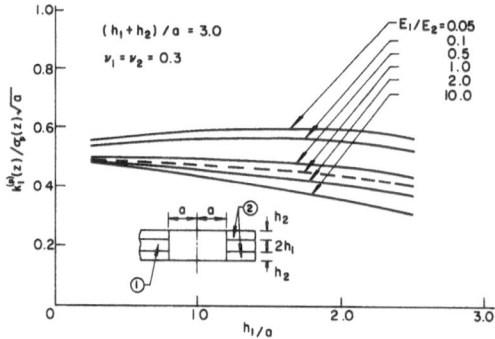

Figure 3.17. Stress intensity factor variations for constant normalizes plate thickness with boundary layers

The approximate theory developed in this chapter can be easily applied to solve problems involving the bending of layered plates. This can be accomplished simply by replacing the symmetric function $f_p(z)$ which governs through-the-thickness variation of the stress field due to in-plane stretching by a skew-symmetric one for the case of bending.

3.7 Appendix: Definition of constants

The various parameters that arise in the formulation of the layered plate theory are given in this Appendix. The quantities I_1, I_2, etc., are integrals that can be computed numerically and they are defined by equations (3.29).

$$b_1 = -2 \frac{a_1}{a_6} \frac{\{[l_1 + (1/a_6)][(a_2/a_1) - l_1] - l_2\}}{\{[l_1 + (1/a_6)]^2 + l_2^2\}} \tag{A3.1a}$$

$$b_2 = 2 \frac{a_1}{a_6} \frac{l_2[(1/a_6) + (a_2/a_1)]}{\{[l_1 + (1/a_6)]^2 + l_2^2\}} \tag{A3.1b}$$

Refer to equations (3.39) for the definitions of a_1, a_2, etc.

$$c_1 = 4a_6 \tag{A3.2a}$$

$$c_2 = I_3 b_1 - 2l_1 a_6 \tag{A3.2b}$$

$$c_3 = \frac{I_3}{2}(l_2 b_2 - l_1 b_1) - \tfrac{1}{2}(l_1^2 - l_2^2)a_6 \tag{A3.2c}$$

$$d_1 = 2l_2 a_6 - I_3 b_2 \tag{A3.3a}$$

$$d_2 = l_1 l_2 a_6 + \frac{I_3}{2}(l_2 b_1 + l_1 b_2) \tag{A3.3b}$$

$$d_3 = I_3[\tfrac{1}{4}l_1 l_2 b_1 + \tfrac{1}{8}(l_1^2 + l_2^2)b_2 - \tfrac{1}{4}l_2(l_2^2 - 3l_2^2)a_6] \tag{A3.3c}$$

$$e_1 = \frac{c_1}{d_1} \tag{A3.4a}$$

$$e_2 = \frac{1}{d_1}\left(c_2 - c_1 \frac{d_2}{d_1}\right) \tag{A3.4b}$$

$$e_3 = \frac{1}{d_1}\left[c_3 - c_2 \frac{d_2}{d_1} - c_1\left(\frac{d_3}{d_1} - \left(\frac{d_2}{d_1}\right)^2\right)\right] \tag{A3.4c}$$

$$m_2 = 2\left(1 - \frac{l_2}{2}e_1\right) \tag{A3.5a}$$

$$m_3 = -\left(l_1 + l_2 e_2 + \frac{l_1 l_2}{2}e_1\right) \tag{A3.5b}$$

$$m_4 = -\left[\frac{(l_1^2 - l_2^2)}{4} + l_2 e_3 + \frac{l_1 l_2}{2}e_2 - \frac{l_2}{8}(l_2^2 - 3l_1^2)e_1\right] \tag{A3.5c}$$

$$m_6 = \frac{2}{(I_1^2 - I_2^2)}\left[(I_1 + I_2) - e_1\left\{I_1 l_2 + \frac{b_1 l_2}{2}\left[\frac{(I_1 + I_2)I_3}{2}\right.\right.\right.$$

$$+ I_4(I_1 - I_2) - I_1 I_3\right] - \frac{b_2}{2}\left[\frac{(I_1 + I_2)I_3}{2}\left(\frac{1}{a_6} l_1\right)\right.$$

$$\left.\left.\left.- (I_4(I_1 - I_2) - I_1 I_3)l_1\right]\right\}\right] \qquad \text{ } \qquad \text{(A3.5d)}$$

$$m_7 = \frac{2}{(I_1^2 - I_2^2)}\left[-I_1(l_1 + l_2 e_2) + \frac{b_1}{2}\left\{-(I_1 + I_2)I_3\left[\frac{4a_6}{1} l_2 e_1\right.\right.\right.$$

$$\left.+ \left(\frac{l_1}{2} + \frac{l_2}{2} e_2 + \frac{l_1 l_2}{4} e_1\right)\right] - [I_4(I_1 - I_2) - I_1 I_3](l_1 + l_2 e_2)\right\}$$

$$+ \frac{b_2}{2}\left\{-(I_1 + I_2)I_3\left[\frac{e_1}{8a_6^2} + \frac{1}{2a_6}\left(\frac{l_1}{2} e_1 - e_2\right) + \frac{(l_1^2 - l_2^2)}{8} e_1\right.\right.$$

$$\left.\left.\left.+ \frac{l_1}{2} e_2 - \frac{l_2}{2}\right] - [I_4(I_1 - I_2) - I_1 I_3](l_1 e_2 - l_2)\right\}\right] \qquad \text{(A3.5e)}$$

$$m_8 = \frac{2}{(I_1^2 - I_2^2)}\left[-I_1 l_2 e_3 + \frac{b_1}{2}\left\{-\frac{(I_1 + I_2)I_3}{2}\left[\left(e_3 - \frac{1}{8a_6^2} e_1\right)l_2\right.\right.\right.$$

$$+ \frac{1}{2a_6}\left(l_1 + l_2 e_2 + \frac{l_1 l_2}{2} e_1\right) + \frac{(l_1^2 - l_2^2)}{4} + \frac{l_1 l_2}{2} e_2$$

$$\left.- \tfrac{1}{8}(l_2^3 - 3l_1^2 l_2)e_1\right] - [I_4(I_1 - I_2) - I_1 I_3](l_2 e_3)\right\}$$

$$+ \frac{b_2}{2}\left\{\frac{(I_1 + I_2)I_3}{2}\left[\frac{1}{8a_6^3} e_1 + \frac{1}{4a_6^2}\left(\frac{l_1 l e_1}{2} - e_2\right)\right.\right.$$

$$- \frac{2}{a_6}\left(\frac{(l_1^2 - l_2^2)}{4} e_1 + l_1 e_2 - l_2 - 2e_3\right) - \tfrac{1}{8}(l_1^2 - 3l_2^2)l_1 e_1$$

$$\left.- \frac{(l_1^2 - l_2^2)}{4} e_2 - l_1 e_3 + \frac{l_1 l_2}{2}\right]$$

$$\left.\left.- [I_4(I_1 - I_2) - I_1 I_3](l_1 e_3)\right\}\right] \qquad \text{(A3.5f)}$$

$$C = \frac{m_6}{m_2} \qquad \text{(A3.6)}$$

References

[1] Reissner, E., On bending of elastic plates, *Quarterly of Applied Math.*, 5, pp. 55–68 (1947).

[2] Hartranft, R. J. and Sih, G. C., An approximate three-dimensional theory of plate with application to crack problems, *Int. J. Engng. Sci.*, 8, pp. 711–729 (1970).

[3] Sih, G. C., A review of the three-dimensional stress problem for a cracked plate, *Int. J. of Fracture Mechanics*, 7, pp. 39–61 (1971).

[4] Sih, G. C., Williams, M. L. and Sewdlow, J. L., Three-dimensional stress distribution near a sharp crack in a plate of finite thickness, *Air Force Materials Laboratory, Wright-Patterson Air Force Base*, AFML-TR-66-242 (1966).

[5] Hartranft, R. J. and Sih, G. C., The use of eigenfunction expansions in the general solution of three-dimensional crack problems, *Journal of Mathematics and Mechanics*, 19, pp. 123–138, (1969).

[6] Copson, E. T., On certain dual integral equations, *Proc. Glasgow Math. Assoc.*, 5, pp. 19–24 (1961).

[7] Villarreal, G., Sih, G. C. and Hartranft, R. J., Photoelastic investigation of a thick plate with a transverse crack, *Journal of Applied Mech.*, 42, pp. 9–14 (1975).

[8] Hilton, P. D. and Sih, G. C., Three-dimensional analysis of laminar composites with through cracks, *American Society of Testing and Materials*, STP 593, pp. 1–35 (1975).

E. S. Folias

4 | *Asymptotic approximations to crack problems in shells*

4.1 Introduction

In nature, shells are the rule rather than the exception. The list of natural shell-like structures is long, and the strength properties of some of them are remarkable. It is logical, therefore, for man to utilize them in man-made structures. But to do this safely, we must understand the fundamental laws which govern the strength and displacement behavior of such structures for they are not immune to failures, particularly in the fracture mode.

It is the intent, therefore, of this chapter to discuss a theoretical method which enables one to determine the stress field that exists in the neighborhood of a crack and furthermore catalog the stress intensity factors for various shell configurations and loads.

In section 2, the author gives a concise summary of the classical shell theory and its limitations. He then goes on to discuss the general character of the equations and subsequently shows that for the two simple geometries, spherical and cylindrical, the equations reduce considerably and in the limit the governing equations of a flat plate are recovered.

Because the solution for a general arbitrary initial curvature presents formidable mathematical complexities, in section 3 he chooses to display the analytical method by specializing it to a spherical shell. In order to preserve unity, he does this in some great length giving sufficient details.

In sections 5 and 6 he gives the stress intensity factors for a cylindrical shell with various crack orientations and for other more complicated shell geometries.

In section 7, he examines what effect, if any, elastic foundations have on the stress intensity factors. Such information can be of great practical value to highway construction and the designing of storage tanks for the oil industry.

4.2   General theory—classical

In the following, we consider bending and stretching of thin shells of revolution, as described by the traditional two-dimensional linear theory and with the additional assumption of shallowness*. In speaking of the formulation of two-dimensional differential equations, we mean the transition from the exact three-dimensional elasticity problem to that of two-dimensional approximate formulation, which is appropriate in view of the 'thinness' of the shell. We shall, furthermore, limit our considerations to homogeneous, isotropic, constant thickness, shallow segments of shells, subjected to small deformations and strains so that the stress-strain relations may be established through Hooke's law.

The basic variables in the theory of shallow shells are the displacement function $w(X, Y)$ in the direction of an axis Z and a stress function $F(X, Y)$ which represents the stress resultatnts tangent to the middle surface of the shell. Following Marguerre [2], the coupled differential equations governing w and F, with X and Y as rectangular cartesian coordinates of the base plane (see Figure 4.1), are given by:

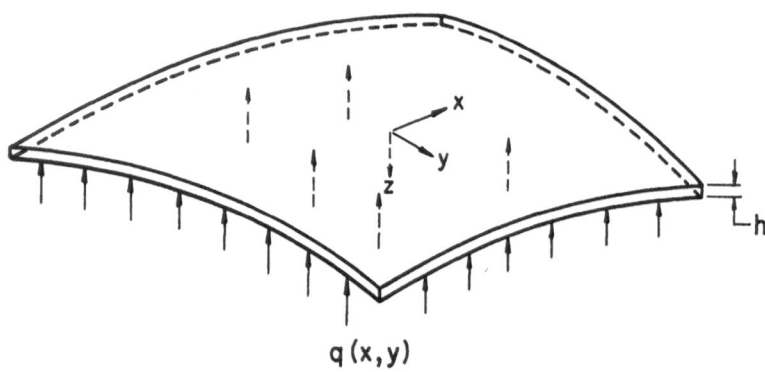

$$q(x,y)$$

Figure 4.1. Initially Curved Sheet

$$\nabla^4 F = Eh \left[2 \frac{\partial^2 w_0}{\partial X \partial Y} \frac{\partial^2 w}{\partial X \partial Y} - \frac{\partial^2 w_0}{\partial X^2} \frac{\partial^2 w}{\partial Y^2} - \frac{\partial^2 w_0}{\partial Y^2} \frac{\partial^2 w}{\partial X^2} \right] \quad (4.1a)$$

---

* According to Ogibalov [1], a shell will be called: shallow if the least radius of curvature is greater by one order of magnitude than the linear dimensions, i.e., $L/R \leqq 0.1$; and thin if $h/R \leqq 0.01$.

$$D\nabla^4 w = -q - 2 \frac{\partial^2 F}{\partial X \partial Y} \frac{\partial^2 w_0}{\partial X \partial Y} + \frac{\partial^2 F}{\partial X^2} \frac{\partial^2 w_0}{\partial Y^2} +$$

$$+ \frac{\partial^2 F}{\partial Y^2} \frac{\partial^2 w_0}{\partial X^2} \tag{4.1b}$$

where ∇^4 is the biharmonic operator, E Young's modulus, h the thickness of the shell, D the flexural rigidity, q the internal pressure and $w_0(X, Y)$ the initial shape of the shell in reference to that of a flat plate.

The usual bending moment components M_x, M_y, M_{xy} are defined in terms of the displacement function w as

$$M_x = -D\left[\frac{\partial^2 w}{\partial X^2} + v \frac{\partial^2 w}{\partial Y^2} \right] \tag{4.2a}$$

$$M_y = -D\left[v \frac{\partial^2 w}{\partial X^2} + \frac{\partial^2 w}{\partial Y^2} \right] \tag{4.2b}$$

$$M_{xy} = -D\left[(1 - v) \frac{\partial^2 w}{\partial X \partial Y} \right] \tag{4.2c}$$

and the membrane forces in terms of the stress function F as

$$N_x = \frac{\partial^2 F}{\partial Y^2} \tag{4.3a}$$

$$N_y = \frac{\partial^2 F}{\partial X^2} \tag{4.3b}$$

$$N_{xy} = -\frac{\partial^2 F}{\partial X \partial Y} \tag{4.3c}$$

Finally, in view of equations (4.2) and (4.3), the bending and extensional stress components become:

$$\sigma_x^{(b)} = -\frac{EZ}{(1 - v^2)}\left[\frac{\partial^2 w}{\partial X^2} + v \frac{\partial^2 w}{\partial Y^2} \right] \tag{4.4a}$$

$$\sigma_y^{(b)} = -\frac{EZ}{(1 - v^2)}\left[v \frac{\partial^2 w}{\partial X^2} + \frac{\partial^2 w}{\partial Y^2} \right] \tag{4.4b}$$

$$\tau_{xy}^{(b)} = -2GZ\left[\frac{\partial^2 w}{\partial X \partial Y} \right] \tag{4.4c}$$

and

$$\sigma_x^{(e)} = \frac{1}{h} \frac{\partial^2 F}{\partial Y^2} \tag{4.5a}$$

$$\sigma_y^{(e)} = \frac{1}{h} \frac{\partial^2 F}{\partial X^2} \tag{4.5b}$$

$$\tau_{xy}^{(e)} = -\frac{1}{h} \frac{\partial^2 F}{\partial X \partial Y} \tag{4.5c}$$

with v being Poisson's ratio.

Because of the coupled nature of the differential equations (4.1), it becomes apparent that there exists an interaction between bending and stretching. That is, a bending load will generally produce both bending and extensional stresses, and similarly a stretching load will also induce both bending and extensional stresses. The subject of eventual concern, therefore, is that of the simultaneous stress fields produced in an initially curved sheet containing a crack.

A theoretical attack of the general problem for an arbitrary initial curvature presents formidable mathematical complexities. However, for the two simple geometries, spherical and cylindrical shells, exact solutions can be obtained in an asymptotic form. On the other hand, for other more complicated shell geometries results can be obtained by a proper superposition of these two solutions.

Spherical shell. For a shallow spherical shell the radius of curvature remains constant in all directions; therefore,

$$\frac{\partial^2 w_0}{\partial X \partial Y} = 0; \frac{\partial^2 w_0}{\partial X^2} = \frac{\partial^2 w_0}{\partial Y^2} = \frac{1}{R} \tag{4.6}$$

Substituting equations (4.6) into (4.1), one recovers Reissner's equations [3]

$$\frac{Eh}{R} \nabla^2 w + \nabla^4 F = 0 \tag{4.7a}$$

$$\nabla^4 w - \frac{1}{RD} \nabla^2 F = -\frac{q}{D} \tag{4.7b}$$

Flat plate. A flat plate represents a degenerative case of a spherical cap when the radius becomes infinite; therefore,

$$\frac{\partial^2 w_0}{\partial X \partial Y} = \frac{\partial^2 w_0}{\partial X^2} = \frac{\partial^2 w_0}{\partial Y^2} = 0 \tag{4.8}$$

Substituting equation (4.8) into (4.1), one recovers the classic equations for a flat plate, i.e.,

$$\nabla^4 F = 0 \tag{4.9a}$$

$$\nabla^4 w = -\frac{q}{D} \tag{4.9b}$$

Cylindrical shell. For a shallow cylindrical shell, one of the principal radii of curvatures is infinite, hile the other one is constant; therefore,

$$\frac{\partial^2 w_0}{\partial X \partial Y} = \frac{\partial^2 w_0}{\partial X^2} = 0; \; \frac{\partial^2 w_0}{\partial Y^2} = \frac{1}{R} \tag{4.10}$$

Substituting equations (4.10) into (4.1), one recovers the equations for a shallow cylindrical shell, i.e.,

$$\frac{Eh}{R} \frac{\partial^2 w}{\partial X^2} + \nabla^4 F = 0 \tag{4.11a}$$

$$\nabla^4 w - \frac{1}{RD} \frac{\partial^2 F}{\partial X^2} = -\frac{q}{D} \tag{4.11b}$$

Other shell geometries. If one chooses the coordinate axes X and Y such that they are parallel to the principal radii of curvature*, then

$$\frac{\partial^2 w_0}{\partial X \partial Y} = 0; \; \frac{\partial^2 w_0}{\partial X^2} = \frac{1}{R_x} \; ; \; \frac{\partial^2 w_0}{\partial Y^2} = \frac{1}{R_y} \tag{4.12}$$

with R_x and R_y being the principal radii of curvatures in the X and Y directions respectively.

Substituting equations (4.12) into (4.1), one finds

$$Eh \left[\frac{1}{R_y} \frac{\partial^2 w}{\partial X^2} + \frac{1}{R_x} \frac{\partial^2 w}{\partial Y^2} \right] + \nabla^4 F = 0 \tag{4.13a}$$

$$\nabla^4 w - \frac{1}{D} \left[\frac{1}{R_y} \frac{\partial^2 F}{\partial X^2} + \frac{1}{R_x} \frac{\partial^2 F}{\partial Y^2} \right] = -\frac{q}{D} \tag{4.13b}$$

---

* In general, when they are not parallel, equations (4.13) will contain additional terms of the form $(\partial^2 w/\partial X \partial Y)$ and $(\partial^2 F/\partial X \partial Y)$.

4.3 The stress field in a cracked spherical shell

Formulation of the Problem. Consider a portion of a thin, shallow spherical shell of constant thickness h and subjected to an internal pressure $q(X, Y)$ (see Figure 4.2). The material of the shell is assumed to be homogeneous and

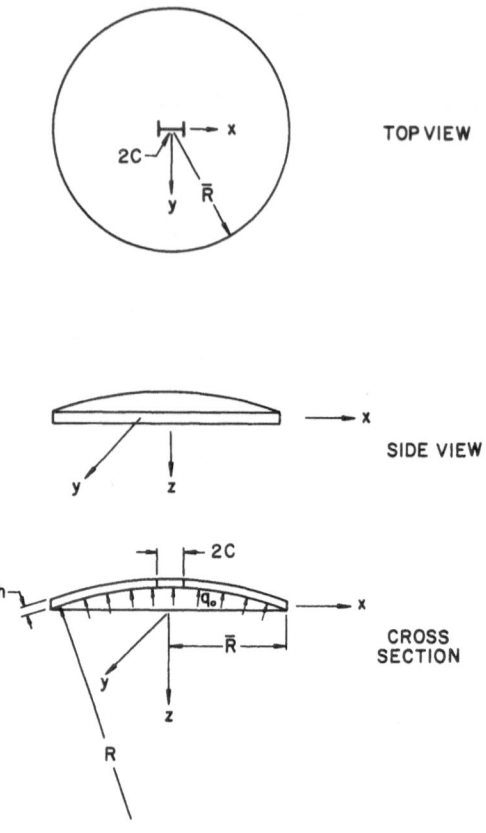

Figure 4.2. Geometrical Configurations of a Pressurized Spherical Cap

isotropic and at the apex there exists a radial cut of length $2c$ with respect to the apex. It is convenient at this point to introduce the dimensionless coordinates

$$x = \frac{X}{c}, \; y = \frac{Y}{c}, \; z = \frac{Z}{c} \tag{4.14}$$

in view of which the coupled differential equations governing the deflection function $w(x, y)$ and the stress function $F(x, y)$ with x and y as dimensionalized rectangular coordinates of the base plane, become

$$-\frac{Ehc^2}{R} \nabla^2 w + \nabla^4 F = 0 \qquad (4.15a)$$

$$\nabla^4 w + \frac{c^2}{RD} \nabla^2 F = -c^4 \frac{q}{D} \qquad (4.15b)$$

As to boundary conditions, we require that (1) on the faces of the crack, the normal moment, equivalent shear, and normal and tangential membrane forces vanish, and (2) away from the crack, the appropriate loading and support condition are satisfied.

In treating this type of problem, it is found convenient to seek the solution into two parts, the 'undisturbed' or 'particular' solution which satisfies equations (4.15) and the loading and support conditions but leaves residual forces along the crack, and the 'complementary' solution which precisely nullifies these residuals and offers no contribution far away from the crack.

However, suppose that one has already found a particular solution satisfying equations (4.15), but that there is a residual normal moment M_y, equivalent vertical shear V_y, normal in-plane stress N_y, and in-plane tangential stress N_{xy}, along the real axis $|x| < 1$, of the form:*

$$M_y^{(P)} = -\frac{D}{c^2} m_0, \; V_y^{(P)} = 0, \; N_y^{(P)} = -\frac{n_0}{c^2}, \; N_{xy}^{(P)} = 0 \qquad (4.16)$$

where, for simplicity, we assume m_0, n_0 to be constants.**

Mathematical Statement of the Problem. Assuming therefore that a particular solution has been found, we need to find two functions of the dimensionless coordinates (x, y), $w(x, y)$ and $F(x, y)$, such that they satisfy the homogeneous part of the differential equations (4.15) and the following boundary conditions. At $y = 0$ and $|x| < 1$:

$$M_y(x, 0) = -\frac{D}{c^2} \left[\frac{\partial^2 w}{\partial y^2} + v \frac{\partial^2 w}{\partial x^2} \right] = \frac{Dm_0}{c^2} = \frac{\bar{\sigma}^{(b)} h^2}{6} \qquad (4.16a)$$

---

* For particular solutions see section 4.8.
** For m_0, n_0 non-constants, see remarks after equations (4.40).

$$V_y(x, 0) = -\frac{D}{c^2}\left[\frac{\partial^2 w}{\partial y^2} + (2 - v)\frac{\partial^2 w}{\partial x^2 \partial y}\right] = 0 \tag{4.16b}$$

$$N_y(x, 0) = \frac{1}{c^2}\frac{\partial^2 F}{\partial x^2} = \frac{n_0}{c^2} = h\bar{\sigma}^{(e)} \tag{4.16c}$$

$$N_{xy}(x, 0) = -\frac{1}{c^2}\frac{\partial^2 F}{\partial x \partial y} = 0 \tag{4.16d}$$

At $y = 0$ and $|x| > 1$ we must satisfy the continuity requirements, i.e.,

$$\lim_{|y| \to 0}\left[\frac{\partial^n}{\partial y^n}(w^+) - \frac{\partial^n}{\partial y^n}(w^-)\right] = 0 \tag{4.17a}$$

$$\lim_{|y| \to 0}\left[\frac{\partial^n}{\partial y^n}(F^+) - \frac{\partial^n}{\partial y^n}(F^-)\right] = 0 \tag{4.17b}$$

for $n = 0, 1, 2, 3$. Furthermore, in order to avoid infinite stresses and infinite displacements we require that the functions w and F with their first derivatives to be finite far away from the crack. These restrictions simplify the mathematical complexities of the problem considerably, and correspond to the usual expectations of the St. Venant Principle. It should be pointed out that the boundary conditions at infinity are not geometrically feasible. However if the crack is small compared to the dimensions of the shell, the approximation is reasonable.

Reduction of the System. Reissner [4] has shown that the homogeneous solution to the system of equation (4.15) can be written in the following form

$$w^{(h)} = \chi + \phi, \quad F^{(h)} = -RDc^{-2}\,\nabla^2\chi + \psi \tag{4.18}$$

where ϕ and ψ are harmonic functions and χ satisfies the same differential equation as the deflection of a plate on an elastic foundation, i.e.,

$$(\nabla^4 + \lambda^4)\chi = 0 \tag{4.19}$$

with

$$\lambda^4 \equiv \frac{Ehc^4}{R^2 D} \equiv \frac{12(1 - v^2)}{(R/h)^2}\left(\frac{c}{h}\right)^4 \tag{4.20}$$

One concludes, therefore, that the effect of the initial curvature is qualitatively equivalent to providing an elastic foundation for an initially flat plate, such that, as the radius of curvature increases, the foundation modulus becomes weaker and weaker. This analogy has also been observed and demonstrated experimentally by Sechler and Williams [5].

Method of Solution. We construct next the following Fourier integral representations with the proper behavior at infinity

$$w(x, y\pm) = \int_0^\infty \{P_1 \exp[-(s^2 - i\lambda^2)^{\frac{1}{2}} |y|]$$
$$+ P_2 \exp[-(s^2 + i\lambda^2)^{\frac{1}{2}} |y|] + P_3 e^{-s|y|}\} \cos xs\, ds \quad (4.21a)$$

$$F(x, y^\pm) = \frac{i\lambda^2 RD}{c^2} \int_0^\infty \{P_1 \exp[-(s^2 - i\lambda^2)^{\frac{1}{2}} |y|]$$
$$- P_2 \exp[-(s^2 + i\lambda^2)^{\frac{1}{2}} |y|] + P_4 e^{-s|y|}\} \cos xs\, ds \quad (4.21b)$$

where the P_i's are arbitrary functions of s to be determined from the boundary conditions, and the $\pm$ signs refer to $y > 0$ and $y < 0$ respectively.

Assuming that one can differentiate under the integral sign, one finds by formally substituting equations (4.21) into the boundary conditions in equations (4.16) that:

$$\lim_{|y| \to 0} \int_0^\infty \{P_1(v_0 s^2 - i\lambda^2) e^{-(s^2 - i\lambda^2)^{1/2}|y|} + P_2(v_0 s^2 + i\lambda^2) e^{-(s^2 + i\lambda^2)^{1/2}|y|}$$
$$+ v_0 s^2 P_3 e^{-s|y|}\} \cos(xs) ds = -m_0; \quad |x| < 1 \quad (4.22a)$$

$$\pm \int_0^\infty \{P_1(s^2 - i\lambda^2)^{\frac{1}{2}}(v_0 s^2 + i\lambda^2) + P_2(s^2 + i\lambda^2)^{\frac{1}{2}}(v_0 s^2 - i\lambda^2)$$
$$+ v_0 s^3 P_3\} \cos(xs) ds = 0; \quad |x| < 1 \quad (4.22b)$$

$$\lim_{|y| \to 0} -\frac{i\lambda^2 RD}{c^2} \int_0^\infty \{P_1 e^{-(s^2 - i\lambda^2)^{1/2}|y|} - P_2 e^{-(s^2 + i\lambda^2)^{1/2}|y|} + P_4 e^{-s|y|}\}$$
$$s^2 \cos(xs) ds = -n_0; \quad |x| < 1 \quad (4.22c)$$

and

$$\mp \frac{i\lambda^2 RD}{c^2} \int_0^\infty \{P_1(s^2 - i\lambda^2)^{\frac{1}{2}} - P_2(s^2 + i\lambda^2)^{\frac{1}{2}} + P_4 s\} s \sin(xs) ds = 0;$$
$$|x| < 1 \quad (422d)$$

where again the $\mp_+$ signs refers to $y > 0$ and $y < 0$ respectively, and $v_0 = 1 - v$. If one, now, chooses

$$v_0 s^3 P_3 = -\{(s^2 - i\lambda^2)^{\frac{1}{2}}(v_0 s^2 + i\lambda^2)P_1 + (s^2 + i\lambda^2)^{\frac{1}{2}}(v_0 s^2 - i\lambda^2)P_2\} \quad (4.23a)$$

$$sP_4 = -\{P_1(s^2 - i\lambda^2)^{\frac{1}{2}} - P_2(s^2 + i\lambda^2)^{\frac{1}{2}})\} \quad (4.23b)$$

then equations (4.22b) and (4.22d) are satisfied automatically and equations (4.22a) and (4.22c) become, respectively

$$\int_0^\infty \{[v_0 s^2 - i\lambda^2 - s^{-1}(s^2 - i\lambda^2)^{\frac{1}{2}}(v_0 s^2 + i\lambda^2)]P_1$$

$$+ [v_0 s^2 + i\lambda^2 - s^{-1}(s^2 + i\lambda^2)^{\frac{1}{2}}(v_0 s^2 - i\lambda^2)]P_2\} \cos (xs) \, ds = - m_0;$$

and $|x| < 1$ (4.24a)

$$- \frac{i\lambda^2 RD}{c^2} \int_0^\infty [(1 - s^{-1}(s^2 - i\lambda^2)^{\frac{1}{2}})P_1$$

$$- (1 - s^{-1}(s^2 + i\lambda^2)^{\frac{1}{2}})P_2] \, s^2 \cos (xs) \, ds = n_0; \quad |x| < 1 \qquad (4.24b)$$

Furthermore, it can easily be shown that all the continuity conditions are satisfied if one considers the following two combinations to vanish

$$\int_0^\infty \frac{P_1}{s^2} (s^2 - i\lambda^2)^{\frac{1}{2}} \cos (xs) \, ds = 0; \quad |x| > 1 \qquad (4.25a)$$

$$\int_0^\infty \frac{P_2}{s^2} (s^2 + i\lambda^2)^{\frac{1}{2}} \cos (xs) \, ds = 0; \quad |x| > 1 \qquad (4.25b)$$

We have reduced, therefore, our problem to that of solving the dual integral equations (4.24) and (4.25) for the unknown functions $P_1(s)$ and $P_2(s)$.

Reduction to Singular Integral Equations. For the determination of the unknown functions $P_1(s)$ and $P_2(s)$, we reduce the problem to a set of coupled singular integral equations of the Cauchy type. This can be accomplished if one lets

$$\int_0^\infty \frac{P_1}{s^2} (s^2 - i\lambda^2)^{\frac{1}{2}} \cos (xs) \, ds = u_1(x); \quad |x| < 1 \qquad (4.26a)$$

$$\int_0^\infty \frac{P_2}{s^2} (s^2 + i\lambda^2)^{\frac{1}{2}} \cos (xs) \, ds = u_2(x); \quad |x| < 1 \qquad (4.26b)$$

which by Fourier inversion give

$$P_1(s^2 - i\lambda^2)^{\frac{1}{2}} = \frac{2s^2}{\pi} \int_0^1 u_1(\xi) \cos (s\xi) \, d\xi \qquad (4.27a)$$

$$P_2(s^2 + i\lambda^2)^{\frac{1}{2}} = \frac{2s^2}{\pi} \int_0^1 u_2(\xi) \cos (s\xi) \, d\xi \qquad (4.27b)$$

where the functions $u_1(\xi)$ and $u_2(\xi)$, due to the symmetry of the problem, are even. Next, substituting formally equations (4.27) into (4.24) one has after changing the order of integration and rearranging

$$N_y = -\frac{2i\lambda^2 RD}{\pi c^2} \int_{-1}^{1} \{u_1(\xi)L_1^* - u_2(\xi)L_2^*\}\, d\xi \tag{4.28a}$$

$$M_y = -\frac{2D}{\pi} \int_{-1}^{1} \{u_1(\xi)L_3^* + u_2(\xi)L_4^*\}\, d\xi \tag{4.28b}$$

where

$$L_1^* = \frac{1}{2}\int_0^\infty \frac{s^4 \exp\left[-(s^2 - i\lambda^2)^{\frac{1}{2}}\,|\,y\,|\right]}{(s^2 - i\lambda^2)^{\frac{1}{2}}}\cos(x - \xi)s\, ds$$
$$-\frac{1}{2}\int_0^\infty s^3 e^{-s\,|\,y\,|}\cos(x - \xi)s\, ds \tag{4.29a}$$

$$L_2^* = \frac{1}{2}\int_0^\infty \frac{s^4 \exp\left[-(s^2 + i\lambda^2)^{\frac{1}{2}}|\,y\,|\right]}{(s^2 + is^2)^{\frac{1}{2}}}\cos(x - \xi)s\, ds$$
$$-\frac{1}{2}\int_0^\infty s^3 e^{s\,|\,y\,|}\cos(x - \xi)s\, ds \tag{4.29b}$$

$$L_3^* = \frac{1}{2}\int_0^\infty \left\{\frac{s^2(v_0 s^2 - i\lambda^2)}{(s^2 - i\lambda^2)^{\frac{1}{2}}}\exp\left[-(s^2 - i\lambda^2)^{\frac{1}{2}}|\,y\,|\right]\right.$$
$$\left.- s(v_0 s^2 + i\lambda^2)\,e^{-s\,|\,y\,|}\right\}\cos(x - \xi)s\, ds \tag{4.29c}$$

$$L_4^* = \frac{1}{2}\int_0^\infty \left\{\frac{s^2(v_0 s^2 + i\lambda^2)}{(s^2 + i\lambda^2)^{\frac{1}{2}}}\exp\left[-(s^2 + i\lambda^2)^{\frac{1}{2}}|\,y\,|\right]\right.$$
$$\left.- s(v_0 s^2 - i\lambda^2)\,e^{-s\,|\,y\,|}\right\}\cos(x - \xi)s\, ds \tag{4.29d}$$

The integration in equations (4.29) may now be carried out explicitly by making use of the Fourier cosine transforms [6]

$$\int_0^\infty e^{-|\,y\,|\,s}\cos(\zeta s)\, ds = \frac{|\,y\,|}{p^2} \tag{4.30a}$$

$$\int_0^\infty \frac{\exp\left[-(s^2 + a^2)^{\frac{1}{2}}|\,y\,|\right]}{(s^2 + a^2)^{\frac{1}{2}}}\cos(\zeta s)\, ds = K_0(ap); \operatorname{Re} a > 0 \tag{4.30b}$$

and similar results obtained by differentiating them with respect to x and y. In these formulas $p^2 = \zeta^2 + |y|^2$, and K_n denotes the modified Bessel function of the third kind of order n.

The expressions in equations (4.29) then become respectively

$$2L_1^* = \frac{\partial}{\partial x}\left\{ -\frac{\lambda^2\beta^2\zeta}{p^4}(\zeta^2 - 3\,|\,y\,|^2)\,K_0(\lambda\beta p) \right.$$

$$\left. -\left[\frac{\lambda^3\beta^3\zeta^3}{p^3} + \frac{2\lambda\beta\zeta}{p^5}(\zeta^2 - 3\,|\,y\,|^2)\right]K_1(\lambda\beta p) + \frac{2\zeta}{p^4} - \frac{8\zeta\,|\,y\,|^2}{p^6}\right\} \quad (4.31a)$$

$$2L_2^* = \frac{\partial}{\partial x}\left\{ -\frac{\lambda^2\alpha^2\zeta}{p^4}(\zeta^2 - 3\,|\,y\,|^2)\,K_0(\lambda\alpha p) \right.$$

$$\left. -\left[\frac{\lambda^3\alpha^3\zeta^3}{p^3} + \frac{2\lambda\alpha\zeta}{p^5}(\zeta^2 - 3\,|\,y\,|^2)\right]K_1(\lambda\alpha p) + \frac{2\zeta}{p^4} - \frac{8\zeta\,|\,y\,|^2}{p^6}\right\} \quad (4.31b)$$

$$2L_3^* = \frac{\partial}{\partial x}\left\{ -\frac{v_0\lambda^2\beta^2\zeta}{p^4}(\zeta^2 - 3\,|\,y\,|^2)\,K_0(\lambda\beta p) \right.$$

$$-v_0\left[\frac{\lambda^3\beta^3\zeta^3}{p^3} + \frac{2\lambda\beta\zeta}{p^5}(\zeta^2 - 3\,|\,y\,|^2)\right]K_1(\lambda\beta p) - \frac{\alpha^2\lambda^3\beta\zeta}{p}K_1(\lambda\beta p)$$

$$\left. + \frac{2v_0\zeta}{p^4} - \frac{8\zeta\,|\,y\,|^2\,v_0}{p^6} - \frac{\alpha^2\lambda^2\zeta}{p^2}\right\} \quad (4.31c)$$

$$2L_4^* = \frac{\partial}{\partial x}\left\{ -\frac{v_0\lambda^2\alpha^2\zeta}{p^4}(\zeta^2 - 3\,|\,y\,|^2)\,K_0(\lambda\alpha p) \right.$$

$$-v_0\left[\frac{\lambda^3\alpha^3\zeta^3}{p^3} + \frac{2\lambda\alpha\zeta}{p^5}(\zeta^2 - 3\,|\,y\,|^2)\right]K_1(\lambda\alpha p) - \frac{\beta^2\lambda^3\alpha\zeta}{p}K_1(\lambda\alpha p)$$

$$\left. + \frac{2v_0\zeta}{p^4} - \frac{8\zeta\,|\,y\,|^2\,v_0}{p^6} - \frac{\beta^2\lambda^2\zeta}{p^2}\right\} \quad (4.31d)$$

where for simplicity we have defined $\alpha^2 = i$ and $\beta^2 = -i$. Thus, the limits, as $|\,y\,| \to 0$, of N_y and M_y are found to be respectively

$$\lim_{|y|\to 0} N_y = -\frac{2i\lambda^2 RD}{\pi c^2}\frac{d}{dx}\int_{-1}^{1}\{u_1(\xi)L_1 - u_2(\xi)L_2\}\,d\xi \quad (4.32a)$$

$$\lim_{|y|\to 0} M_y = -\frac{2D}{\pi}\frac{d}{dx}\int_{-1}^{1}\{u_1(\xi)L_3 + u_2(\xi)L_4\}\,d\xi \quad (4.32b)$$

where the integrals are understood to be of Cauchy principal value and

$$2L_1 \equiv -\frac{\lambda^2\beta^2}{\zeta}K_0(\lambda\beta\,|\,\zeta\,|) - \left(\lambda^3\beta^3\frac{\zeta}{|\,\zeta\,|} + \frac{2\lambda\beta}{\zeta\,|\zeta|}\right)K_1(\lambda\beta\,|\,\zeta\,|) + \frac{2}{\zeta^3}$$

$$(4.33a)$$

$$2L_2 \equiv -\frac{\lambda^2\alpha^2}{\zeta} K_0(\lambda\alpha\,|\,\zeta\,|) - \left(\lambda^3\alpha^3 \frac{\zeta}{|\,\zeta\,|} + \frac{2\lambda\alpha}{\zeta\,|\,\zeta\,|}\right) K_1(\lambda\alpha\,|\,\zeta\,|) + \frac{2}{\zeta^3}$$

(4.33b)

$$2L_3 \equiv -\frac{v_0\lambda^2\beta^2}{\zeta} K_0(\lambda\beta\,|\,\zeta\,|) - v_0\left(\lambda^3\beta^3 \frac{\zeta}{|\,\zeta\,|} + \frac{2\lambda\beta}{\zeta\,|\,\zeta\,|}\right) K_1(\lambda\beta\,|\,\zeta\,|)$$

$$- \lambda^3\alpha^2\beta \frac{\zeta}{|\,\zeta\,|} K_1(\lambda\beta\,|\,\zeta\,|) + \frac{2v_0}{\zeta^3} - \frac{\alpha^2\lambda^2}{\zeta}$$

(4.33c)

$$2L_4 \equiv -\frac{v_0\lambda^2\alpha^2}{\zeta} K_0(\lambda\alpha\,|\,\zeta\,|) - v_0\left(\lambda^3\alpha^3 \frac{\zeta}{|\,\zeta\,|} + \frac{2\lambda\alpha}{\zeta\,|\,\zeta\,|}\right) K_1(\lambda\alpha\,|\,\zeta\,|)$$

$$- \beta^2\lambda^3\alpha \frac{\zeta}{|\,\zeta\,|} K_1(\lambda\alpha\,|\,\zeta\,|) + \frac{2v_0}{\zeta^3} - \frac{\beta^2\lambda^2}{\zeta}$$

(4.33d)

If we set N_y, M_y, in the limit as $|y| \to 0$, equal to $-n_0$ and $-m_0$ respectively, integrate with respect to x, then we find that they must satisfy the integral equations

$$\int_{-1}^{1} \{u_1(\xi)2L_1 - u_2(\xi)2L_2\}\,\mathrm{d}\xi = -\frac{\pi n_0 c^2}{i\lambda^2 RD}\,x; \quad |x| < 1$$

(4.34a)

$$\int_{-1}^{1} \{u_1(\xi)2L_3 + u_2(\xi)2L_4\}\,\mathrm{d}\xi = -\pi m_0 x; \quad |x| < 1$$

(4.34b)

where the kernels L_1, L_2, L_3, L_4, have singularities of the order $1/\zeta = 1/(x-\xi)$, as can easily be seen by observing their behavior for small arguments:

$$2L_1 = -\frac{\lambda^2\beta^2}{2(x-\xi)} + \lambda^4\beta^4(x-\xi)\left[\frac{5}{32} - \frac{3\gamma}{8} - \frac{3}{8}\ln\lambda\beta\frac{|x-\xi|}{2}\right]$$

$$+ O(\lambda^6(x-\xi)^3 \ln\lambda\,|x-\xi|)$$

(4.35a)

$$2L_2 = -\frac{\lambda^2\alpha^2}{2(x-\xi)} + \lambda^4\alpha^4(x-\xi)\left[\frac{5}{32} - \frac{3\gamma}{8} - \frac{3}{8}\ln\lambda\alpha\frac{|x-\xi|}{2}\right]$$

$$+ O(\lambda^6(x-\xi)^3 \ln\lambda\,|x-\xi|)$$

(4.35b)

$$2L_3 = -\frac{\lambda^2\alpha^2(4-v_0)}{2(x-\xi)} + \lambda^4\beta^4(x-\xi)\left[\frac{5v_0-8}{32} + \frac{4-3v_0}{8}\right.$$

$$\left.\left(\gamma + \ln\lambda\beta\frac{|x-\xi|}{2}\right)\right] + O(\lambda^6(x-\xi)^3 \ln\lambda\,|x-\xi|)$$

(4.35c)

$$2L_4 = -\frac{\lambda^2\beta^2(4-v_0)}{2(x-\xi)} + \lambda^4\alpha^4(x-\xi)\left[\frac{5v_0-8}{32} + \frac{4-3v_0}{8}\right.$$

$$\left(\gamma + \ln \frac{\lambda\alpha\,|\,x - \xi\,|}{2}\right)\right] + O(\lambda^6(x - \xi)^3 \ln \lambda\,|\,x - \xi\,|) \qquad (4.35d)$$

We require, therefore, that the solutions $u_1(x)$, $u_2(x)$ be Hölder continuous for some positive Hölder indices μ_1 and μ_2 for all x in the closed interval $[-1, 1]$. Thus in particular, $u_1(x)$, $u_2(x)$ are to be bounded near the ends of the crack.

However, because of the complicated nature of the kernels L_i, an exact solution for the unknown functions $u_1(x)$ and $u_2(x)$ is extremely difficult. On the other hand, for most practical applications the parameter λ attains small values as follows from the definition of λ, namely

$$\lambda = \frac{[12(1 - v^2)]^{\frac{1}{4}}}{(R/h)^{\frac{1}{2}}}(c/h) = [12(1 - v^2)]^{\frac{1}{4}}(c/R)(R/h)^{\frac{1}{2}} \qquad (4.36)$$

It is clear that λ is small for large ratios of R/h and small crack lengths. As a practical matter, if we consider crack lengths less than one tenth of the periphery, i.e., $2c < 2\pi R/10$, and for $R/h < 10^3$ a corresponding upper bound for λ can be obtained, namely $\lambda < 20$. Thus the range of λ becomes $0 < \lambda < 20$ and for most practical cases is between 0 and 2, depending upon the size of the crack.

Solution for Small λ. For the simple case $\lambda = 0$, the problem reduces to that of a flat sheet under applied bending and stretching loads, the solution of which has been investigated by many authors. For example, the problem for both bending and stretching for an orthotropic plate, containing a finite crack, was investigated by Ang and Williams [7] and a solution was obtained by means of dual integral equations. It can easily be shown* that the dual integral equations can be transformed to two singular integral equations of the type (4.34) with simpler kernels. Furthermore, these are not coupled and the solutions can easily be obtained as in § 47 of [9]. Without going into the details they are found to be of the form $A(1 - \xi^2)^{\frac{1}{2}}$, where A is a constant.

Similarly, the solution for an initially curved sheet must, in the limit, check the above result and because $u_1(\xi)$ and $u_2(\xi)$ are in particular to be bounded near the ends of the crack, it is reasonable to assume solutions of the form**

$$u_1(\xi) = (1 - \xi^2)^{\frac{1}{2}}[A_0 + \lambda^2 A_1(1 - \xi^2) + \ldots]; \quad |\xi| < 1 \qquad (4.37a)$$

* See Noble [8].
** In fact, one can show [10] that this is precisely the form of the asymptotic solution for small λ.

$$u_2(\xi) = (1 - \xi^2)^{\frac{1}{2}} [B_0 + \lambda^2 B_1 (1 - \xi^2) + \ldots]; \quad |\xi| < 1 \qquad (4.37b)$$

where the coefficients $A_0, A_1, \ldots, B_0, B_1, \ldots$ can be functions of λ but not of ξ.

Substituting equations (4.31) into (4.34), and making use of the relation

$$\int_{-1}^{1} (1 - \xi^2)^{\frac{1}{2}} (x - \xi) \ln \frac{\lambda \alpha |x - \xi|}{2} d\xi = \frac{\pi}{4} \left(1 + \ln \frac{\lambda^2 \alpha^2}{16}\right) x + \frac{\pi}{6} x^3 \qquad (4.38)$$

we find by equating coefficients that*

$$A_0 = \frac{n_0 c^2}{\lambda^4 RD} \left\{ 1 + \frac{\pi \lambda^2}{16} \frac{8 - 3v_0}{4 - v_0} + \frac{\lambda^2 \alpha^2}{4 - v_0} \left(\frac{8 - 7v_0}{32} + \frac{4 - 3v_0}{8} \gamma\right) \right. $$
$$+ \frac{\lambda^2 \alpha^2}{16} \frac{4 - 3v_0}{4 - v_0} \ln \frac{\lambda^2 \alpha^2}{16} \left. \right\} + \frac{m_0}{\lambda^2 \alpha^2 (4 - v_0)}$$
$$\left\{1 + \frac{\pi \lambda^2}{16} \frac{8 - 3v_0}{4 - v_0} + \lambda^2 \alpha^2 \left(\frac{7}{32} + \frac{3\gamma}{8}\right) + \frac{3}{16} \lambda^2 \alpha^2 \left(1 + \ln \frac{\lambda^2 \alpha^2}{16}\right)\right\}$$
$$+ O(\lambda^2 \ln \lambda) \qquad (4.39a)$$

$$B_0 = \frac{n_0 c^2}{\lambda^4 RD} \left\{ 1 + \frac{\lambda^2 \pi}{16} \frac{8 - 3v_0}{4 - v_0} + \frac{\lambda^2 \beta^2}{4 - v_0} \left(\frac{8 - 7v_0}{32} + \frac{4 - 3v_0}{8} \lambda\right) \right.$$
$$+ \frac{\lambda^2 \beta^2}{16} \frac{4 - 3v_0}{4 - v_0} \ln \frac{\lambda^2 \beta^2}{16} \left. \right\} + \frac{m_0}{\lambda^2 \beta^2 (4 - v_0)} \left\{1 + \frac{\lambda^2 \pi}{16} \frac{8 - 3v_0}{4 - v_0}\right.$$
$$+ \lambda^2 \beta^2 \left(\frac{7}{32} + \frac{3\gamma}{8}\right) + \frac{3}{16} \lambda^2 \beta^2 \left(1 + \ln \frac{\lambda^2 \beta^2}{16}\right) \left. \right\} + O(\lambda^2 \ln \lambda) \qquad (4.39b)$$

It should be pointed out that if coefficients A_0, B_0 of higher accuracy are desired, say up to order λ^{2n}, then it is necessary to solve an $n \times n$ algebraic system. In effect, this is a method of successive approximations for which the question of convergence is investigated in Reference [10].

It thus appears that for $\lambda < \lambda^*$ the power series solutions of the form

$$u_1^{(N)}(\xi) = (1 - \xi^2)^{\frac{1}{2}} \sum_{n=0}^{N} A_n \lambda^{2n} (1 - \xi^2)^n, \qquad (4.40a)$$

$$u_2^{(N)}(\xi) = (1 - \xi^2)^{\frac{1}{2}} \sum_{n=0}^{N} B_n \lambda^{2n} (1 - \xi^2)^n, \qquad (4.40b)$$

---

* For brevity, in this analysis we will restrict ourselves to terms up to $O(\lambda^2)$.

in the limit as $N \to \infty$, will converge to the exact solutions* $u_1(\xi)$ and $u_2(\xi)$ of the integral equations (4.34). However, since most particular solutions will give us a non-uniform residual moment and normal membrane stress along the crack, it is only natural to ask how the solution changes. Suppose for $|x| < 1$, we expand m_0 and n_0 in the form $\sum_n a_n x^{2n}$ (even powers because of the symmetry of the problem), then our previous method of solution will still be applicable. And as can easily be seen from equations (4.34), although the coefficients A_n, B_n in this case may change, the character of the solution will still remain the same. Finally, because we desire to focus our attention upon the singular stresses around the neighborhood of the crack point, we need only to compute coefficients A_0 and B_0.

Alternate Method of Solution. It is also possible to solve the coupled dual integral equations directly by using a method which was developed by the author some time ago [11] and is parallel to the previous method. Thus motivated by equations (4.25) and (4.26), one assumes the unknown functions P_1 and P_2 in the form

$$\frac{P_1(s^2 - i\lambda^2)^{\frac{1}{4}}}{s^2} = \sum_{k=0}^{\infty} A_k \frac{J_{k+1}(s)}{(s)^{k+1}} \tag{4.41a}$$

and

$$\frac{P_2(s^2 + i\lambda^2)^{\frac{1}{4}}}{s^2} = \sum_{k=0}^{\infty} B_k \frac{J_{k+1}(s)}{(s)^{k+1}} \tag{4.41b}$$

where the coefficients A_k and B_k are constants to be determined.

The advantage of such a form is that equations (4.25) are automatically satisfied and furthermore,

$$u_1(x) = \int_0^\infty \frac{P_1}{s^2}(s^2 - i\lambda^2)^{\frac{1}{4}} \cos(xs)ds = \sum_{k=0}^{\infty} A_k \frac{\sqrt{\pi}}{2^{k+1}\Gamma(k + \frac{3}{2})}(1 - x^2)^{k+\frac{1}{2}} \tag{4.42a}$$

and

$$u_2(x) = \int_0^\infty \frac{P_2}{s^2}(s^2 + i\lambda^2)^{\frac{1}{4}} \cos(xs)ds = \sum_{k=0}^{\infty} B_k \frac{\sqrt{\pi}}{2^{k+1}\Gamma(k + \frac{3}{2})}(1 - x^2)^{k+\frac{1}{2}} \tag{4.42b}$$

In general, the functions u_i, have some physical meaning. For example, in this case their algebraic combination represents the crack opening displace-

---

* This matter is discussed at same length in [10].

ment and as a result such an expansion is plausible.

It follows that equations (4.24) take the forms

$$\sum_{k=0}^{\infty} \int_0^{\infty} \left\{ \left[\frac{v_0 s^4 - i\lambda^2 s^2}{(s^2 - i\lambda^2)^{\frac{1}{2}}} - s(v_0 s^2 + i\lambda^2) \right] A_k \right.$$

$$+ \left[\frac{v_0 s^4 + i\lambda^2 s^2}{(s^2 + i\lambda^2)^{\frac{1}{2}}} - s(v_0 s^2 - i\lambda^2) \right] B_k \left. \right\} \frac{J_{k+1}(s)}{(s)^{k+1}} \cos (xs) ds$$

$$= - m_0; \quad |x| < 1 \tag{4.43a}$$

and

$$\sum_{k=0}^{\infty} \int_0^{\infty} \left\{ \left[\frac{s^2}{(s^2 - i\lambda^2)^{\frac{1}{2}}} - s \right] A_k - \left[\frac{s^2}{(s^2 + i\lambda^2)^{\frac{1}{2}}} - s \right] B_k \right\}$$

$$s^2 \frac{J_{k+1}(s)}{(s)^{k+1}} \cos (xs) ds = n_0; \quad |x| < 1 \tag{4.43b}$$

Multiplying, subsequently, both sides by

$$\frac{(1 - x^2)^{j+\frac{1}{2}}}{2^j \Gamma(j + \frac{3}{2}) \sqrt{\pi}}$$

and integrating with respect to x from 0 to 1 one finds

$$\sum_{k=0}^{\infty} \{ A_k G_{k,j}(\lambda; i) + B_k G_{k,j}(\lambda; -i) \} = - m_0 H_j; \quad j = 0, 1, 2, \ldots \tag{4.44a}$$

and

$$\sum_{k=0}^{\infty} \{ A_k F_{k,j}(\lambda; i) - B_k F_{k,j}(\lambda; -i) \} = n_0 H_j; \quad j = 0, 1, 2, \ldots \tag{4.44b}$$

where for simplicity we have made the following definitions

$$G_{k,j}(\lambda; i) = \int_0^{\infty} \left\{ \frac{v_0 s^4 - i\lambda^2 s^2}{(s^2 - i\lambda^2)^{\frac{1}{2}}} - s(v_0 s^2 + i\lambda^2) \right\} \frac{J_{k+1}(s)}{(s)^{k+1}} \frac{J_{j+1}(s)}{(s)^{j+1}} ds \tag{4.45a}$$

$$F_{k,j}(\lambda; i) = - \int_0^{\infty} s^3 \left(1 - \frac{s}{(s^2 - i\lambda^2)^{\frac{1}{2}}} \right) \frac{J_{k+1}(s)}{(s)^{k+1}} \frac{J_{j+1}(s)}{(s)^{j+1}} ds \tag{4.45b}$$

and

$$H_j = - \frac{1}{2^{j+1} (j + 1)!} \tag{4.46}$$

Equations (4.44) now represent two infinite systems of algebraic equations which are to be solved for the unknown coefficients* A_k and B_k. Such systems have been studied extensively and the questions of existence and uniqueness of the solution are discussed in reference [13].

As a practical matter now, if one addresses himself to the major contribution of the solution, which comes primarily from the terms with $k = j = 0$, and uses the following integral approximation

$$\int_0^\infty s \left\{1 - \frac{s}{(s^2 + \beta^2)^{\frac{1}{2}}}\right\} \left\{\frac{J_1(s)}{(s)}\right\}^2 ds \simeq \int_0^\infty s \; \frac{(\beta^2/2)}{s^2 + (\beta^2/2)} \left\{\frac{J_1(s)}{(s)}\right\}^2 ds$$

$$= \left(\frac{1}{2}\right)\{1 - 2I_1(\beta/\sqrt{2}) K_1(\beta/\sqrt{2})\}, \tag{4.47}$$

then

$$A_0 \simeq - \frac{m_0}{2\Delta} F_{0,0}(\lambda; -i) + \frac{n_0}{2\Delta} G_{0,0}(\lambda; -i) \tag{4.48a}$$

$$B_0 \simeq - \frac{m_0}{2\Delta} F_{0,0}(\lambda; i) - \frac{n_0}{2\Delta} G_{0,0}(\lambda; i) \tag{4.48b}$$

in which Δ is given by

$$\Delta \simeq F_{0,0}(\lambda; -i) G_{0,0}(\lambda; i) + F_{0,0}(\lambda; i) G_{0,0}(\lambda; -i) \tag{4.49}$$

Furthermore, in view of the approximation (4.47) ,one has

$$F_{0,0}(\lambda; i) = + \frac{i\lambda^2}{2} \; I_1(e^{-i\pi/4} \lambda/\sqrt{2})K_1 (e^{-i\pi/4} \lambda/\sqrt{2}) \tag{4.50a}$$

$$G_{0,0}(\lambda; +i) = v_0 F_{0,0}(\lambda; +i) + \frac{i\lambda^2}{2} \{1 - 2I_1(e^{-i\pi/4} \lambda/\sqrt{2})$$

$$K_1(e^{-i\pi/4} \lambda/\sqrt{2})\} - i\lambda^2 \tag{4.50b}$$

It is clear now from the above, that the general expressions for the coefficients A_0 and B_0 are complicated series expressions involving the modified Bessel functions I_n and K_n. As complicated as they may seem, the use of an electronic computer makes the work a routine.

---

* In the field of fracture mechanics it is only necessary to compute the first coefficients A_0 and B_0 for only the first term of the series in equations (4.41) lead to the well known [12] $1/\sqrt{r}$ stress singular behavior ahead of the crack tip.

Determination of w and F.   In view of equations (4.27), (4.37) to (4.39) and the relation

$$\int_0^\infty s^{-\mu} J_\mu(as) \cos(xs)\, ds$$
$$= \begin{cases} \sqrt{\pi}(2a)^{-\mu} [\Gamma(\mu + \tfrac{1}{2})]^{-1} (a^2 - x^2)^{\mu - \frac{1}{2}}; & 0 < x < a \;\; ; \mathrm{Re}\,\mu > -\tfrac{1}{2} \\ 0; & a < x < \infty; \mathrm{Re}\,\mu > -\tfrac{1}{2} \end{cases}$$

$$(4.51)$$

which can be found on page 44 of [6], we have:

$$P_1(s) = \frac{s}{(s^2 - i\lambda^2)^{\frac{1}{2}}} \left\{ A_0 J_1(s) + 3\lambda^2 A_1 \frac{J_2(s)}{s} + O(\lambda^4) \right\} \tag{4.52a}$$

$$P_2(s) = \frac{s}{(s^2 + i\lambda^2)^{\frac{1}{2}}} \left\{ B_0 J_1(s) + 3\lambda^2 B_1 \frac{J_2(s)}{s} + O(\lambda^4) \right\} \tag{4.52b}$$

$$P_3(s) = -(A_0 + B_0)J_1(s) - 3\lambda^2(A_1 + B_1) \frac{J_2(s)}{s}$$

$$- \frac{i\lambda^2}{v_0 s^2} (A_0 - B_0)J_1(s) + O(\lambda^4) \tag{4.52c}$$

$$P_4(s) = -(A_0 - B_0)J_1(s) - 3\lambda^2(A_1 - B_1)s^{-1} J_2(s) + O(\lambda^4) \tag{4.52d}$$

Therefore a substitution of the above relations into equations (4.21) will determine the bending deflection w and membrane stress function F. Furthermore, the corresponding integrals will converge and the differentiations under the integral sign are also justified at least for $y \neq 0$. The values of the derivatives at $y = 0$ and $|x| < 1$ can be obtained by a proper limiting process.

The Stress Field.   Without going into the details, the stress distribution around the crack tip for a symmetrical loading* is found to be:

Extensional stresses: through the thickness

$$\sigma_x^{(e)} = \frac{k_1^{(e)}}{(2r)^{\frac{1}{2}}} \cos(\theta/2) \left[1 - \sin(\theta/2) \sin(3\theta/2) \right] \tag{4.53a}$$

$$\sigma_x^{(e)} = \frac{k_1^{(e)}}{(2r)^{\frac{1}{2}}} \cos(\theta/2) \left[1 + \sin(\theta/2) \sin(3\theta/2) \right] \tag{4.53b}$$

---

* The antisymmetric loading case will be discussed later.

$$\sigma_{xy}^{(e)} = \frac{k_1^{(e)}}{(2r)^{\frac{1}{2}}} \cos{(\theta/2)} \sin{(\theta/2)} \cos{(3\theta/2)} \tag{4.53c}$$

Bending stresses: on the 'tension side' of the shell

$$\sigma_x^{(b)} = -\frac{k_1^{(b)}}{(2r)^{\frac{1}{2}}} \kappa[3\cos{(\theta/2)} + \cos{(5\theta/2)}] \tag{4.54a}$$

$$\sigma_y^{(b)} = \frac{k_1^{(b)}}{(2r)^{\frac{1}{2}}} \kappa\left[\cos{(5\theta/2)} + \frac{11 + 5v}{1 - v}\cos{(\theta/2)}\right] \tag{4.54b}$$

$$\sigma_{xy}^{(b)} = \frac{k_1^{(b)}}{(2r)^{\frac{1}{2}}} \kappa\left[\sin{(5\theta/2)} + \left(\frac{7 + v}{1 + v}\right)\sin{(\theta/2)}\right] \tag{4.54c}$$

where v is Poisson's ratio and (r, θ) are the polar coordinates around the crack tip. In general, the stress intensity factors $k_1^{(e)}$ and $k_1^{(b)}$ are functions of crack size, geometry of the shell, material properties and loading characteristics. In this case, they are related to the coefficients A_0 and B_0 by the expressions

$$k_1^{(e)} = \frac{\lambda^4 RD}{2hc^4}(A_0 + B_0)\sqrt{c} \tag{4.55a}$$

$$k_1^{(b)} = \frac{iEh\lambda^2}{4(1 - v^2)c^2}(A_0 - B_0)(3 + v)\sqrt{c} \tag{4.55b}$$

Thus, in view of equations (4.39), the stress intensity factors

$$k_1^{(e)} = \bar{\sigma}^{(e)}\sqrt{c}\left\{1 + \frac{3\pi}{32}\lambda^2\right\} + \bar{\sigma}^{(b)}\sqrt{c}\frac{(\lambda^2)(1 - v^2)^{\frac{1}{2}}}{3^{\frac{1}{2}}}\left\{\frac{7}{32} + \frac{3}{8}\left(\lambda + \ln\frac{\lambda}{4}\right)\right\}$$
$$+ O(\lambda^4 \ln \lambda) \tag{4.56}$$

and

$$k_1^{(b)} = -\bar{\sigma}^{(e)}\frac{\lambda^2(3^{\frac{1}{2}})\sqrt{c}}{(1 - v^2)^{\frac{1}{2}}(3 + v)}\left\{\frac{1 + 7v}{32} + \frac{1 + 3v}{8}\left(\gamma + \ln\frac{\lambda}{4}\right)\right\}$$
$$- \bar{\sigma}^{(b)}\sqrt{c}\left\{1 + \frac{1 + 3v}{3 + v}\frac{\pi\lambda^2}{32}\right\} + O(\lambda^4 \ln \lambda) \tag{4.57}$$

are obtained.

The reader should be cautioned for equations (4.56) and (4.57) represent exact asymptotic expansions up to $0(\lambda^2)$ terms. Consequently, they are good approximations for $0 \leq \lambda < 1$. For larger values of λ, one must also include higher order terms in order to guarantee convergence. This can be done by either of the two methods discussed previously and the aid of a computer. For example, the numerical solution of the system of equation (4.44) leads to a numerical result that may well be approximated within a 5% error by the simple relation*

$$k_1^{(e)} \simeq \sigma^{(e)} \sqrt{c} \, (1 + 0.466 \, \lambda^2)^{\frac{1}{4}} \qquad (4.58)$$

which is valid for all values of λ.

In view of the above, one may conjecture that in an initially curved sheet,
1. the stresses are proportional to $(c/r)^{\frac{1}{2}}$,
2. the stresses have the same angular distribution as that of a flat plate,
3. the stress intensity factors are functions of the shell geometry and, in the limit, we recover the flat plate,
4. the stresses include interaction terms for bending and strectching.

A typical term for a spherical shell is

$$\frac{\sigma_{\text{shell}}}{\sigma_{\text{plate}}} \simeq 1 + \frac{c^2}{Rh} \left(a_1 + b_1 \ln \frac{c}{(Rh)^{\frac{1}{2}}} \right) + O\left(\frac{1}{R^2}\right) \qquad (4.59)$$

where the expression inside the parenthesis is a positive quantity. One concludes, therefore, that a spherical initial curvature, in reference to that of a flat sheet, is to increase the stresses in the neighborhood of the crack tip and, as a result, reduce its resistance to fracture initiation.

It should be emphasized that classical bending theory has been used in deducing the foregoing results. Hence it is inherent that only the Kirchoff equivalent shear free condition is satisfied along the crack [15], and not the vanishing of both individual shearing stresses. While outside the local region the stress distribution should be accurate, one might expect the same type of discrepancy to exist near the crack point as that found by Knowles and Wang [16] in comparing Kirchoff and Reissner bending results for a flat plate. In this case the order of the stress singularity remained unchanged but the angular distribution around the crack changed so as to precisely be the same as that due to solely extensional loading.

Recently, Sih and Hagendorf [17] investigated this matter further by deriving an improved theory of shallow shells which incorporates the effect of a transverse shear deformation. As expected, their results** showed that

---

* For the prediction of failures in pressurized vessels [14] the contributions of $k_1^{(b)}$ are negligible in comparison to those of $k_1^{(e)}$.
** See chapter 6.

classic theory cannot adequately predict the exact angular dependence of the bending stresses in the vicinity of a crack. However, in general, these bending stresses are so small when compared to the extensional stresses that can be neglected. On the other hand, for very long cracks such contributions become significant and consequently may no longer be neglected. Unfortunately, in such cases bulging effects become extremely important and any theory, whether classic or shear, is inadequate.

In-plane Shear Load.   If on the other hand, the residual loads, i.e., equation (4.16), are of the form

$$M_y^{(P)} = 0, \ V_y^{(P)} = 0, \ N_y^{(P)} = 0, \ N_{xy}^{(P)} = -\frac{t_0}{c^2} \tag{4.60}$$

then the solution can be constructed in a similar manner [10] and the results are

Extensional stresses: through the thickness:

$$\sigma_x^{(e)} = -\frac{k_2^{(e)}}{(2r)^{\frac{1}{2}}} \sin(\theta/2) \left[2 + \cos(\theta/2)\cos(3\theta/2)\right] \tag{4.61a}$$

$$\sigma_y^{(e)} = \frac{k_2^{(e)}}{(2r)^{\frac{1}{2}}} \sin(\theta/2)\cos(\theta/2)\cos(3\theta/2) \tag{4.61b}$$

$$\sigma_{xy}^{(e)} = \frac{k_2^{(e)}}{(2r)^{\frac{1}{2}}} \cos(\theta/2) \left[1 - \sin(\theta/2)\sin(3\theta/2)\right] \tag{4.61c}$$

Bending stresses: on the 'tension side' of the shell:

$$\sigma_x^{(b)} = \frac{k_2^{(b)}}{(2r)^{\frac{1}{2}}} \kappa \left[\sin(5\theta/2) - \left(\frac{9+7\nu}{1-\nu}\right)\sin(\theta/2)\right] \tag{4.62a}$$

$$\sigma_y^{(b)} = \frac{k_2^{(b)}}{(2r)^{\frac{1}{2}}} \kappa \left[\sin(\theta/2) - \sin(5\theta/2)\right] \tag{4.62b}$$

$$\sigma_{xy}^{(b)} = \frac{k_2^{(b)}}{(2r)^{\frac{1}{2}}} \kappa \left[\left(\frac{3\nu+5}{1-\nu}\right)\cos(\theta/2) - \cos(5\theta/2)\right] \tag{4.62c}$$

where the stress intensity factors $k_2^{(e)}$ and $k_2^{(b)}$ are given by

$$k_2^{(e)} = -\tau^{(e)} \sqrt{c} \left\{1 + \frac{\pi\lambda^2}{32} + O(\lambda^4 \ln \lambda)\right\} \tag{4.63a}$$

$$k_2^{(b)} = -\bar{\tau}^{(e)} \frac{\sqrt{3}(3+v)}{2(1-v^2)^{\frac{1}{2}}} \lambda^2 \left\{ \frac{7+v-4(5-v)\gamma}{16(1-v)} - \frac{5-v}{4(1-v)} \ln \frac{\lambda}{4} \right.$$

$$\left. + O(\lambda^2 \ln \lambda) \right\} \tag{4.63b}$$

with

$$\bar{\tau}^{(e)} = \frac{t_0}{hc^2} \tag{4.64}$$

In the limit, as $\lambda \to 0$ one recovers precisely the results of references [18] and [15]. Here again the results are correct up to $O(\lambda^4)$ terms and as a result they are only good approximations for $0 \le \lambda < 1$.

Effect of Transverse Vibrations. In addition to the usual external applied loads, pressure vessels are frequently also subjected to vibrations. Consequently, an investigation was carried out in order to assess analytically what effect, if any, do vibrations have on the mechanism of fracture. The analysis has shown [19] that in general, transverse vibrations reduce the stress intensity factor. However, when the forcing frequency ω approaches the natural frequency of the uncracked shell, the stress intensity factor increases without bound. This phenomenon, coupled with the usual $1/\sqrt{r}$ singular behavior, causes the pressure vessel to fail at nominal values even lower than the yield stress.

Thus, without going into the mathematical details [19], the stress intensity factors for a residual load* of the form

$$M_y^{(P)} = -\frac{D}{c^2} m_0 \cos(\omega t + \phi), \quad V_y^{(P)} = 0, \tag{4.65a}$$

$$N_y^{(P)} = -\frac{n_0}{c^2} \cos(\omega t + \phi), \quad N_{xy}^{(P)} = 0, \tag{4.65b}$$

are

case (i) $\lambda^4 \ge 0$:

$$k_1^{(e)} = \sqrt{c} \left\{ \frac{n_0}{hc^2} + \frac{m_0 E}{(3+v)R} \left[\frac{7}{32} + \frac{3}{8} \left(\gamma + \ln \frac{\lambda}{4} \right) \right] + O(\lambda^4 \ln \lambda) \right\} \cos(\omega t + \phi) \tag{4.66a}$$

---

* See equation (4.16).

$$k_1^{(b)} = \left\{ -\frac{\lambda^4 R n_0 \sqrt{c}}{2(1-v^2)c^4} \left[\frac{1+7v}{32} + \frac{1+3v}{8}\left(\gamma + \ln\frac{\lambda}{4} \right) \right] + \right.$$

$$\left. + \frac{m_0 E h \sqrt{c}}{2(1-v^2)c^4} + O(\lambda^4 \ln\lambda) \right\} \cos(\omega t + \phi) \tag{4.66b}$$

case (ii) $\lambda^4 < 0$:

$$k_1^{(e)} = \sqrt{c}\left\{ \frac{n_0}{hc}\left[1 + \frac{3\pi}{32}\lambda^2 \right] + \frac{m_0 E}{R(3+v)}\left[\frac{7}{32} + \frac{3}{8}\left(\gamma + \ln\frac{\lambda}{4} \right) \right] \right.$$

$$\left. + O(\lambda^4 \ln\lambda) \right\} \cos(\omega t + \phi) \tag{4.67a}$$

$$k_1^{(b)} = \left\{ -\frac{\lambda^4 R n_0 \sqrt{c}}{2(1-v^2)c^4}\left[\frac{1+7v}{32} + \frac{1+3v}{8}\left(\gamma + \ln\frac{\lambda}{4} \right) \right] + \right.$$

$$\left. + \frac{m_0 E h \sqrt{c}}{2(1-v^2)c^4}\left[1 + \frac{\pi\lambda^2}{32}\frac{1+3v}{3+v} \right] + O(\lambda^4 \ln\lambda) \right\} \cos(\omega t + \phi) \tag{4.67b}$$

where

$$\lambda^4 \equiv \frac{\rho h c^4}{D}\omega^2 - \frac{E h c^4}{R^2 D}, \tag{4.68}$$

ω the forcing frequency and ρ the density of the material.

From these results, the following special limiting cases of practical interest can be examined:

1. If $\omega \to 0$ and $R \neq \infty$, the stresses of a non-vibrating cracked spherical shell are recovered and coincide with those obtained in [10].
2. If $\omega \neq 0$ and $R \to \infty$, we recover the vibrating cracked plate expressions in [20].
3. If $\omega = 0$ and $R \to \infty$, the stresses of a flat sheet are recovered and coincide with those obtained previously for bending [15] and extension [18].
4. If $\lambda \to 0$, i.e., when the forcing frequency reaches the natural frequency $(E/\rho)^{\frac{1}{2}}(1/R)$ of the uncracked shell, the extensional stress intensity factor becomes infinite.

As a practical matter, it is of some value to compare the dynamic with the static stress along the line of crack prolongation. For $c = 1$ in., $h = 0.1$ in., $R = 32.6$ in., $v = 1/3$, $E = 16 \times 10^6$ psi, $\rho = 0.315$ lbf/in.^{3}.

(i) for $n_0 \neq 0$, $m_0 = 0$:

$$\frac{\sigma_{y_{\text{dynamic}}}}{\sigma_{y_{\text{static}}}} = \begin{cases} \left[\left[0.67(1 + 0.29\lambda^2) - 1.24\lambda^4 \left(0.25 + 0.13 \ln \frac{\lambda^2}{16} \right) \right] \\ \cos(\omega t + \phi); \lambda^4 < 0 \\ \left[0.67 - 1.24\lambda^2 \left(0.25 + 0.13 \ln \frac{\lambda^2}{16} \right) \right] \cos(\omega t + \phi); \\ \lambda^4 > 0 \end{cases}$$

(ii)   for $n_0 = 0$, $m_0 \neq 0$:

$$\frac{\sigma_{y_{\text{dynamic}}}}{\sigma_{y_{\text{static}}}} = \begin{cases} \left[\left[0.87(1 + 0.18\lambda^2) + 0.14 \left(0.43 + 0.19 \ln \frac{\lambda^2}{16} \right) \right] \\ \cos(\omega t + \phi); \lambda^4 < 0 \\ \left[0.87 + 0.14 \left(0.43 + 0.19 \ln \frac{\lambda^2}{16} \right) \right] \cos(\omega t + \phi); \lambda^4 > 0 \end{cases}$$

where

$$\lambda^4 = 2.1 \times 10^{-5} \omega^2 - 1$$

The plots of the ratio

$$I = \frac{\sigma_{y_{\text{dynamic}}}}{\sigma_{y_{\text{static}}}} \cos(\omega t + \phi)$$

for various values of ω are given in Figures 4.3 and 4.4.

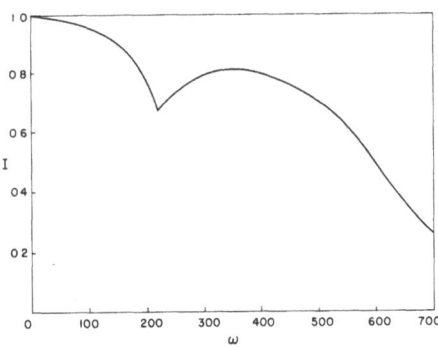

Figure 4.3. Ratio of Dynamic and Static Stresses *vs* ω for $m_0 = 0$

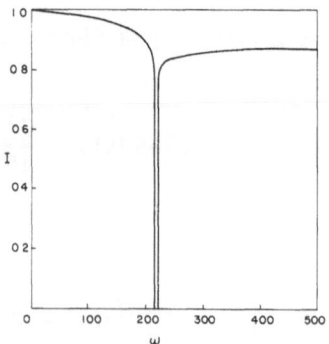

Figure 4.4. Ratio of Dynamic and Static Stresses vs ω for $n_0 = 0$

4.4 The stress field in a cracked plate

The problem of a flat plate containing a finite crack has been investigated by many authors for various types of loadings and the results are reported in other chapters of this volume. The solution, however, for an infinite plate (see Figure 4.5) may also be obtained from that of a spherical cap by simply

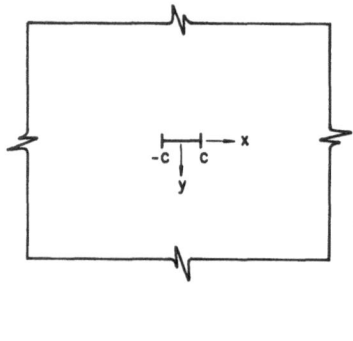

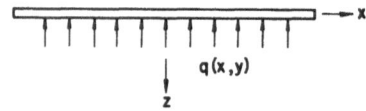

Figure 4.5. Cracked Plate Subjected to a Lateral Load q

letting $R \to \infty$ or $\lambda \to 0$. Thus the stress field around the crack tip is given by equations (4.53) and (4.54) where the stress intensity factors now are

$$k_1^{(e)} = \bar{\sigma}^{(e)} \sqrt{c} \qquad (4.69a)$$

$$k_1^{(b)} = - \bar{\sigma}^{(b)} \sqrt{c} \qquad (4.69b)$$

4.5 The stress field in a cracked cylindrical shell

For a cylindrical shell, one of the principal radii of curvature is infinite and the other constant. It appears therefore that this geometric simplicity leads to rather straightforward analytical solutions. However, the fact that the curvature varies between zero and a constant as one considers different angular positions – say around the point of a crack which is aligned parallel to the cylinder axis – more than obviates the initial geometric simplification and therefore increases the mathematical complexities considerably. For this reason, Sechler and Williams [5] suggested an approximate equation, based upon the behavior of a beam on an elastic foundation, and were able

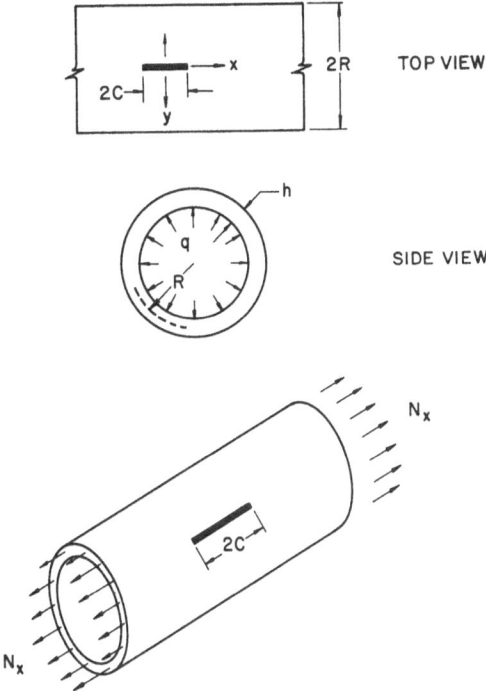

Figure 4.6. Geometry and Coordinates of an Axially Cracked Cylindrical Shell Under Uniform Axial Extension N_x and Internal Pressure q_0

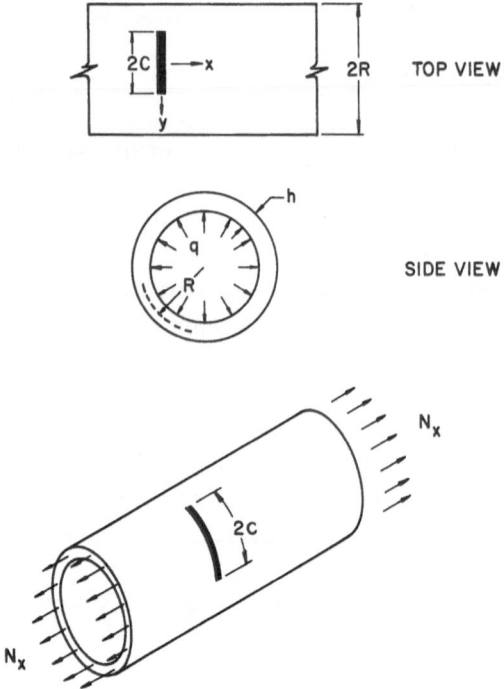

Figure 4.7. Geometry and Coordinates of a Peripherally Cracked Cylindrical Shell Under Uniform Axial Extension N_x and Internal Pressure q_0

to obtain a reasonable agreement with the experimental results. Subsequently, using the method of section 2, the author investigated this problem in a more sophisticated manner and the details for an axial and a peripheral crack (see Figures 4.6 and 4.7) can be found in references [21] and [22], respectively.

Again, omitting the mathematical details, the stresses around the crack tip are given also by equations (4.53) and (4.54), where the stress intensity factors are:*

* It should be emphasized that these results are only valid for small λ and that for large λ one must consider more terms of the asymptotic expansions. Using the method described previously on the alternate method of solution, the stress intensity factors have been determined numerically for $\nu = \frac{1}{3}$ and $\sigma^{(b)} = 0$ and may well be approximated within a 6% error and for all λ by the simple relations:

$$k_1^{(e)} = \sqrt{c}\,(1 + 0.317\,\lambda^2)^{\frac{1}{2}}$$

and

$$k_1^{(e)} = \sqrt{c}\,(1 + 0.05\,\lambda^2)^{\frac{1}{2}}\,.$$

(i) *for an axial crack* (see Figure 6)

$$k_1^{(e)} = \bar{\sigma}^{(e)} \sqrt{c} \left\{ 1 + \frac{5\pi\lambda^2}{64} \right\} + \bar{\sigma}^{(b)} \frac{(1-v^2)^{\frac{1}{4}}\lambda^2 \sqrt{c}}{\sqrt{3}(3+v)}$$

$$\left\{ \frac{5+37v}{96(1-v)} + \frac{1+5v}{16(1-v)} \cdot \left(\gamma + \ln\frac{\lambda}{8} \right) \right\} + O(\lambda^4 \ln \lambda); \quad \lambda < 1 \qquad (4.70a)$$

$$k_1^{(b)} = -\bar{\sigma}^{(e)} \sqrt{c} \frac{\sqrt{3}\,\lambda^2}{(1-v^2)^{\frac{1}{4}}} \left\{ \frac{5+37v}{96} + \frac{1+5v}{16}\left(\lambda + \ln\frac{\lambda}{8} \right) \right\}$$

$$- \bar{\sigma}^{(b)} \sqrt{c} \left\{ 1 - \frac{1+2v+5v^2}{(3+v)\,(1-v)} \frac{\pi\lambda^2}{64} \right\} + O(\lambda^4 \ln \lambda); \quad \lambda < 1 \qquad (4.70b)$$

(ii) *for a peripheral crack* (see Figure 4.7)

$$k_1^{(e)} = \bar{\sigma}^{(e)} \sqrt{c} \left\{ 1 + \frac{\pi\lambda^2}{64} \right\} + \bar{\sigma}^{(b)} \frac{(1+v^2)^{\frac{1}{4}}\,\lambda^2 \sqrt{c}}{\sqrt{3}\,(3+v)}$$

$$\left\{ \frac{(1+v)}{32(1-v)} + \frac{(1+v)}{16(1-v)}\left(\lambda + \ln\frac{\lambda}{8} \right) \right\} + O(\lambda^4 \ln \lambda); \quad \lambda < 2.5 \qquad (4.71a)$$

$$k_1^{(b)} = -\bar{\sigma}^{(e)} \frac{\sqrt{3}\,\lambda^2 \sqrt{c}}{(1-v^2)^{\frac{1}{4}}} \left\{ \frac{1+v}{32} + \frac{1+v}{16}\left(\gamma + \ln\frac{\lambda}{8} \right) \right\} - \frac{\bar{\sigma}^{(b)}}{3+v}$$

$$\left\{ 1 - \frac{5+2v+v^2}{(3+v)(1-v)} \frac{\pi\lambda^2}{64} \right\} + O(\lambda^4 \ln \lambda); \quad \lambda < 2.5 \qquad (4.71b)$$

(iii) *for an arbitrary orientation crack** (see Figure 4.8)

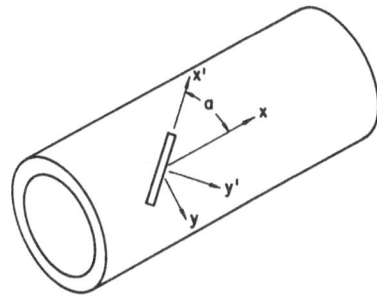

Figure 4.8

$$k_1^{(e)} = \bar{\sigma}^{(e)} \sqrt{c} \left\{ 1 + \frac{(5\cos^2\alpha + \sin^2\alpha)\pi\lambda^2}{64} \right\} + \bar{\sigma}^{(b)} \frac{(1-v^2)^{\frac{1}{4}}\lambda^2 \sqrt{c}}{\sqrt{3}(3+v)}$$

$$\left\{\left[\frac{5+37v}{96(1-v)}+\frac{1+5v}{16(1-v)}\left(\gamma+\ln\frac{\lambda\cos\alpha}{8}\right)\right]\cos^2\alpha\right.$$

$$\left.+\left[\frac{1+v}{32(1-v)}+\frac{1+v}{16(1-v)}\left(\gamma+\ln\frac{\lambda\sin\alpha}{8}\right)\right]\sin^2\alpha\right\}+O(\lambda^4\ln\lambda);\quad\lambda<1$$

$$\tag{4.72a}$$

$$k_1^{(b)}=-\bar{\sigma}^{(e)}\frac{\sqrt{3}\,\lambda^2\,\sqrt{c}}{(1-v^2)^{\frac{1}{4}}}\left\{\left[\frac{5+37v}{96}+\frac{1+5v}{16}\left(\gamma+\ln\frac{\lambda\cos\alpha}{8}\right)\right]\cos^2\alpha\right.$$

$$\left.+\left[\frac{1+v}{32}+\frac{1+v}{16}\left(\gamma+\ln\frac{\lambda\sin\alpha}{8}\right)\right]\sin^2\alpha\right\}$$

$$-\bar{\sigma}^{(b)}\sqrt{c}\left\{1-\frac{(1+2v+5v^2)\cos^2\alpha+(5+2v+v^2)\sin^2\alpha}{(3+v)(1-v)}\frac{\pi\lambda^2}{64}\right\}$$

$$+O(\lambda^4\ln\lambda);\quad\lambda<1\tag{4.72b}$$

and*

$$k_2^{(e)}=\bar{\tau}^{(e)}\sqrt{c}\left\{1+\sqrt{5}\frac{\pi\lambda^2}{64}\sin2\alpha\right\}$$

$$+\bar{\tau}^{(b)}\sqrt{c}\frac{(1-v^2)^{\frac{1}{4}}\lambda^2}{\sqrt{3}(3+v)}\left[\frac{5+37v}{96(1-v)}+\frac{1+5v}{16(1-v)}\left(\gamma+\ln\frac{\lambda\cos\alpha}{8}\right)\right]^{\frac{1}{2}}$$

$$\left[\frac{(1+v)}{32(1-v)}+\frac{(1+v)}{16(1-v)}\left(\gamma+\ln\frac{\lambda\sin\alpha}{8}\right)\right]^{\frac{1}{2}}\sin2\alpha+O(\lambda^4\ln\lambda);\quad\lambda<1\tag{4.73a}$$

$$k_2^{(b)}=\bar{\tau}^{(e)}\frac{\sqrt{3}\,\lambda^2\,\sqrt{c}}{(1-v^2)^{\frac{1}{4}}}\left\{\left[\frac{5+37v}{96}+\frac{1+5v}{16}\left(\gamma+\ln\frac{\lambda\cos\alpha}{8}\right)\right]^{\frac{1}{2}}\right.$$

$$\left.\left[\frac{1+v}{32}+\frac{1+v}{16}\left(\gamma+\ln\frac{\lambda\sin\alpha}{8}\right)\right]^{\frac{1}{2}}\right\}\sin2\alpha$$

$$+\bar{\tau}^{(b)}\sqrt{c}\left\{1+\frac{(5v_0^2-12v_0+8)^{\frac{1}{4}}(v^2+2v+5)^{\frac{1}{4}}}{(4-v_0)v_0}\frac{\pi\lambda^2}{64}\sin2\alpha\right\}$$

$$+O(\lambda^4\ln\lambda);\quad\lambda<1\tag{4.73b}$$

4.6 Approximate stress intensity factors for other shell geometries

Because the complementary or perturbed solution presents contributions only in the immediate vicinity of the crack tip, one may consider—at least

* The reader should note that the angular distribution here is given by equations (4.61) and (4.62).

locally — the principal radii or curvatures constant. Thus, assuming that the crack is parallel to one of the principal axes, say along the x-axis, one may hypothesize that the stress intensity factors depend primarily on the curvatures that one observes as he travels along and perpendicular to the crack prolongation. Consequently, one may estimate the stress intensity factors by a proper superposition of the results of an axial and a peripheral crack in a cylindrical shell. In particular, for $\bar{\sigma}^{(b)} = 0$

$$k_1^{(e)} \simeq \bar{\sigma}^{(e)} \sqrt{c} \left\{ 1 + \frac{\pi \lambda_x^2}{64} + \frac{5\pi \lambda_y}{64}^2 \right\} + O(\lambda^4 \ln \lambda); \quad \lambda < 1 \tag{4.74a}$$

$$k_1^{(b)} \simeq \bar{\sigma}^{(e)} \frac{\sqrt{3} \sqrt{c}}{(1 - v^2)^{\frac{1}{2}}} \left\{ \frac{5 + 37v}{96} \lambda_x^2 + \frac{1 + 5v}{16} \lambda_y^2 \left(\gamma + \ln \frac{\lambda_y}{8} \right) \right.$$
$$\left. + \frac{1 + v}{32} \lambda_x^2 + \frac{1 + v}{16} \lambda_x^2 \left(\gamma + \ln \frac{\lambda_x}{8} \right) \right\} + O(\lambda^4 \ln \lambda); \quad \lambda < 1 \tag{4.74b}$$

In order to check the validity of such a superposition we will consider as our first example* a spherical cap the stress intensity factors for which we know exactly.

Example 1: Sphere. For this shell the curvature is constant in all directions; therefore, in view of equations (4.70a) and (4.71a), one has

$$k_1^{(e)} \simeq \bar{\sigma}^{(e)} \sqrt{c} \left\{ 1 + \frac{\pi \lambda^2}{64} + \frac{5\pi \lambda^2}{64} \right\} = \bar{\sigma}^{(e)} \left\{ 1 + \frac{3\pi \lambda^2}{32} \right\}; \quad \lambda < 1 \tag{4.75}$$

which is identical to equation (4.56). Similarly,

$$k_1^{(b)} \simeq -\bar{\sigma}^{(e)} \frac{\sqrt{3} \lambda^2 \sqrt{c}}{(1 - v^2)^{\frac{1}{2}}} \left\{ \frac{5 + 37v}{96} + \frac{1 + 5v}{96} \left(\gamma + \ln \frac{\lambda}{8} \right) \right.$$
$$\left. + \frac{1 + v}{32} + \frac{1 + v}{16} \left(\gamma + \ln \frac{\lambda}{8} \right) \right\}$$
$$= -\bar{\sigma}^{(e)} \frac{\sqrt{3} \lambda^2 \sqrt{c}}{(1 - v^2)^{\frac{1}{2}}} \left\{ \frac{-0.1 + 5v}{32} + \frac{1 + 3v}{8} \left(\gamma + \ln \frac{\lambda}{4} \right) \right\}; \quad \lambda < 1 \tag{4.76}$$

which agrees fairly well with equation (4.57). One may conclude, therefore, that such a hypothesis may not be unreasonable.

---

* In the following examples, we have assumed $\sigma^{(b)} = 0$.

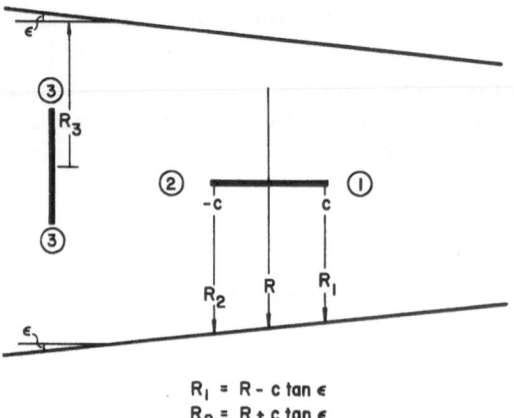

$$R_1 = R - c \tan \epsilon$$
$$R_2 = R + c \tan \epsilon$$

Figure 4.9. Conical Circular Shell

Example 2: Circular conical shell (see Figure 4.9). In this case, one curvature is infinite, the other finite; therefore,

(*i*) *for an axial crack:*

$$k_1^{(e)} \simeq \bar{\sigma}^{(e)} \sqrt{c} \left\{ 1 + \frac{5\pi}{64} \lambda_1^2 \right\}; \quad \lambda_1 < 1 \tag{4.77a}$$

$$k_1^{(e)} \simeq \bar{\sigma}^{(e)} \sqrt{c} \left\{ 1 + \frac{5\pi}{64} \lambda_2^2 \right\}; \quad \lambda_2 < 1 \tag{4.77b}$$

(*ii*) *for a peripheral crack:*

$$k_1^{(e)} \simeq \bar{\sigma}^{(e)} \sqrt{c} \left\{ 1 + \frac{\pi}{64} \lambda_3^2 \right\}; \quad \lambda_3 < 1 \tag{4.78}$$

where

$$\lambda_1^2 \equiv [12(1 - v^2)]^{\frac{1}{2}} \frac{c^2}{(R - c \tan \varepsilon)h}$$

$$\lambda_2^2 \equiv [12(1 - v^2)]^{\frac{1}{2}} \frac{c^2}{(R + c \tan \varepsilon)h}$$

$$\lambda_3^2 \equiv [12(1 - v^2)]^{\frac{1}{2}} \frac{c^2}{R_3 h} \tag{4.79}$$

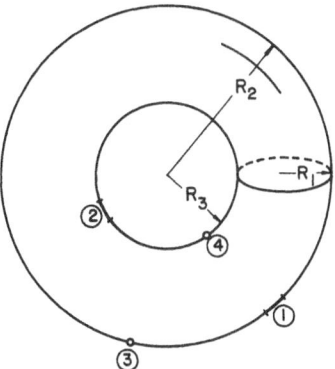

Figure 4.10. Toroidal Shell

Example 3: Toroidal shell (see Figure 4.10). For an axial crack in the outer surface

$$k_1^{(e)} \simeq \bar{\sigma}^{(e)} \sqrt{c} \left\{ 1 + \frac{5\pi}{64} \lambda_1^2 + \frac{\pi}{64} \lambda_2^2 \right\}; \quad \lambda_{1,2} < 1 \tag{4.80}$$

for an axial crack in the inner surface

$$k_1^{(e)} \simeq \bar{\sigma}^{(e)} \sqrt{c} \left\{ 1 + \frac{5\pi}{64} \lambda_1^2 - \frac{\pi}{64} \lambda_3^2 \right\}; \quad \lambda_{1,3} < 1 \tag{4.81}$$

for a peripheral crack in the outer surface

$$k_1^{(e)} \simeq \bar{\sigma}^{(e)} \sqrt{c} \left\{ 1 + \frac{5\pi}{64} \lambda_2^2 + \frac{\pi}{64} \lambda_1^2 \right\}; \quad \lambda_{1,2} < 1 \tag{4.82}$$

and for a peripheral crack in the inner surface

$$k_1^{(e)} \simeq \bar{\sigma}^{(e)} \sqrt{c} \left\{ 1 - \frac{5\pi}{64} \lambda_3^2 + \frac{\pi}{64} \lambda_1^2 \right\}; \quad \lambda_{1,3} < 1 \tag{4.83}$$

4.7 Plates on elastic foundations

Analyses of plates resting on foundations usually fall into two groups. The first group follows the well known theory of Winkler and Zimmerman [23] in which the elastic foundation is considered as a system of separate unconnected springs. Such a hypothesis simplifies considerably the analysis of structures on elastic foundations and leads frequently to incorrect results. The second group follows the theory in which one describes the physical properties of the natural foundation more accurately by the hypothesis that the foundation is an elastic isotropic semi-infinite space [24]. Here again, such a hypothesis leads to cumbersome calculations and therefore the method becomes impractical.

Recently a new theory based on Vlasov's general variational method [25] has been proposed [26]. This theory considers the elastic foundation as a single or double layer model whose properties are described by two or more generalized elastic characteristics. The advantage of this theory is that it is more accurate than the theory of Winkler and Zimmermann and simpler than the theory of the elastic semi-infinite space.

Winkler-Zimmermann Foundation. The characteristics of the fracture of plates resting on a Winkler-Zimmermann foundation have been investigated and the results are reported in references [27] and [28]. In this case, the governing differential equation for the displacement function $w(x, y)$, with x and y as dimensionless coordinates (see Figure 4.11), is given in the classical theory by

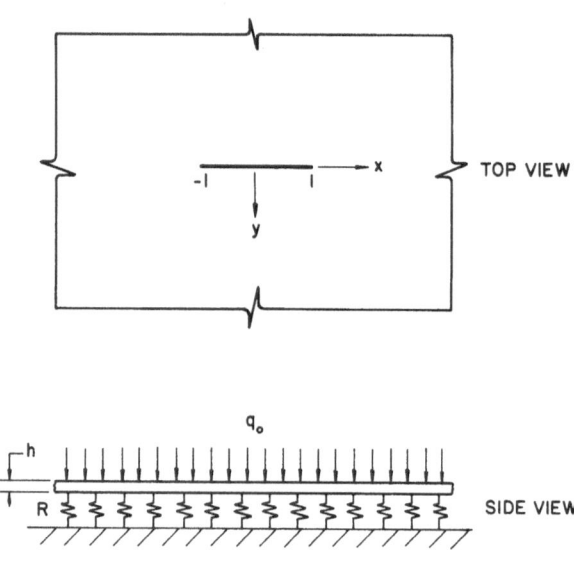

Figure 4.11

$$(\nabla^4 + \lambda^4) \, w(x, y) = \frac{q(x, y)c^4}{D},$$ (4.84)

where

$$\lambda^4 \equiv \frac{k}{D} \, c^4$$ (4.85)

and k the elastic foundation spring constant. In reference [27], the author, using the method described in section 4.3, was able to obtain an asymptotic expansion of the solution for small values of the parameter λ. Thus without going into the details, the stresses at the surface $z = h/2c$ are given by equations (4.54) where the bending stress intensity factor now is given by

$$k_1^{(b)} = \bar{\sigma}^{(b)} \sqrt{c} \left\{ 1 + \frac{3 + 2v + 3v^2}{(3 + v)(1 - v)} \frac{\pi\lambda^2}{32} + \left[\frac{3 + 6v + 15v^2}{(3 + v)(1 - v)} \left(\gamma + \ln\frac{\lambda}{4} \right) \right. \right.$$
$$\left. \left. + \frac{3v^2 - 1}{(3 + v)(1 - v)} \right] \frac{\lambda^4}{128} \right\}^{-1} + O(\lambda^6 \ln \lambda); \quad \lambda < 2$$ (4.86)

On the other hand, for large values of the parameter λ reference [28] gives

$$k_1^{(b)} = \bar{\sigma}^{(b)} \sqrt{c} \, \frac{1}{\sqrt{\lambda}} + O(\lambda^{-3/2}); \quad \lambda > 4$$ (4.87)

For example, along the crack prolongation and for $v = 1/3$,

$$\frac{\sigma_y^{(b)}(\varepsilon, 0)}{\bar{\sigma}^{(b)}} \simeq \begin{cases} \dfrac{\sqrt{c}}{(2r)^{\frac{1}{2}}} \dfrac{1}{\left[1 + \dfrac{9}{5} \dfrac{\pi}{32} \lambda^2 \right]} & ; \quad \lambda < 1 \\[4mm] \dfrac{\sqrt{c}}{(2r)^{\frac{1}{2}}} \dfrac{1}{\sqrt{\lambda}} & ; \quad \lambda > 4 \end{cases}$$ (4.88)

Since the behavior of the stress intensity factor at the two extremes is known, one may construct a curve with the proper asymptotes. Such a plot is given in Figure 4.12. One concludes, therefore, that the general effect of an elastic foundation is to decrease the stresses in magnitude by a factor which depends on the type of foundation, the crack length, and the material properties.

Single Layer Foundation. Following [26], the differential equation governing the displacement function is $w(X, Y)$ of a plate resting on single layered elastic foundation (see Figure 4.13).

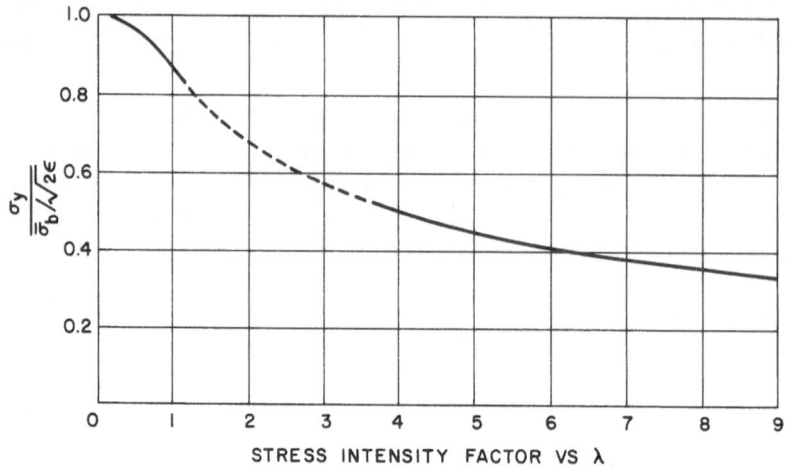

STRESS INTENSITY FACTOR VS λ

Figure 4.12

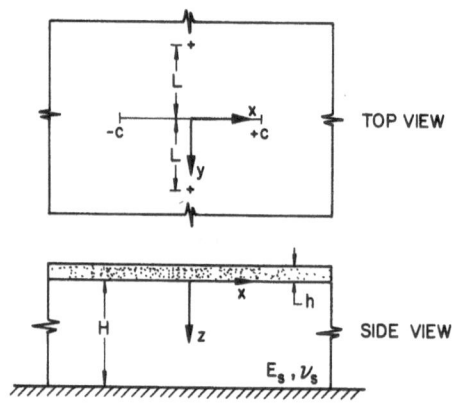

Figure 4.13. A Cracked Plate on a Single Layered Foundation

$$\nabla^4 w - 2r^{*2}\nabla^2 w + \delta^{*4}w + m^*\frac{\partial^2 w}{\partial t^2} = \frac{q}{D} \tag{4.89}$$

with the quantities $r*$, $\delta*$ and $m*$ as constants defined by*

$$r*^2 = \frac{E_s^*}{(1 + v_s^*) D} \int_0^H \psi^2(z) \, dz$$

$$\delta*^4 = \frac{E_s^*}{(1 - v_s^{*2})D} \int_0^H \{\psi'(z)\}^2 \, dz$$

$$m* = \left(\frac{\gamma_p h}{g} + \frac{\gamma_s}{g} \int_0^H \psi^2(z) \, dz \right) \frac{1}{D} \tag{4.90}$$

and

$$E_s^* = \frac{E_s}{1 - v_s^2}$$

$$v_s^* = \frac{v_s}{1 - v_s} \tag{4.91}$$

D the flexural rigidity of the plate, γ_p and γ_s the specific weights of the plate and elastic foundation and g the gravitational acceleration.

The problem of a finite crack, of length $2c$, in the plate has been investigated and the results are reported in reference [29]. Thus without going through the details, the stress intensity factor, in view of the definitions

$$r = cr*, \quad \delta = c\delta*, \quad k = r/\delta, \tag{4.92}$$

become

(i) for $\delta < r < 1$

$$k_1^{(b)} = \bar{\sigma}_b \sqrt{c(12 - \tfrac{3}{2}r^2)} \left\{ 12 + \{ - \tfrac{3}{4}[v(2 - v)(5 - 12\gamma + 12 \ln 2) + \right.$$

$$- 2(3 - 4\gamma + 4 \ln 2) + 16v(1 - 2\gamma + 2 \ln 2)] +$$

$$+ \tfrac{3}{2}(3v + 1)(1 - v)(1 - \ln 2 + 2 \ln r)\} \frac{r^2}{(1 - v)(3 + v)} +$$

$$+ \{\tfrac{3}{64}[v(2 - v)(5 - 12\gamma + 12 \ln 2)(3 - 4\gamma + 4 \ln 2) - 8v(1 - 2\gamma + 2 \ln 2)]$$

---

* It is assumed that no horizontal displacements occur in the elastic foundation and that the vertical displacement is given by a single function $\psi(z)$. From reference [26] a typical function is

$$\psi(z) = \frac{\sinh [r_*(H - z)]}{\sinh [r_* H]}$$

where r_* is a coefficient determining the variation with depth of the displacement.

$$- \tfrac{3}{32}(3v + 1)(1 - v)(\tfrac{3}{2} - \ln 2 + 2 \ln r)\} \frac{r^4}{(1 - v)(3 + v)} + \dots\Big\}^{-1}$$

$$(4.93)$$

Notice that δ does not appear in equation (4.93) for it is negligible. A plot of this is given in Figure 4.14.

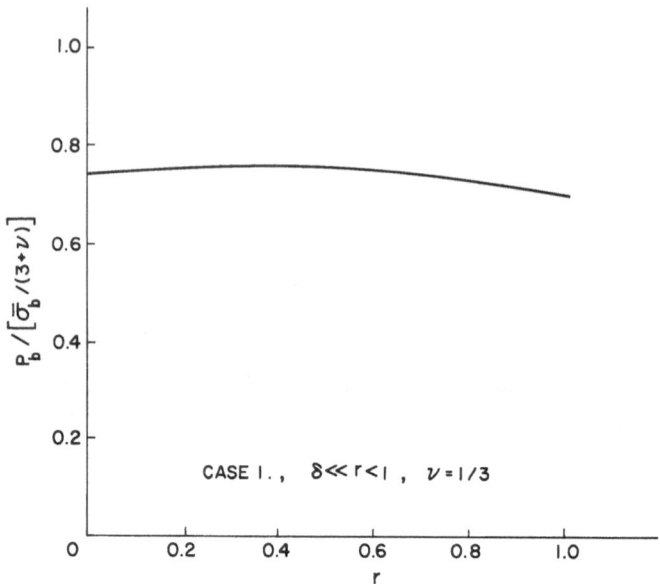

Figure 4.14. Stress Coefficient Versus r

(ii) for $r = k\delta < 1$, where k is a real constant

$$k_1^{(b)} = \bar{\sigma}_b \sqrt{c}\Big\{1 + \frac{1}{8(k^4 - 1)^{\frac{1}{4}}} \ln\Big[\frac{k^2 + (k^4 - 1)^{\frac{1}{2}}}{k^2 - (k^4 - 1)^{\frac{1}{2}}}\Big]$$

$$\frac{(1 + v) + \tfrac{1}{4}(3v + 1)(1 - v)(2k^4 - 1)}{(1 - v)(3 + v)}\delta^2$$

$$- \frac{1}{16(1 - v)(3 + v)}\big[v(2 - v)(5 - 12\gamma + 12 \ln 2) + (3 - 4\gamma + 4 \ln 2) +$$

$$- 8v(1 - 2\gamma + 2 \ln 2) - 2(1 + 3v)(1 - v)(1 - 2 \ln 2)]k^2\delta^2 +$$

$$+ \tfrac{1}{4}k^2(\ln \delta)\delta^2 + \frac{3k^2\delta^2}{4 - 3k^2\delta^2} - \frac{3}{32(1 - v)(3 + v)}\big[v(2 - v)$$

$$(5 - 12\gamma + 12\ln 2) + (3 - 4\gamma + 4\ln 2) - 8v(1 - 2\gamma + 2\ln 2) - 2(1 + 3v)$$

$$(\tfrac{1}{2} - 2\ln 2)]\frac{k^4\delta^4}{4 - 3k^2\delta^2} + \frac{3}{16(k^4-1)^{\frac{1}{2}}}\ln\frac{k^2 + (k^4 - 1)^{\frac{1}{2}}}{k^2 - (k^4 - 1)^{\frac{1}{2}}}$$

$$\frac{(1 + v) + \tfrac{1}{4}(3v + 1)(1 - v)(2k^4 - 1)}{(1 + v)(3 + v)}\quad \frac{k^2\delta^4}{4 - 3k^2\delta^2} + \frac{3k^4\delta^4}{32 - 24k^2\delta^2}$$

$$\left.\ln\delta + \ldots\right\}^{-1} \tag{4.94}$$

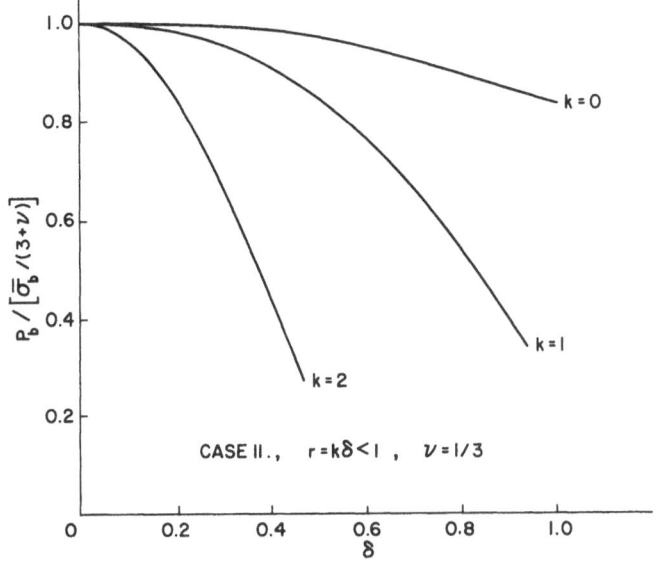

Figure 4.15. Stress Coefficient Versus δ

A plot is given in Figure 4.15. Furthermore, in the limit as $k \to 0$ one recovers the results for a Winkler and Zimmermann foundation.

4.8 Particular solutions

In general, the actual stress fields will depend upon the contributions of the particular solutions reflecting the magnitude and distribution of the applied load. On the other hand, the singular part of the solution, that is the terms producing infinite elastic stresses at the crack tip, will depend upon the local stresses existing along the locus of the crack before it is cut, which of course are precisely the stresses which must be removed by the particular solutions described above in order to obtain the stress-free edges as required physically.

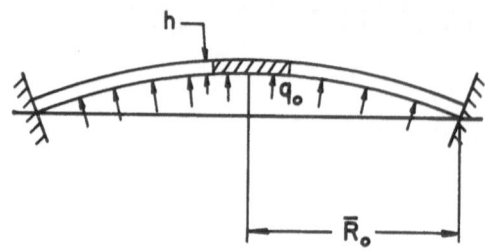

Figure 4.16. Pressurized Spherical Cap With Fixed Ends

Clamped Spherical Shell. Consider a clamped segment of a shallow spherical shell of base radius $\bar{R}_0$ and containing at the apex a finite radial crack of length $2c$ (see Figure 4.16). The shell is subjected to a uniform internal pressure q_0 with radial extension $N_r = (q_0/2)R$, and because it is clamped we require that the displacement and slope vanish at $\bar{R} = \bar{R}_0$. For this problem the residual 'applied bending' and 'applied stretching' loads at the crack are:*

$$\bar{\sigma}^{(e)} = q_0 R/2h, \quad \bar{\sigma}^{(b)} = 0 \text{ and } \bar{\tau}^{(e)} = 0 \tag{4.95}$$

Closed Cylindrical Tank. Consider a shallow cylindrical shell containing a crack of length $2c$. The shell is subjected to a uniform internal pressure q_0 with an axial extension $N_x = (q_0 R/2)$, $M_y = 0$, far away from crack. For this problem, if the crack is parallel to the axis of the cylinder, then

$$\bar{\sigma}^{(e)} = (q_0 R/h), \quad \bar{\sigma}^{(b)} = 0 \text{ and } \bar{\tau}^{(e)} = 0 \tag{4.96}$$

If the crack is perpendicular to the axis of the cylinder, then

$$\bar{\sigma}^{(e)} = q_0 R/2h, \quad \bar{\sigma}^{(b)} = 0 \text{ and } \bar{\tau}^{(e)} = 0 \tag{4.97}$$

In the event that the crack makes an angle α with the axis of the cylinder, then

---

* For more details see reference [30].

$$\bar{\sigma}^{(e)} = (q_0R/4h)\,(3 + \cos 2\alpha),\ \bar{\sigma}^{(b)} = 0,\ \text{and}\ \bar{\tau}^{(e)} = (q_0R/4h)\sin 2\alpha \qquad (4.98)$$

Infinite Plate. Consider an infinite thin plate containing a crack of length $2c$. At infinity the plate is subjected to a uniform extensional load σ_∞ and in-plane shear load τ_∞, then

$$\bar{\sigma}^{(e)} = \sigma_\infty,\ \ \bar{\sigma}^{(b)} = 0\ \text{and}\ \bar{\tau}^{(e)} = \tau_\infty. \qquad (4.99)$$

Rectangular Strip on a Spring Foundation. Consider a rectangular strip, infinitely long in the x-direction and of finite width b in the y-direction. Furthermore, let the strip be subjected to a constant moment M_0 and zero shear at $y = \pm b$, and simultaneously subject to a uniform normal loading q_0. Then*

$$\bar{\sigma}^{(e)} = \bar{\tau}^{(e)} = 0$$

and

$$\bar{\sigma}^{(b)} = -\ \frac{6M_0}{h^2c^2}\ \frac{\cos\,(\lambda b/\sqrt{2})\sinh\,(\lambda b/\sqrt{2}) + \sin\,(\lambda b/\sqrt{2})\cosh\,(\lambda b/\sqrt{2})}{\cosh\,(\lambda b/\sqrt{2})\sinh\,(\lambda b/\sqrt{2}) + \sin\,(\lambda b/\sqrt{2})\cos\,(\lambda b/\sqrt{2})} \qquad (4.100)$$

where

$$\lambda^4 \equiv \frac{k}{D}\,c^4 \qquad (4.101)$$

Plate on a Single Layered Foundation. Consider an infinite elastic plate which rests on a single-layered elastic foundation and contains a finite, through the thickness, crack of length $2c$. The plate is subjected to two equal concentrated lateral static loads of intensity P_0 with corresponding points of application $(0, L, -h)$ and $(0, -L, -h)$ (see Figure 4.13). Furthermore, it will be assumed that $L > > c$. Then**

$$\bar{\sigma}^{(e)} = 0\ \text{and}\ \bar{\sigma}^{(b)} = -\ \frac{6D}{h^2c^2}\ \frac{P_0c^4}{2\pi D(\lambda_+^2 - \lambda_-^2)}\left\{\frac{(1-v)}{(x^2+l^2)^{\frac{1}{2}}}\left[1 - \frac{2x^2}{x^2+l^2}\right]\right.$$

$$\{(\lambda_+K_1[\lambda_+(x^2+l^2)^{\frac{1}{2}}] - \lambda_-K_1[\lambda_-(x^2+l^2)^{\frac{1}{2}}]\} +$$

$$\left. + \frac{l^2+vx^2}{x^2+l^2}\,\{\lambda_+^2K_0[\lambda_+(x^2+l^2)^{\frac{1}{2}}] - \lambda_-^2K_0[\lambda_-(x^2+l^2)^{\frac{1}{2}}]\}\right\} \qquad (4.102)$$

---

* For more details see reference [28].
** See reference [29].

where

$$l = \frac{L}{c}, \quad \lambda_{\pm}^2 = r^2 \pm (r^4 - \delta^4)^{\frac{1}{2}} \tag{4.103}$$

Now, since we have already assumed that $1 \ll l$, it is easy to see that the above bending moment (along the crack) is approximately a constant, i.e., $-Dm_0/c^2$. Alternately, as an engineering approximation, one may think of the quantity $(-Dm_0/c^2)$ as an upper bound, or lower bound, or even a mean value of the precise bending moment along the crack in order to obtain an estimate of the stresses in the vicinity of the crack.

4.9 Discussion

From the above analysis it becomes evident that in an initially curved sheet the stresses near the crack tip possess the usual $1/\sqrt{r}$ singular behavior which is characteristic to two-dimensional linear elastic crack problems. Furthermore, the angular distribution around the crack tip is precisely the same as that of a flat plate and that the initial curvature appears only in the stress intensity factors and it appears in such a way that in the limit as $R \to \infty$ one recovers the flat sheet behavior.

A typical term is of the form

$$\frac{\sigma_{\text{shell}}}{\sigma_{\text{plate}}} \simeq 1 + \left\{ \frac{a_1}{R_1} + \frac{a_2}{R_2} + \frac{b_1}{R_1} \ln \frac{c}{(R_1 h)^{\frac{1}{2}}} + \frac{b_2}{R_2} \ln \frac{c}{(R_2 h)^{\frac{1}{2}}} \right\} \cdot \frac{c^2}{h}$$
$$+ O\left(\frac{1}{R_1^2}, \frac{1}{R_2^2} \right) \tag{4.104}$$

Thus the general effect of a positive (negative) initial curvature, in reference to that of a flat sheet, is to increase (decrease) the stresses in the neighborhood of the crack point and reduce (increase) its resistance to fracture initiation. For a cylindrical shell with an axial crack, for example, equation (4.178) reads

$$\frac{\sigma_{\text{hoop}}}{\sigma_{\text{plate}}} \simeq \frac{1}{(1 + 0.317\lambda^2)^{\frac{1}{2}}}, \tag{4.105}$$

which correlates flat sheet behavior with that of initially curved specimens.

In a similar manner, the general effect of an elastic foundation is to decrease the magnitude of the stress intensity factor in the neighborhood of the crack tip and as a result prevent further fracture. This decrease clearly depends on the values of the parameters which characterize the elastic foundation.

References

[1] Ogibalov, P. M., *Dynamics and strength of shells*, translated from Russian, published for NASA and NSF by the Israel Program for Scientific Translations.

[2] Marguerre, K., Zur theorie der gekrummten platte grosser formanderung, *Proc. 5th Int. Congr. Appl. Mech.* pp. 93–101 (1938).

[3] Reissner, E., On some problems in shell theory, Structural Mechanics, *Proceedings of the First Symposium on Naval Structural Mechanics*, pp. 11–14, pp. 74–113 (1958).

[4] Reissner, E., A note on membrane and bending stresses in spherical shells, *Soc. Industr. Appl. Math.* 4, pp. 230–240 (1956).

[5] Sechler, E. E. and Williams, M. L., The critical crack length in pressurized monocoque cylinders, Final Report, GALCIT 96, Calif. Inst. Tech., September 1959. See also: M. L. Williams, *Proc. Crack Propagation Symp.*, Cranfield, 1, pp. 130 (1961).

[6] Erdelyi, A et al., Tables of integral transforms, *Bateman Manuscript Project*, McGraw-Hill, N. Y. (1954).

[7] Ang, D. D. and Williams, M. L., Combined stresses in an orthotropic plate having a finite crack, *J. Appl. Mech.*, pp. 372–378 (1961).

[8] Noble, B., The approximate solution of dual integral equations by Variational Methods, *Proceedings Edinburgh Mathematical Society*, 11, pp. 113–126 (1958–59).

[9] Muskheleshvili, N. I., *Singular integral equations*, English translation by J. R. M. Radok, P. Noordhoff, Ltd., The Netherlands (1953).

[10] Folias, E. S., *The stresses in a spherical shell containing a crack*, ARL 64–23, Aerospace Research Laboratories (1964).

[11] Folias, E. S., *The effect of initial curvature on cracked flat sheets*, UTEC DO 68–070, University of Utah (1968).

[12] Sneddon, I. N. and Lowengrub, M., *Crack problems in the classical theory of elasticity*, John Wiley & Sons, Inc., New York (1969).

[13] Kantorovich, L. V. and Krylov, V. I., *Approximate methods of higher analysis*, P. Noordhoff, The Netherlands (1964).

[14] Folias, E. S., On the predictions of catastrophic failures in pressurized vessels, *Prospects of Fracture Mechanics*, edited by G. C. Sih, H. C. van Elst and D. Broek, Noordhoff International Publishing (1974).

[15] Williams, M. L., The bending stress distribution at the base of a stationary crack, *J. Appl. Mech.*, pp. 78–82 (1961).

[16] Knowles, J. K. and Wang, N. M., On the bending of an elastic plate containing a crack, *J. Math. and Phys.* 39, pp. 223–236 (1960).

[17] Sih, G. C. and Hagendorf, H. C., A new theory of spherical shells with cracks, *Thin Shell Structures*, edited by Y. C. Fung and E. E. Sechler, Prentice-Hall, pp. 519–545 (1974).

[18] Williams, M. L., On the stress distribution at the base of a stationary crack, *J. Appl. Mech.* (1957).

[19] Do, S. H. and Folias, E. S., On the steady-state transverse vibrations of a cracked spherical shell, *International Journal of Fracture Mechanics*, 7, No. 1, pp. 23–37 (1971).

[20] Folias, E. S., On the steady-state transverse vibrations of a cracked plate, *Engineering Fracture Mechanics Journal*, Vol. 1, No. 2, pp. 363–368 (1968).

[21] Folias, E. S., An axial crack in a pressurized cylindrical shell, *International Journal of Fracture Mechanics*, 1, 2, pp. 104–113 (1965), or The stresses in a cylindrical shell containing an axial crack, ARL–64–174, Aerospace Research Laboratories, United States Air Force, Dayton, Ohio, pp. 1–42 (1964).

[22] Folias, E. C., A circumferential crack in a pressurized cylindrical shell, *International Journal of Fracture Mechanics*, 3, 1, pp. 1–11 (1967).

[23] Zimmermann, H., *Die berechnung des eisenbahn oberbaues*, Berlin: W. Ernst U. Sohn (1888).

[24] Boussinesq, J., *Application des potentiels à l'étude de l'équilibre et du mouvement des solides élastiques*, Paris: Gauthier-Villars (1888).

[25] Vlasov, V. Z., *General theory of shells and its application in engineering*, Moskva Leningrad, Gostekhizdat (1949).

[26] Vlasov, V. Z. and Leont'ev, U. N., *Beams, plates and shells on elastic foundations*, NASA Technical Translation No. TTF–357 (Acc. No. TT65–50135).

[27] Folias, E. S., On a plate supported by an elastic foundation and containing a finite crack, *International Journal of Fracture Mechanics*, 6, 3 pp. 257–263 (1970).

[28] Ang, D. D., Folias, E. S. and Williams, M. L., The bending stress in a cracked plate on an elastic foundation, *J. Appl. Mech.*, 30, pp. 245–251 (1963).

[29] Lin, Si-Tsai, and Folias, E. S., On the fracture of highway pavement, *International Journal of Fracture*, 11, pp. 93–106 (1975).

[30] Folias, E. S., A finite line crack in a pressurized spherical shell, *International Journal of Fracture Mechanics*, 1, 1, pp. 20–46 (1966).

F. Erdogan

5 | *Crack problems in cylindrical and spherical shells*

5.1 Introduction

Particularly within the past decade the so-called linear fracture mechanics has established itself as a highly satisfactory working tool in studying the phenomena of brittle fracture and fatigue crack propagation in structural solids. The technique appears to be most effective when 'plane strain' conditions prevail along the existing crack front. It has also been shown that the stress intensity factor, which is the basic element in the linear fracture mechanics, is the most appropriate correlation parameter in fatigue crack propagation studies of relatively thin-walled plates under membrane loading where the crack is a through crack, and 'generalized plane stress' conditions are assumed to exist. The plane assumption here, of course, is an approximation in which the three-dimensional effects resulting from the intersection of the crack plane with the stress-free surfaces of the plate are neglected.*
From the view point of practical applications this boundary layer or thickness effect does not appear to be very significant. Therefore, one may be justified in using standard plate or shell theories for studying the fracture problems in thin-walled structures provided the plane of the crack is perpendicular to the surface of the sheet.

   With the assumption of linearity, it is known that the relevant information in crack problems may be obtained from a local perturbation problem in which the only external loads are the crack surface tractions. In 'thin-walled' structures this would mean that after solving the plate or shell problem under given external loads by ignoring the crack, the stress intensity factors may

---

* See Chapter 2 of this volume for the effect of plate thickness and related approximate techniques.

be found by using the equal and opposite of the membrane and bending resultants at the location of the crack as the crack surface tractions. Since at the present time linear problems are the only tractable crack problems, the geometry of the particular thin-walled structure must then be such that locally, small deformation plate or linearized shallow shell theories are applicable. At first sight it may appear that in such cases it is sufficient to approximate the structure locally by a flat plate. However, recent studies have shown that local shell curvatures may have a rather considerable effect on the stress intensity factors. Hence, in thin-walled curved structures the crack problem must be considered in conjunction with a shell rather than a plate theory.

Because of the peculiarity of the crack problems in shells, there are analytical limitations regarding the type of problems which can be solved by the existing techniques. Aside from the considerations regarding the linearity of the problem, the two major limitations arise from the geometry and material behavior of the shell. The geometrical factors include the relative size of the crack with respect to the radii of curvature and dimensions of the shell, spatial variation of the curvatures, and the shape and orientation of the crack. The material factors are primarily the anisotropy and nonhomogeneity. In addition to linear elasticity, in the existing solutions [e.g., 1-13] it is assumed that the shell is 'infinitely' large, the curvatures are constant (i.e., the shell is a circular cylinder or a sphere), the crack is along a principal plane of curvature, and the material is isotropic and homogeneous. If these assumptions are disregarded, mathematically the problem does not seem to be tractable. Further remarks will be made in this chapter regarding this point. If the material is isotropic and homogeneous, in applications one could obtain approximate solutions with an acceptable accuracy by approximating local shell-crack geometry with an ideal shell which has a solution, namely a spherical shell with a meridional crack, a cylindrical shell with an axial crack, or a cylindrical shell with a circumferential crack.

From a practical view point the assumption of homogeneity of the material in shells does not seem to be a critical restriction. Even in thin-walled structures made of composites one may easily assume that the gross behavior of the material is homogeneous. However, in practice a mild or strong anisotropy in shells appears to be a rule rather than an exception. Most metallic shells are manufactured through rolling or extrusion process, and hence, are generally mildly anisotropic. Shells which are made of composites such as fiberglass, boron-epoxy, graphite-epoxy, etc., are of course, strongly anisotropic. Since the treatment of anisotropic or, even orthotropic shells is not tractable, it is therefore desirable to have a technique for approximately evaluating the effect of material anisotropy on the critical fracture parameter, namely, the stress intensity factor.

This chapter describes a method of solution for the specially orthotropic shells containing a crack. The method is described by considering symmetric and skew-symmetric problems in cylindrical shells with an axial crack (for details see [14–16]). Its extension to the other two ideal geometries seems to be straightforward. Most of the numerical results given in the chapter, which includes the effect of Poisson's ratio and interaction of two cracks, is, however, on the isotropic shells. The analysis and the results given in this chapter are based on an 8th order linearized shallow shell theory in which Kirchhoff assumption is made with regard to the transverse shear and the twisting moment on the crack surface. Since there are five traction components on the boundary, to satisfy all the boundary conditions individually a 10th order theory should be used.* Also, since any bending theory is necessarily approximate, one would expect that the shell thickness will have a slight effect on the membrane component and a considerable effect on the bending component of the stress intensity factor.**

5.2 Formulation of the specially orthotropic cylindrical shell problem

The linearized bending theory of anisotropic shallow shells dates back to a paper by Ambartsumyan [18] and the detailed treatment of the subject may be found in [19–21]. Referring to Figure 5.1, let an infinitely long orthotropic

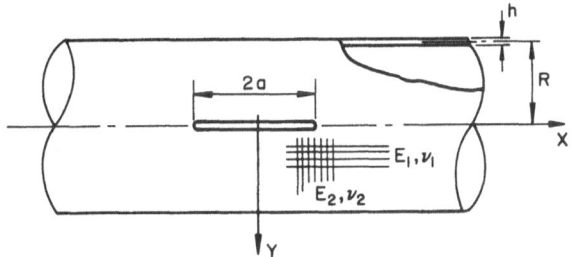

Figure 5.1. Geometry of a cylindrical shell with an axial crack

* See Chapter 6 of this volume.
** See Chapter 2 of this volume. See also [17] for cracked plate under bending.

circular cylindrical shell of elastic constants E_1, v_1, E_2, v_2, G_{12}, thickness h, and mean radius R contain an axial through crack of length $2a$. Assume that through a proper superposition the problem has been reduced to a local perturbation problem in which the crack surface tractions are the only external loads. Defining the nondimensional orthogonal coordinates in the tangent plane by

$$x_1 = X/a \quad , \quad y_1 = Y/a \tag{5.1}$$

The differential equations for the orthotropic cylindrical shell may be written as

$$D_1\nabla_1^4 w(x_1, x_2) - \frac{a^2}{R} \frac{\partial^2}{\partial x_1^2} F(x_1, x_2) = 0, \tag{5.2a}$$

$$\nabla_2^4 F(x_1, x_2) + \frac{hE_2a^2}{R} \frac{\partial^2}{\partial x_1^2} w(x_1, x_2) = 0, \tag{5.2b}$$

where x_1 and x_2 are the principal directions of orthotropy taken respectively along axial and circumferential directions, F is a stress function, w is the displacement component normal to the shell surface, and the operators ∇_1^4 and ∇_2^4 are defined by

$$\nabla_1^4 = \frac{\partial^4}{\partial x_1^4} + 2\left[v_2 + 2(1 - v_1 v_2) \frac{G_{12}}{E_1}\right] \frac{\partial^4}{\partial x_1^2 \partial x_2^2} + \frac{E_2}{E_1} \frac{\partial^4}{\partial x_2^4}, \tag{5.3a}$$

$$\nabla_2^4 = \frac{\partial^4}{\partial x_1^4} + 2\left(\frac{E_2}{2G_{12}} - v_2\right) \frac{\partial^4}{\partial x_1^2 \partial x_2^2} + \frac{E_2}{E_1} \frac{\partial^4}{\partial x_2^4}. \tag{5.3b}$$

The notation for the orthotropic elastic constants are defined by the following stress strain relations:

$$\varepsilon_{11} = \frac{1}{E_1} (\sigma_{11} - v_1\sigma_{22}), \tag{5.4a}$$

$$\varepsilon_{22} = \frac{1}{E_2} (\sigma_{22} - v_2\sigma_{11}), \tag{5.4b}$$

$$\varepsilon_{12} = \frac{1}{2G_{12}} \sigma_{12}, \tag{5.4c}$$

$$\frac{v_1}{E_1} = \frac{v_2}{E_2}. \tag{5.4d}$$

The stress and moment resultants are related to F and w through the following expressions:

$$N_{11} = \frac{1}{a^2} \frac{\partial^2 F}{\partial x_2^2}, \tag{5.5a}$$

$$N_{22} = \frac{1}{a^2} \frac{\partial^2 F}{\partial x_1^2}, \tag{5.5b}$$

$$N_{12} = -\frac{1}{a^2} \frac{\partial^2 F}{\partial x_1 \partial x_2}, \tag{5.5c.}$$

$$M_{11} = -\frac{D_1}{a^2} \left(\frac{\partial^2 w}{\partial x_1^2} + v_2 \frac{\partial^2 w}{\partial x_2^2} \right), \tag{5.5d}$$

$$M_{22} = -\frac{D_2}{a^2} \left(\frac{\partial^2 w}{\partial x_2^2} + v_1 \frac{\partial^2 w}{\partial x_1^2} \right), \tag{5.5e}$$

$$V_1 = Q_1 + \frac{\partial M_{12}}{\partial x_2} = -\frac{D_1}{a^3} \left[\frac{\partial^3 w}{\partial x_1^3} + \left(v_2 + \frac{h^3 G_{12}}{3 D_1} \right) \frac{\partial^3 w}{\partial x_1 \partial x_2^2} \right], \tag{5.5f}$$

$$V_2 = Q_2 + \frac{\partial M_{12}}{\partial x_1} = -\frac{D_2}{a^3} \left[\frac{\partial^3 w}{\partial x_2^3} + \left(v_1 + \frac{h^3 G_{12}}{3 D_2} \right) \frac{\partial^3 w}{\partial x_1^2 \partial x_2} \right], \tag{5.5g}$$

where

$$D_k = E_k h^3 / 12(1 - v_1 v_2), \quad (k = 1, 2). \tag{5.6}$$

The membrane and bending stress components are obtained from relations of the form

$$\sigma_{11}^m = N_{11}/h, \ldots, \sigma_{11}^b = 12 M_{11} Z / h^3, \ldots \tag{5.7}$$

In solving the problem, for example by expressing the function F and w in terms of appropriate Fourier integrals, (2) may be reduced to a system of two fourth order linear ordinary differential equations. The characteristic function of this system will be an 8th degree polynomial the coefficients of which will be functions of the transform variable. For the problem to be analytically tractable, it is essential that the roots of the characteristic equation be obtainable in closed form. For anisotropic shells in general and for orthotropic shells in particular this does not seem to be possible. In order to express the roots in closed form the operators ∇_1^4 and ∇_2^4 must be properly factorized. From equations (5.3) it may be seen that these operators can indeed be factorized and expressed in the following form

$$\nabla_1^4 = \left(\frac{\partial^2}{\partial x_1^2} + (E_2/E_1)^{\frac{1}{2}} \frac{\partial^2}{\partial x_2^2} \right) = \nabla_2^4 \tag{5.8}$$

if the elastic constants satisfy the following conditions:

$$[v_2 + 2(1 - v_1 v_2)G_{12}/E_1](E_1/E_2)^{\frac{1}{2}} = 1, \tag{5.9a}$$

$$\left(\frac{E_2}{2G_{12}} - v_2 \right)(E_1/E_2)^{\frac{1}{2}} = 1. \tag{5.9b}$$

Now, by direct substitution it may easily be shown that the conditions (9) are satisfied provided the elastic constants are related by

$$G_{12} = \frac{(E_1 E_2)^{\frac{1}{2}}}{2[1 + (v_1 v_2)^{\frac{1}{2}}]}. \tag{5.10}$$

Considering also the relation in equation (5.4d), this means that the sheet material has only three independent constants. Such a material is said to be 'specially orthotropic'. The analysis and certain results given in this chapter will then be valid only for those sheet materials in which the measured value of G_{12} and that calculated from equation (5.10) in terms of measured E_i and v_i, $(i = 1, 2)$ are in reasonably good agreement.

If the variables are changed once more as follows

$$x_1 = x, \quad (E_1/E_2)^{\frac{1}{4}}x_2 = y, \tag{5.11}$$

the operators ∇_1^4 and ∇_2^4 become

$$\nabla_1^4 = \nabla_2^4 = \left(\frac{\partial^2}{\partial x^2} + \frac{\partial^2}{\partial y^2} \right)^2 = \nabla^4. \tag{5.12}$$

With equation (5.12), equations (2) become identical to the differential equations for isotropic shells in which E and $D = Eh^3/[12(1 - v^2)]$ are replaced by E_2 and D_1, respectively, i.e.,

$$D_1\nabla^4 w(x, y) - \frac{a^2}{R} \frac{\partial^2}{\partial x^2} F(x, y) = 0, \tag{5.13a}$$

$$\nabla^4 F(x, y) + \frac{h^2 E_2 a^2}{R} \frac{\partial^2}{\partial x^2} w(x, y) = 0. \tag{5.13b}$$

Let the stress and moment resultants on $y = 0$, $-a < X < a$ obtained from the solution of the shell under given external loads by ignoring the crack be

$$N_Y(X, 0) = N_Y^0(X) = n_0(x) \tag{5.14a}$$

$$N_{XY}(X, 0) = N_{XY}^0(X) = t_0(x) \tag{5.14b}$$

$$M_Y(X, 0) = M_Y^0(X) = m_0(x) \tag{5.14c}$$

$$V_Y(X, 0) = V_Y^0(X) = v_0(x) \tag{5.14d}$$

Considering the perturbation problem and referring to (5.1), (5.5), and (5.11), the system of differential equations (5.13) must then be solved under the following boundary conditions specified on the crack surface:

$$\lim_{y \to \pm 0} \frac{D_2}{a^2}\left(c^2 \frac{\partial^2 w}{\partial y^2} + v_1 \frac{\partial^2 w}{\partial x^2}\right) = m_0(x), \tag{5.15a}$$

$$\lim_{y \to \pm 0} \frac{1}{a^2}\frac{\partial^2 F}{\partial x^2} = -n_0(x), \tag{5.15b}$$

$$\lim_{y \to \pm 0} \frac{c}{a^2}\frac{\partial^2 F}{\partial x \partial y} = t_0(x), \tag{5.15c}$$

$$\lim_{y \to \pm 0} \frac{D_2}{a^3}\left[c^3 \frac{\partial^3 w}{\partial y^3} + c\left(v_1 + \frac{h^3 G_{12}}{3 D_2}\right)\frac{\partial^3 w}{\partial x^2 \partial y}\right] = v_0(x), \, (-1 < x < 1) \tag{5.15d}$$

where

$$c = (E_1/E_2)^{\frac{1}{4}} \tag{5.16}$$

It is now clear that by properly decomposing the input functions given by equations (5.15) into even and odd components, the solution of the general problem may be expressed as the sum of a 'symmetric' and a 'skew-symmetric' solution. In the following two sections the solutions of these problems will be presented in some detail.

5.3   The skew-symmetric problem

From a practical view point the important skew-symmetric problem is that having the following crack surface tractions:

$$m_0(x) = 0, \, n_0(x) = 0, \, t_0(x) = t_0(-x), \, v_0(x) = -v_0(-x), \, (-1 < x < 1) \tag{5.17}$$

Outside the crack, the antisymmetry of the problem and the conditions of continuity require that

$$M_y(X, 0) = 0, \quad N_y(X, 0) = 0 \tag{5.18a}$$

$$\lim_{y \to +0} \frac{\partial^n}{\partial y^n} w(x, y) = \lim_{y \to -0} \frac{\partial^n}{\partial y^n} w(x, y), \quad (n = 0, 1, 2, 3), \tag{5.18b}$$

$$\lim_{y \to +0} \frac{\partial^n}{\partial y^n} F(x, y) = \lim_{y \to -0} \frac{\partial^n}{\partial y^n} F(x, y), \quad (n = 0, 1, 2, 3), \quad |x| > 1 \tag{5.18c}$$

Since the external loads in equations (5.17) are self-equilibrating local tractions, the functions F and w satisfy the regularity conditions at $x = \mp \infty$ and hence may be expressed in terms of Fourier integrals. Thus using the symmetry considerations, after some routine manipulations the solution of equations (5.13) may be expressed as

$$w(x,y) = \text{sgn}(y) \int_0^\infty \sum_1^4 Q_j(\alpha) e^{m_j |y|} \sin \alpha x \, d\alpha \tag{5.19a}$$

$$F(x,y) = \text{sgn}(y) \int_0^\infty \sum_1^4 K_j Q_j(\alpha) e^{m_j |y|} \sin \alpha x \, d\alpha, \tag{5.19b}$$

where the functions $Q_j(\alpha)$, $(j = 1, \ldots, 4)$ are unknown and

$$K_1 = K_2 = -i(E_2 h D_1)^{\frac{1}{2}}, \quad K_3 = K_4 = i(E_2 h D_1)^{\frac{1}{2}}$$

$$m_1 = -(\alpha^2 + i_1 \lambda \alpha)^{\frac{1}{2}}, \qquad m_2 = -(\alpha^2 - i_1 \lambda \alpha)^{\frac{1}{2}}$$

$$m_3 = -(\alpha^2 + i_2 \lambda \alpha)^{\frac{1}{2}}, \qquad m_4 = -(\alpha^2 - i_2 \lambda \alpha)^{\frac{1}{2}}$$

$$i_1 = e^{\pi i/4}, \ i_2 = e^{-\pi i/4}, \ \lambda^4 = 12(E_2/E_1)(1 - v_1 v_2) \frac{a^4}{R^2 h^2} \tag{5.20}$$

Substituting from equations (5.19) into equations (5.15a), (5.15b), (5.17a), (5.17b), and (5.18a) it is found that

$$Q_3 = \left[\frac{\alpha(v_1 - c^2)}{i_2 \lambda c^2} + \frac{1}{2} \right] (Q_1 + Q_2) - \frac{i}{2} (Q_1 - Q_2) \tag{5.21a}$$

$$Q_4 = -\left[\frac{\alpha(v_1 - c^2)}{i_2 \lambda c^2} - \frac{1}{2} \right] (Q_1 + Q_2) + \frac{i}{2} (Q_1 - Q_2) \tag{5.21b}$$

The two remaining equations to determine Q_j $(j = 1, \ldots, 4)$ are obtained from the mixed boundary conditions in equations (5.15c), (5.15d), (5.18c) and (5.18d). Since w and F are odd functions of y, equations (5.18c) and (5.18d) are automatically satisfied for $n = 1$ and $n = 3$. Using equations (5.21) and (5.19) it may be shown that the conditions in equations (5.18c) and (5.18d) for $n = 0$ and $n = 2$ will be satisfied if

$$\int_0^\infty (Q_1 + Q_2) \sin \alpha x \, d\alpha = 0 \tag{5.22a}$$

$$\int_0^\infty (Q_1 + Q_2)\alpha^2 \sin \alpha x \, d\alpha = 0 \tag{5.22b}$$

$$\int_0^\infty \frac{v_1 - c^2}{c^2}(Q_1 + Q_2)\alpha^2 \sin \alpha x \, d\alpha + \int_0^\infty i_1\lambda(Q_1 - Q_2)\alpha \sin \alpha x \, d\alpha = 0,$$
$$(|x| > 1) \tag{5.22c}$$

Here equations (5.22a) and (5.22b) refer to the conditions that w and $\partial^2 w/\partial y^2$ vanish on $y = 0$, $|x| > 1$. Since analytically equation (5.22b) follows from equation (5.22a), equations (5.22) is actually equivalent to only two independent conditions. For dimensional consistency these conditions will be selected as follows:

$$\int_0^\infty (Q_1 + Q_2)\alpha^2 \sin \alpha x \, d\alpha = 0 \tag{5.23a}$$

$$\int_0^\infty i_1\lambda(Q_1 - Q_2)\alpha \sin \alpha x \, d\alpha = 0, \quad |x| > 1 \tag{5.23b}$$

With the selection of equation (5.23) as the conditions for $|x| > 1$ it should be noted that a single-valuedness condition i.e., $w = 0$ for $y = 0$, $|x| > 1$ still remains to be satisfied. This condition will be necessary to obtain a unique solution for the resulting integral equations.

Substituting from equations (5.19) into equation (5.15c) and (5.15d), and again, for dimensional consistency, integrating equation (5.15d), it is found that

$$\lim_{y \to +0}\left[-\frac{c}{a^2}\int_0^\infty \sum_1^4 K_j m_j Q_j e^{m_j y}\alpha \cos \alpha x \, d\alpha \right] = -t_0(x), \tag{5.24a}$$

$$\lim_{y \to +0}\int_0^x \left\{ -\frac{D_2}{a^3}\int_0^\infty \sum_1^4 \left[c^3 m_j^3 - \alpha^2 cm_j\left(v_1 + \frac{h^3 G_{12}}{3D_2} \right) \right] Q_j e^{m_j y} \sin \alpha x \, d\alpha \right\}dx$$
$$= -\int_0^x v_0(x) \, dx, \quad (|x| < 1) \tag{5.24b}$$

With equations (5.21), (5.23) and (5.24) give a system of dual integral equations to determine Q_1 and Q_2. Define now the following auxiliary functions:

$$u_1(x) = \int_0^\infty i_1\lambda\alpha(Q_1 - Q_2) \sin \alpha x \, d\alpha \tag{5.25a}$$

$$u_2(x) = \int_0^\infty \alpha^2(Q_1 + Q_2) \sin \alpha x \, d\alpha, \quad 0 \le x < \infty \tag{5.25b}$$

Note that u_1 and u_2 are related to the second derivatives of w and F and hence

are expected to have the same type of singularity as N_{ij} and M_{ij} at the crack
tips ($x = \mp 1, y = 0$). From equations (5.25) and (5.23) it follows that

$$Q_1 - Q_2 = \frac{2}{\pi i_1 \lambda \alpha} \int_0^1 u_1(t) \sin \alpha t \, dt \qquad (5.26a)$$

$$Q_1 + Q_2 = \frac{2}{\pi \alpha^2} \int_0^1 u_2(t) \sin \alpha t \, dt \qquad (5.26b)$$

Substituting now from equations (5.26) and (5.21) into equations (5.24) and
observing that u_1 and u_2 are odd functions, the following integral equations
are obtained to determine u_1 and u_2:

$$\lim_{y \to +0} \frac{1}{\pi} \int_{-1}^1 \sum_1^2 h_{ij}(x, t, y) u_j(t) dt = f_i(x), \ (i = 1, 2, |x| < 1), \qquad (5.27)$$

where

$$f_1(x) \frac{ia^2 t_0(x)}{c(E_2 h D_1)^{\frac{1}{4}}}, \ f_2(x) = \frac{a^3}{D_2} \int_0^x v_0(x) \, dx \qquad (5.28)$$

$$h_{1j}(x, t, y) = \int_0^\infty F_{1j}(\alpha, y) \sin \alpha(t - x) d\alpha, \quad (j = 1, 2) \qquad (5.29a)$$

$$F_{11}(\alpha, y) = \frac{1}{2}\left[-\frac{\alpha^2}{i_1 \lambda} n_1 - \alpha n_2 - \frac{\alpha^2}{i_2 \lambda} n_3 - \alpha n_4 \right] \qquad (5.29b)$$

$$F_{12}(\alpha, y) = \frac{1}{2}\left[-i_1 \lambda n_1 - \alpha n_2 + \left(\frac{2\alpha^2(v_1 - c^2)}{i_2 \lambda c^2} + i_2 \lambda \right) n_3 \right.$$

$$\left. + \left(\frac{2\alpha(v_1 - c^2)}{c^2} + \alpha \right) n_4 \right] \qquad (5.29c)$$

$$h_{2j}(x, t, \alpha) = \int_0^\infty F_{2j}(\alpha, y) \left[\sin \alpha(t - x) - \sin \alpha t \right] d\alpha, \quad (j = 1, 2) \qquad (5.30a)$$

$$F_{21}(\alpha, y) = \frac{1}{2}\left[\left(c^2 - v_1 - \frac{h^3 G_{12}}{3D_2} \right)\left(\frac{\alpha^2}{i_1 \lambda} n_1 - \frac{\alpha^2}{i_2 \lambda} n_3 \right) \right.$$

$$\left. + \left(2c^2 - v_1 - \frac{h^3 G_{12}}{3D_2} \right) \alpha(n_2 - n_4) + i_1 \lambda c^2 n_1 - i_2 \lambda c^2 n_3 \right] \qquad (5.30b)$$

$$F_{22}(\alpha, y) = \frac{1}{2}\left[\left(c^2 - v_1 - \frac{h^3 G_{12}}{3D_2} \right)\left(\alpha n_2 + \frac{2\alpha^2(v_1 - c^2)}{i_2 \lambda c^2} n_3 + \alpha n_4 \right) \right.$$

$$\left. + \left(2c^2 - v_1 - \frac{h^3 G_{12}}{3D_2} \right)\left(i_1 \lambda n_1 + \frac{2\alpha(v_1 - c^2)}{c^2} n_4 + i_2 \lambda n_3 \right) \right.$$

$$+ \frac{i\lambda^2 c^2}{\alpha} n_2 + 2i_2\lambda(v_1 - c^2)n_3 - \frac{i\lambda^2 c^2}{\alpha} n_4 \bigg]. \qquad (5.30c)$$

$$n_1 = \frac{e^{m_1 y}}{m_1} - \frac{e^{m_2 y}}{m_2}, \quad n_2 = \frac{e^{m_1 y}}{m_1} + \frac{e^{m_2 y}}{m_2},$$

$$n_3 = \frac{e^{m_3 y}}{m_3} - \frac{e^{m_4 y}}{m_4}, \quad n_4 = \frac{e^{m_3 y}}{m_3} + \frac{e^{m_4 y}}{m_4}, \qquad (5.31)$$

By examining the asymptotic behavior of the integrands for large and small values of α in equations (5.29) and (5.30) for the kernels h_{ij}, it may be seen that some of the integrals are uniformly convergent. In these integrals the limit can be put under the integral sign and the resulting kernels are simple Fredholm kernels. In the expression $\sin \alpha p$ ($p = t - x$ or $p = t$) as $p \to 0$ it may also be seen that the remaining integrals become divergent, meaning that the kernels contain singularities. These singular parts of the kernels may be separated in a standard way by adding and subtracting the asymptotic value of the integrand for large values of α. For example, noting that for large α

$$\alpha^2 n_1(\alpha, y) = i_1 \lambda e^{-\alpha y} + e^{-\alpha y} 0(\alpha^{-2}), \quad y > 0, \qquad (5.32)$$

it is seen that

$$\int_0^\infty \alpha^2 n_1(\alpha, y) \sin \alpha p \, d\alpha$$

$$= i_1 \lambda \int_0^\infty e^{-\alpha y} \sin \alpha p \, d\alpha + \int_0^\infty [\alpha^2 n_1(\alpha, y) - i_1 \lambda e^{-\alpha y}] \sin \alpha p \, d\alpha$$

$$= i_1 \lambda \frac{p}{p^2 + y^2} + \int_0^\infty \left[\left(\frac{e^{m_1 y}}{m_1} - \frac{e^{m_2 y}}{m_2} \right) \alpha^2 - i_1 \lambda e^{-\alpha y} \right] \sin \alpha p \, d\alpha \qquad (5.33)$$

where the last integral is uniformly convergent for all p and $y \to 0$, and hence, the limit $y = 0$ can be put under the integral sign, whereas the integrated term gives a Cauchy type singularity $1/p$. Similarly

$$\int_0^\infty \alpha n_2(\alpha, y) \sin \alpha p \, d\alpha$$

$$= \frac{2p}{p^2 + y^2} + \int_0^\infty \left[\left(\frac{e^{m_1 y}}{m_1} + \frac{e^{m_2 y}}{m_2} \right) \alpha - 2e^{-\alpha y} \right] \sin \alpha p \, d\alpha. \qquad (5.34)$$

Thus, after separating the singular parts of the kernels and going to limit, equation (5.27) becomes

$$\int_{-1}^1 \sum_1^2 a_{ij} u_j(t) \frac{dt}{t - x} + \int_{-1}^1 \sum_1^2 k_{ij}(x, t) u_j(t) dt = \pi f_i(x) \quad (|x| < 1), \qquad (5.35)$$

where

$$a_{11} = 1, \quad a_{12} = 1 - \frac{v_1}{c_2}, \quad a_{21} = 0,$$

$$a_{22} = -3v_1 + c^2 + \left(1 + \frac{v_1}{c^2}\right)\left(v_1 + \frac{h^3 G_{12}}{3D_2}\right), \tag{5.36}$$

and the Fredholm kernels $k_{ij}(x, t)$ are given by

$$k_{1j}(x, t) = \int_0^\infty [F_{1j}(\alpha, 0) - a_{1j}] \sin \alpha(t - x) d\alpha, \quad (j = 1, 2), \tag{5.37}$$

and

$$k_{21}(x, t) = \int_0^\infty F_{21}(\alpha, 0) [\sin \alpha(t - x) - \sin \alpha t] d\alpha, \tag{5.38a}$$

$$k_{22}(x, t) = -\frac{a_{22}}{t} + \int_0^\infty [F_{22}(\alpha, 0) - a_{22}] \sin \alpha(t - x) - \sin \alpha t] d\alpha. \tag{5.38b}$$

Since u_1 and u_2 are related to the second derivatives of F and w, the elements of the fundamental matrix of the singular integral equations (5.35) will be $(1 - x^2)^{-\frac{1}{2}}$ and the solution will be of the form

$$u_j(x) = G_j(x) (1 - x^2)^{-\frac{1}{2}}, \quad (j = 1, 2), \quad (|x| < 1) \tag{5.39}$$

where the functions G_j are bounded in $(-1 \le x \le 1)$. Thus the index of the system in equation (5.35) is $\kappa = 1$, and hence theoretically the solution is not unique and will contain two arbitrary constants [22]. These constants are determined by using the single-valuedness condition mentioned earlier, namely that $w(x, 0) = 0$ for $|x| > 1$. Referring to equations (5.19), (5.21), (5.22) and (5.25) it may be shown that $2u_2(x) = -(\partial^2/\partial x^2)w(x, 0)$. Since $u_2(x) = 0$ for $|x| > 1$, it then follows that for $w(x, 0) = 0$ for $|x| > 1 \, u_2(x)$ must satisfy the following conditions:

$$\int_{-1}^1 u_2(x) \, dx = 0, \quad \int_{-1}^1 dx \int_{-1}^x u_2(t) \, dt = 0. \tag{5.40}$$

The unknown functions G_1 and G_2 defined by equation (5.39) may be obtained from equations (5.35) and (5.40) in a straightforward way by using the technique outlined in [23].

To examine the asymptotic behavior of the stresses around the crack tips let us assume that the bounded functions G_1 and G_2 are expressed in terms of the following infinite series:

$$G_1(x) = \sum_1^\infty A_n T_{2n-1}(x), \quad G_2(x) = \sum_1^\infty B_n T_{2n-1}(x), \tag{5.41}$$

where $T_k(x)$ is the Chebyshev polynomial of the first kind and the symmetry property of $u_j(x) = -u_j(-x)$ $(j = 1, 2)$ has been used. From equations (5.26), (5.39) and (5.41) by using the relation [24]

$$\int_0^1 T_{2n+1}(x)(1 - x^2)^{-\frac{1}{2}} \sin \alpha x \, d\alpha = (-1)^n \frac{\pi}{2} J_{2n+1}(\alpha), \quad (n = 0, 1, 2 \ldots)$$
$$\tag{5.42}$$

it may be shown that

$$i_1 \lambda \alpha (Q_1 - Q_2) = \sum_1^\infty (-1)^{n-1} A_n J_{2n-1}(\alpha), \tag{5.43a}$$

$$\alpha^2 (Q_1 + Q_2) = \sum_1^\infty (-1)^{n-1} B_n J_{2n-1}(\alpha). \tag{5.43b}$$

The expressions for the stresses may then be obtained by substituting from equations (5.43), (5.21), (5.19) into (5.5) and (5.7). For example,

$$\sigma_{xy}^m = -\frac{c}{ha^2} \frac{\partial^2 F}{\partial x \partial y} = \frac{c}{ha^2} \int_0^\infty \sum_1^4 K_j Q_j(\alpha) m_j \alpha e^{m_j y} \cos \alpha x \, d\alpha, (y \geq 0).$$
$$\tag{5.44}$$

At $(y = 0, x = \pm 1)$, the integrals in (5.44) are divergent, meaning that the stresses will have a singularity at the crack tips. Noting that the integrand in equation (5.44) is integrable around $\alpha = 0$ and is bounded and continuous elsewhere in the domain, the divergent behavior of the integral must be due to the asymptotic behavior of the integrand for large values of α. Thus, substituting from equation (5.43) and (5.21), equation (5.44) may be expressed as

$$\sigma_{xy}^m = \frac{c}{ha^2} i(E_2 hD_1)^{\frac{1}{4}} \sum_1^\infty (-1)^{n-1} \left(A_n - \frac{v_1 - c^2}{c^2} B_n \right)$$

$$\times \int_0^\infty J_{2n-1}(\alpha) [-1 + \alpha y + O(\alpha^{-1})] e^{-\alpha y} \cos \alpha x \, d\alpha \quad (y \geq 0). \tag{5.45}$$

Noting that for large values of α [24]

$$J_{2n-1}(\alpha) \cong (-1)^{n-1} J_1(\alpha) \cong \left(\frac{2}{\pi\alpha} \right)^{\frac{1}{4}} (-1)^{n-1} \times$$

$$\times \left[\cos \left(\alpha - \frac{3\pi}{4} \right) + O(\alpha^{-1}) \right] (n = 1, 2, \ldots), \tag{5.46}$$

and using the results in [24] to evaluate the integrals, we obtain the leading term in (5.45) as follows:

$$\sigma_{xy}^m(r, \theta) = \frac{ci}{ha^2} (E_2 h D_1)^{\frac{1}{2}} \sum_1^\infty \left(- A_n + \frac{v_1 - c^2}{c^2} B_n \right)$$

$$\times \frac{1}{4(2r)^{\frac{1}{2}}} \left(3 \cos \frac{\theta}{2} + \cos \frac{5\theta}{2} \right) + O(r^{\frac{1}{2}}),$$ (5.47)

where (r, θ) are the polar coordinates measured from the crack tip,

$$(x = 1, y = 0), r^2 = (x - 1)^2 + y^2, \tan \theta = y/(x - 1).$$

For example, if

$$t_0(x) = N_0, \quad v_0(x) = 0,$$

defining the following normalized functions (see equation (5.28)):

$$G_j^*(x) = G_j(x)/u_0, \quad (j = 1, 2),$$ (5.48a)

$$u_0 = \frac{a^2 N_0}{c(E_2 h D_1)^{\frac{1}{2}}} = \frac{\lambda^2 R N_0 c}{h(E_1 E_2)^{\frac{1}{2}}},$$ (5.48b)

$$G_1^*(x) = \sum_1^\infty a_n T_{2n-1}(x),$$ (5.48c)

$$G_2^*(x) = \sum_1^\infty b_n T_{2n-1}(x),$$ (5.48d)

equation (5.46) may be expressed as

$$\sigma_{xy}^m = \left(\frac{N_0 \sqrt{a}}{h} \right) \left[i \sum_1^\infty \left(- a_n + \frac{v_1 - c^2}{c^2} b_n \right) \right] \frac{1}{4(2ra)^{\frac{1}{2}}} \left(3 \cos \frac{\theta}{2} + \cos \frac{5\theta}{2} \right)$$

$$+ O(r^{\frac{1}{2}}).$$ (5.49)

Observing that the stress intensity factor in a flat plate under uniform shear stress N_0/h and that in a shell are defined by

$$k_p = N_0 a^{\frac{1}{2}}/h, \quad k_s^m = \lim_{x \to 1} [2(x - 1)a]^{\frac{1}{2}} \sigma_{xy}^m(x, 0),$$ (5.50)

from equation (5.49) the membrane component of the stress intensity factor ratio for the shell is found to be

$$C_m = \frac{k_s^m}{k_p} = i \sum_1^\infty \left(- a_n + \frac{v_1 - c^2}{c^2} b_n \right).$$ (5.51)

Further, noting that $T_n(1) = 1$, $(n = 0, 1, 2, \ldots)$ from equation (5.48) and

(5.50) it follows that

$$C_m = \frac{k_s^m}{k_p} = \left[-G_1^*(1) + \frac{v_1 - c^2}{c^2} G_2^*(1) \right] i. \tag{5.52}$$

The remaining membrane stresses may be obtained in a similar way. Thus, for small values of r the membrane stresses in the shell may be expressed as

$$\sigma_{xx}^m(r, \theta) = \frac{C_m k_p}{4c(2ra)^{\frac{1}{2}}} \left(7 \sin \frac{\theta}{2} + \sin \frac{5\theta}{2} \right) + O(r^{\frac{1}{2}}), \tag{5.53a}$$

$$\sigma_{yy}^m(r, \theta) = \frac{C_m k_p}{4(2ra)^{\frac{1}{2}}} \left(- \sin \frac{\theta}{2} + \sin \frac{5\theta}{2} \right) + O(r^{\frac{1}{2}}), \tag{5.53b}$$

$$\sigma_{xy}^m(r, \theta) = \frac{C_m k_p}{4(2ra)^{\frac{1}{2}}} \left(3 \cos \frac{\theta}{2} + \cos \frac{5\theta}{2} \right) + O(r^{\frac{1}{2}}). \tag{5.53c}$$

Defining now the bending component of the stress intensity factor by

$$k_s^b = \lim_{x \to 1} \left[2(x - 1)a \right]^{\frac{1}{2}} \sigma_{xy}^b(x, 0, h) = C_b k_p, \tag{5.54}$$

in a similar way the asymptotic expressions for the bending stresses around the crack tip may be obtained as follows:

$$\sigma_{xy}^b(r, \theta, Z) = \frac{C_b k_p}{(2ra)^{\frac{1}{2}}} \frac{2Z}{h} \frac{1}{4[2 + (v_1 - c^2)/c^2]} \left[8 + \frac{3(v_1 - c^2)}{c^2} \right] \times$$

$$\times \cos \frac{\theta}{2} + \frac{v_1 - c^2}{c^2} \cos \frac{5\theta}{2} \right] + O(r^{\frac{1}{2}}), \tag{5.55a}$$

$$\sigma_{xx}^b(r, \theta, Z) = \frac{C_b k_p}{(2ra)^{\frac{1}{2}}} \frac{2Z}{h} \frac{c}{4[2 + (v_1 - c^2)/c^2][1 - (v_1 v_2)^{\frac{1}{2}}]}$$

$$\times \left\{ \left[8(1 - v_2 c^2) - \frac{v_1 - c^2}{c^2} (1 + 7v_2 c^2) \right] \sin \frac{\theta}{2} \right.$$

$$+ \frac{v_1 - c^2}{c^2} (1 - v_2 c^2) \sin \frac{5\theta}{2} \right\} + O(r^{\frac{1}{2}}), \tag{5.55b}$$

$$\sigma_{yy}^b(r, \theta, Z) = \frac{C_n k_p}{(2ra)^{\frac{1}{2}}} \frac{2Z}{h} \frac{c^2 - v_1}{4[2 + (v_1 - c^2)/c^2][1 - (v_1 v_2)^{\frac{1}{2}}] c^3}$$

$$\times \left[\left(\frac{v_1 - c^2}{c^2} + 8c^2 - 8 \right) \sin \frac{\theta}{2} - \frac{v_1 - c^2}{c^2} \sin \frac{5\theta}{2} \right] + O(r^{\frac{1}{2}}), \tag{5.55c}$$

where the bending component of the stress intensity ratio is found to be*:

$$C_b = -\frac{\{3[1-(v_1v_2)^{\frac{1}{2}}]\}^{\frac{1}{2}}}{1+(v_1v_2)^{\frac{1}{2}}}\left(2+\frac{v_1-c^2}{c^2}\right)G_2^*(1). \tag{5.56}$$

Thus, once the singular integral equations (5.35) are solved after normalization described by equations (5.48), the stress intensity factors can be obtained without further analysis. The analysis in this section remains valid for the isotropic shell with $E_1 = E_2 = E$, $v_1 = v_2 = v$, $G_{12} = G$, and $c = (E_1/E_2)^{\frac{1}{4}}$ $= 1$.

5.4 The symmetric problem

Consider now the symmetric problem in which the only external loads are the following crack surface tractions (see equations (5.14)):

$$m_0(x) = m_0(-x), \quad n_0(x) = n_0(-x), \quad t_0(x) = 0$$

$$v_0(x) = 0, \quad (-1 < x < 1) \tag{5.57}$$

In addition to the boundary conditions specified by equations (5.14), (5.15), and (5.57) on the crack surface $(-1 < x < 1, y = 0)$, outside the crack $(|x| > 1, y = 0)$ the symmetry and continuity considerations require that (18c, d) and the following conditions be satisfied:

$$N_{XY}(X, 0) = n_{xy}(x, 0) = 0, \quad V_Y(X, 0) = v_y(x, 0) = 0 \quad (|x| > 1) \tag{5.58}$$

In this case, using again the Fourier transforms, the solution of equation (5.13) may be expressed as

$$w(x, y) = \int_0^\infty \sum_1^4 R_j(\alpha)e^{m_j|y|}\cos\alpha x\, d\alpha, \tag{5.59a}$$

$$F(x, y) = \int_0^\infty \sum_1^4 K_j R_j(\alpha)e^{m_j|y|}\cos\alpha x\, d\alpha, \tag{5.59b}$$

where K_j and m_j are defined by equations (5.20). Substituting from equations (5.59) into (5.15), the homogeneous conditions in equations (5.57c), (5.57d), and (5.58) give the following two algebraic relations:

$$m_3 R_3 = \left(\frac{c_0\alpha}{i_2\lambda c^2}+\frac{1}{2}\right)(m_1 R_1 + m_2 R_2) + \frac{i}{2}(m_2 R_2 - m_1 R_1), \tag{5.60a}$$

* See [16] for details.

$$m_4 R_4 = \left(\tfrac{1}{2} - \frac{c_0 \alpha}{i_2 \lambda c^2} \right) (m_1 R_1 + m_2 R_2) - \frac{i}{2} (m_2 R_2 - m_1 R_1), \quad (5.60b)$$

$$c_0 = v_1 + c^2 [1 - 2(v_1 v_2)^{\frac{1}{2}}], \quad (5.60c)$$

After some manipulations it can be shown that the continuity conditions are satisfied if

$$\int_0^\infty (m_1 R_1 + m_2 R_2) \cos \alpha x \, d\alpha = 0, \quad (5.61a)$$

$$\int_0^\infty \frac{i_1 \lambda}{\alpha} (m_1 R_1 - m_2 R_2) \cos \alpha x \, d\alpha = 0, \quad (|x| > 1). \quad (5.61b)$$

The remaining boundary conditions in equations (5.57a), (5.57b) with (5.14), (5.15), and (5.59) may be expressed as

$$\lim_{y \to +0} \int_0^x \left[-\frac{1}{a^2} \int_0^\infty \sum_1^4 (c^2 m_j^2 - v_1 \alpha^2) R_j e^{m_j y} \cos \alpha x \, d\alpha \right] dx$$

$$= -\frac{1}{D_2} \int_0^x m_0(x) dx, \quad (5.62a)$$

$$\lim_{y \to +0} \int_0^x \left[\frac{1}{a^2} \int_0^\infty \sum_1^4 K_j R_j e^{m_j y} \alpha^2 \cos \alpha x \, d\alpha \right] dx =$$

$$= -\int_0^x n_0(x) dx, \quad (-1, < x < 1) \quad (5.62b)$$

With equations (5.6), (5.61) and (5.62) give a system of dual integral equations to determine R_1 and R_2. In equation (5.62) the integral equations are written in integrated form to make them dimensionally consistent with equations (5.61) (i.e., the quantities which appear in equations (5.61) and (5.62) now represent the first derivatives of F and w). Defining

$$\int_0^\infty (m_1 R_1 + m_2 R_2) \cos \alpha x \, d\alpha = v_1(x), \quad (5.63a)$$

$$\int_0^\infty \frac{i_1 \lambda}{\alpha} (m_1 R_1 - m_2 R_2) \cos \alpha \, d\alpha = v_2(x), \quad 0 \le x < \infty \quad (5.63b)$$

and using equation (5.60), and the symmetry conditions $v_j(x) = v_j(-x)$, $(j = 1,2)$, equations (5.61) and (5.62) may be reduced to:

$$\lim_{y \to +0} \frac{1}{\pi} \int_{-1}^1 \sum_1^2 g_{ij}(x, t, y) v_j(t) dt = p_i(x), \quad (i = 1, 2), \quad (|x| < 1) \quad (5.64)$$

Following a procedure similar to that of the previous section to separate the singular kernels and going to limit, equation (5.64) may be put into the following standard form:

$$\int_{-1}^{1} \sum_{1}^{2} \left[\frac{b_{ij}}{t-x} + l_{ij}(x, t) \right] v_j(t) dt = \pi p_i(x), \quad (i = 1, 2, |x| < 1) \tag{5.65}$$

where

$$p_1(x) = \frac{a^2}{i(E_2 h D_1)} \int_0^x n_0(x)\, dx, \quad p_2(x) = \frac{a^2}{D_2} \int_0^x m_0(x)\, dx; \tag{5.66}$$

$$b_{11} = -\frac{2c_0}{c^2}, \quad b_{12} = 2, \quad b_{21} = 4(c^2 - v_1) + \frac{2c_0}{c^2}(2 + v_1 - c^2),$$

$$b_{22} = 2\lambda c^2 (1 - i_2); \tag{5.67}$$

$$l_{11}(x, t) = \int_0^\infty \left[\left(\frac{\alpha}{m_1} + \frac{\alpha}{m_2} + 2 \right) - \frac{2c_0}{i_2 \lambda c^2} \left(\frac{\alpha^2}{m_3} - \frac{\alpha^2}{m_4} - i_2 \lambda \right) \right.$$

$$\left. - \left(\frac{\alpha}{m_3} + \frac{\alpha}{m_4} + 2 \right) \right] \sin \alpha(t - x)\, d\alpha, \tag{5.68a}$$

$$l_{12}(x, t) = \int_0^\infty \left[\frac{\alpha^2}{i_1 \lambda} \left(\frac{1}{m_1} - \frac{1}{m_2} \right) + \frac{\alpha^2}{i_2 \lambda} \left(\frac{1}{m_3} - \frac{1}{m_4} \right) - 2 \right]$$

$$\times \sin \alpha(t - x)\, d\alpha, \tag{5.68b}$$

$$l_{21}(x, t) = \int_0^\infty \left[(v_1 - c^2) \left(\frac{\alpha}{m_1} + \frac{\alpha}{m_2} + \frac{\alpha}{m_3} + \frac{\alpha}{m_4} + 4 \right) \right.$$

$$- i_1 \lambda c \left(\frac{1}{m_1} + \frac{1}{m_2} \right) - i_2 \lambda c \left(\frac{1}{m_3} + \frac{1}{m_4} \right) - (c^2 - v_1) \frac{2c_0}{i_2 \lambda c}$$

$$\left. \times \left(\frac{\alpha^2}{m_3} - \frac{\alpha^2}{m_4} - i_2 \lambda \right) \right] \sin \alpha(t - x)\, d\alpha, \tag{5.68c}$$

$$l_{22}(x, t) = \int_0^\infty \left[\frac{v_1 - c^2}{i_1 \lambda} \left(\frac{\alpha^2}{m_1} - \frac{\alpha^2}{m_2} - i_1 \lambda \right) - \lambda c^2 \left(\frac{\alpha}{m_1} + \frac{\alpha}{m_2} + 2 \right) \right.$$

$$+ \frac{c^2 - v_1}{i_2 \lambda} \left(\frac{\alpha^2}{m_3} - \frac{\alpha^2}{m_4} - i_2 \lambda \right) + i_2 \lambda c^2 \left(\frac{\alpha}{m_3} + \frac{\alpha}{m_4} + 2 \right) \Bigg] \sin \alpha(t - x)\, d\alpha. \tag{5.68d}$$

Here m_j and constants i_1, i_2, and λ are defined by equations (5.20) and c_0 is given by equation (5.60c).

From equations (5.63) and (5.59) it may be seen that physically the quantities v_1 and v_2 correspond to the first derivatives of F and w. Therefore, the elements of the fundamental matrix of the system of singular integral equations (5.65) will be $(1 - x^2)^{\frac{1}{2}}$ and the solution will be of the following form:

$$v_j(x) = H_j(x)(1 - x^2)^{\frac{1}{2}}, \quad (j = 1, 2, |x| < 1), \tag{5.69}$$

where H_j is bounded in $-1 \leq x \leq 1$. Thus the index of the system in equations (5.65) is $\kappa = -1$, and there are no additional conditions (other than the consistency conditions of the singular integral equations) necessary for a unique solution [22]. Note, again, that equations (5.65) are complex and are equivalent to four real integral equations which may be solved numerically in a simple way by using the technique outlined in [23].

To examine the stress state around the crack tips let the functions H_j be expressed in terms of the following infinite series:

$$H_1(x) = \sum_0^\infty A_n U_{2n}(x), \quad H_2(x) = \sum_0^\infty B_n U_{2n}(x), \tag{5.70}$$

where $U_k(x)$ is the Chebyshev polynomial of the second kind and the symmetry property $v_j(x) = v_j(-x)$ has been used. By using the relation [24]

$$\int_0^1 U_{2n}(t)(1 - t^2)^{\frac{1}{2}} \cos \alpha t \, dt = (-1)^n \frac{\pi}{2} (2n + 1) \frac{1}{\alpha} J_{2n+1}(\alpha),$$
$$(n = 0, 1, 2, \ldots), \tag{5.71}$$

from equations (5.63), (5.69), and (5.70) it follows that

$$m_1 R_1 + m_2 R_2 = \sum_0^\infty (-1)^n (2n + 1) A_n \frac{1}{\alpha} J_{2n+1}(\alpha), \tag{5.72a}$$

$$m_1 R_1 - m_2 R_2 = \frac{1}{i_1 \lambda} \sum_0^\infty (-1)^n (2n + 1) B_n J_{2n+1}(\alpha). \tag{5.72b}$$

Substituting into the stress expressions from equations (5.72), (5.60), and (5.59) and omitting the details it may now be shown that*

$$\sigma_{yy}^m(r, \theta) = \frac{i(E_2 h D_1)^{\frac{1}{2}}}{4ha^2} \sum_0^\infty (2n + 1) \left(\frac{c_0}{c^2} A_n - B_n \right)$$

$$\times \frac{1}{(2r)^{\frac{1}{2}}} \left(5 \cos \frac{\theta}{2} - \cos \frac{5\theta}{2} \right) + O(r^{\frac{1}{2}}) \tag{5.73}$$

* See [16] for the evaluation of the related integrals.

where (r, θ) are the polar coordinates at the crack tip defined by

$$r^2 = (x - 1)^2 + y^2, \quad \tan \theta = y/(x - 1).$$

Defining the membrane component of the stress intensity factor in the shell by

$$k_s^m = \lim_{x \to 1} [2(x - 1)a]^{\frac{1}{2}} \sigma_{yy}^m(x, 0), \tag{5.74}$$

and observing that $U_{2n}(1) = 2n + 1$, k_s^m is found to be

$$k_s^m = \frac{i(E_2 hD_1 a)^{\frac{1}{2}}}{ha^2} \left[\frac{c_0}{c^2} H_1(1) - H_2(1) \right]. \tag{5.75}$$

For example, if $n_0(x) = N_0 = $ constant, $m_0(x) = 0$, and the corresponding plate stress intensity factor is defined by $k_p = (N_0\sqrt{a})/h$, the membrane component of the stress intensity factor ratio becomes

$$A_m = \frac{k_s^m}{k_p} = i \left[\frac{c_0}{c^2} H_1^*(1) - H_2^*(1) \right] \tag{5.76}$$

where

$$H_j^*(x) = H_j(x) \frac{(E_2 hD_1)^{\frac{1}{2}}}{N_0 a^2}, \quad (j = 1, 2) \tag{5.77}$$

the functions H_j^* are obtained from equations (5.65) after the normalization given by equation (5.77).

The reamining stress components may be obtained in a similar way. The asymptotic stress state in the neighborhood of the crack tip may then be expressed as

$$\sigma_{yy}^m(r, \theta) = A_m \frac{k_p}{(2ra)^{\frac{1}{2}}} \frac{1}{4} \left(5 \cos \frac{\theta}{2} - \cos \frac{5\theta}{2} \right) + O(r^{\frac{1}{2}}), \tag{5.78a}$$

$$\sigma_{xx}^m(r, \theta) = A_m \frac{k_p}{(2ra)^{\frac{1}{2}}} \frac{c^2}{4} \left(3 \cos \frac{\theta}{2} + \cos \frac{5\theta}{2} \right) + O(r^{\frac{1}{2}}), \tag{5.78b}$$

$$\sigma_{xy}^m(r, \theta) = A_m \frac{k_p}{(2ra)^{\frac{1}{2}}} \frac{c}{4} \left(\sin \frac{\theta}{2} - \sin \frac{5\theta}{2} \right) + O(r^{\frac{1}{2}}), \tag{5.78c}$$

$$\sigma_{yy}^b(r, \theta, Z) = A_b \frac{k_p}{(2ra)^{\frac{1}{2}}} \frac{Z}{2h} \left[\left(8v_0 - 8v_c + \frac{5v_c v_0}{c^2} \right) \cos \frac{\theta}{2} \right.$$

$$- \frac{v_0 v_c}{c^2} \cos \frac{5\theta}{2} \Bigg] + O(r^{\frac{1}{2}}), \tag{5.79a}$$

$$\sigma_{xx}^b(r, \theta, Z) = A_b \frac{k_p}{(2ra)^{\frac{1}{2}}} \frac{2Z}{h} \left[\left(8 + \frac{5v_c}{c^2} - 8v_c c^2 + 3v_2 v_c \right) \cos \frac{\theta}{2} \right.$$

$$\left. - \frac{v_c}{4c^2} (1 - v_2 c^2) \cos \frac{5\theta}{2} \right] + O(r^{\frac{1}{2}}), \tag{5.79b}$$

$$\sigma_{xy}^b(r, \theta, Z) = A_b \frac{k_p}{(2ra)^{\frac{1}{2}}} \frac{Z}{h} \frac{a^2 h^3 G_{12}}{12 D_1} \left[\frac{v_c}{c^2} \sin \frac{5\theta}{2} \right.$$

$$\left. - \left(8 + \frac{v_c}{c^2} \right) \sin \frac{\theta}{2} \right] + O(r^{\frac{1}{2}}), \tag{5.79c}$$

where

$$v_0 = c^2 - v_1, \ c = (E_1/E_2)^{\frac{1}{4}}, \ v_c = c^2 - \left(v_1 + \frac{h^3 G_{12}}{3 D_2} \right). \tag{5.80}$$

and the bending component of the stress intensity ratio is found to be

$$A_b = k_s^b/k_p = \frac{1}{k_p} \lim_{x \to 1} [2(x - 1)a]^{\frac{1}{2}} \sigma_{yy}^b(x, 0, h)$$

$$= \frac{1}{2c^2} \frac{[12(1 - v_1 v_2)]^{\frac{1}{2}}}{1 - v_1 v_2} \left[\left(\frac{c_0}{c^2} - 2 \right)(c^2 - v_1) + 2c_0 \right] H_1^*(1). \tag{5.81}$$

By letting $E_1 = E_2 = E$, $v_1 = v_2 = v$, $G_{12} = G$, and $c = 1$ the results of this section too reduce to the solution of isotropic cylindrical shell.

The 'bulging' of the shell in the neighborhood of the crack, i.e., w may be directly evaluated in terms of the solution given in this section. Also, in the present as well in the two other ideal shell geometries (that is, in the cylindrical shell with an axial or a circumferential crack, and in the spherical shell with a meridional crack) it can be shown that the auxiliary functions defined to reduce the problem to singular integral equations are directly related to the crack surface displacements. For the isotropic shells these displacements are obtained and presented for various values of the shell parameter λ in [8].

The θ-dependence in the asymptotic stress expressions given by equations (5.53), (5.55), (5.78), and (5.79) is identical to the expressions for isotropic shells. However, note that the dimensionless coordinates r, θ, x, and y in the specially orthotropic shells are defined by

$$r^2 = (x - 1)^2 + y^2, \ \tan \theta = y/(x - 1), \ x = X/a, \ y = c \ Y/a \tag{5.82}$$

where X, Y, Z are the actual rectangular coordinates and the actual geometric angle Θ in the shell is given by

$$\tan \Theta = Y/(X - a) = \frac{y}{c(x - 1)} = \frac{1}{c} \tan \theta. \tag{5.83}$$

Therefore, because of equation (5.83), the angular variation of the asymptotic stresses in the specially orthotropic shells is different and a good deal more complicated than that in isotropic shells.

The analysis given in this and the previous sections indicates that, since the roots of the characteristic equation m_j, $(j = 1, \ldots, 4)$, shown in equations (5.19) and (5.20) are functions of the transform variable α, $(0 < \alpha < \infty)$ mathematically the problem would have been intractable if $m_j(\alpha)$ were not evaluated in closed form. This is essential for extracting the singular parts of the kernels of the resulting integral equations as well as for studying and obtaining the correct singular behavior of the solution. The analysis also shows that this critical aspect of the problem relating to the singular nature of the integral equations and their solution is entirely dependent on the asymptotic behavior of certain functions for large values of α (see, for example, equations (5.29) to (5.34)). The variable α appears in these functions explicitly as well as through $m_j(\alpha)$. In the equation which determines m_j, the coefficients of the characteristic function, which is an 8th degree polynomial, are functions of α. Therefore, for the problems in which the roots $m_j(\alpha)$ cannot be expressed in closed form, it appears that if the asymptotic solution of the characteristic equation giving $m_j(\alpha)$ for large values of α can be obtained correctly in closed form, then the singular parts of the kernels can be separated and the singular nature of the solution can be studied. Furthermore, by also evaluating $m_j(\alpha)$ for small values of α in closed form and for intermediate discrete values of α numerically, at least in principle, it is possible to evaluate the Fredholm kernels in the integral equations numerically and, at the cost of a rather high computational effort, to obtain a meaningful approximate solution.

5.5 Results for a specially orthotropic cylindrical shell

In order to give an idea about the effect the material orthotropy may have on the stress intensity factors in a cylindrical shell containing a longitudinal through crack, in this section some numerical results on cylinders made of three different materials will be presented. These are an isotropic cylinder, a titanium cylinder which is mildly orthotropic, and a graphite cylinder which is strongly orthotropic. The measured elastic constants of the orthotropic materials are shown in Table 5.1. The table also shows the 'average shear

TABLE 5.1

Elastic constants of the orthotropic materials

	Titanium	Graphite
E_1(psi)	1.507×10^7	1.5×10^6
E_2(psi)	2.08×10^7	40×10^6
v_1	0.1966	0.0075
v_2	0.2714	0.2000
G_{12}	6.78×10^6	4.0×10^6
$G_{av.}$	7.15×10^6	3.73×10^6

modulus' calculated from (see equation (5.10))

$$G_{av.} = \frac{(E_1 E_2)^{\frac{1}{2}}}{2[1 + (v_1 v_2)^{\frac{1}{2}}]}, \tag{5.84}$$

where E_1 is the modulus in the axial direction and the notation is given by equations (5.4). If the measured shear modulus G_{12} were equal to the calculated modulus $G_{av.}$, then the material would be specially orthotropic and the analysis given in the previous sections would be valid without any approximations. The table indicates that these two values are sufficiently close so that the special orthotropy assumption may be used to study the effect of material orthotropy on the stress intensity factors.

Figures 5.2 to 5.5 show the results for a pressurized shell with an axial crack. The membrane and bending components of the stress intensity factor ratio A_m and A_b shown in the figures are defined by equations (5.74), (5.76) and (5.81). For the pressurized shell the corresponding flat plate stress intensity factor is

$$k_p = \frac{p_0 R}{h} \sqrt{(a)}, \tag{5.85}$$

where p_0 is the internal pressure and the dimensions R, h, a are shown in Figure 5.1. Generally the results in cylindrical as well as spherical shells are presented in terms of the dimensionless 'shell parameter' λ defined by

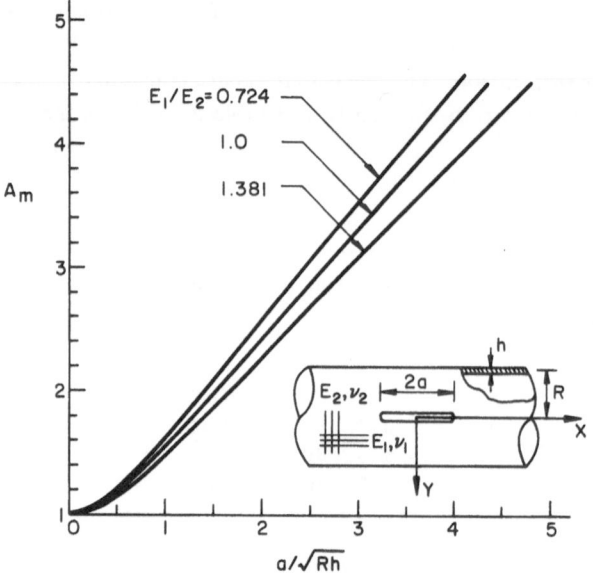

Figure 5.2. Membrane component of the stress intensity factor ratio A_m for a pressurized Titanium and for an isotropic ($\nu = 1/3$) cylinder

$$\lambda = [(12(1 - \nu^2)]^{\frac{1}{4}}\, a/\sqrt{(Rh)} \tag{5.86}$$

in isotropic shells, and

$$\lambda = [12(1 - \nu_1\nu_2)E_2/E_1]^{\frac{1}{4}}\, a/\sqrt{(Rh)} \tag{5.87}$$

in orthotropic shells (see equation (5.20)). It is seen that the parameter λ in the specially orthotropic shells depends on two elastic constants and, therefore, is not an appropriate correlation coefficient to be used for the purpose of comparing the results in two different shells with the same geometry and different materials. Thus, in Figures 5.2 to 5.5 a purely geo-metrical parameter, namely $a/\sqrt{(Rh)}$ is used as the independent variable.

Also, from the analysis given in the previous sections it is clear that the dependence of the results on elastic constants is not through λ only. Hence, the orthotropic results shown in the figures are for the specific material constants given in Table 5.1. Similarly, for the isotropic shells the Poisson's ratio ν appears in the analysis through λ as well as elsewhere. The isotropic shell shown in Figures 5.2 to 5.5 under the designation $(E_1/E_2) = 1$ are thus

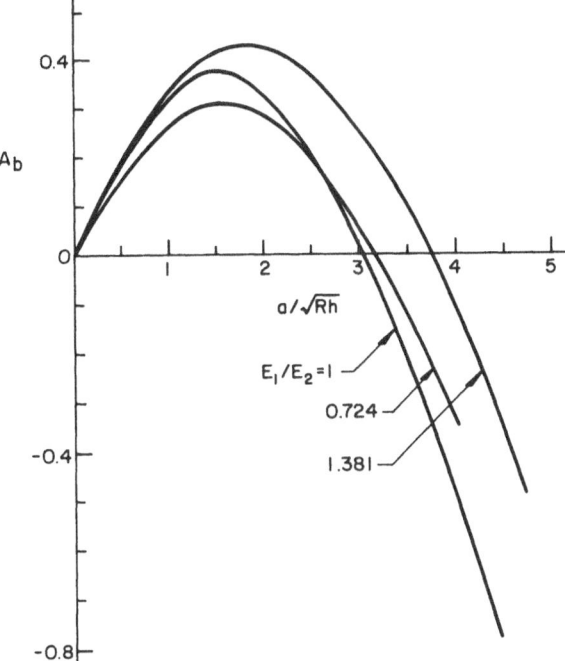

Figure 5.3. Bending component of the stress intensity factor ratio A_b for a pressurized Titanium and for an isotropic cylinder

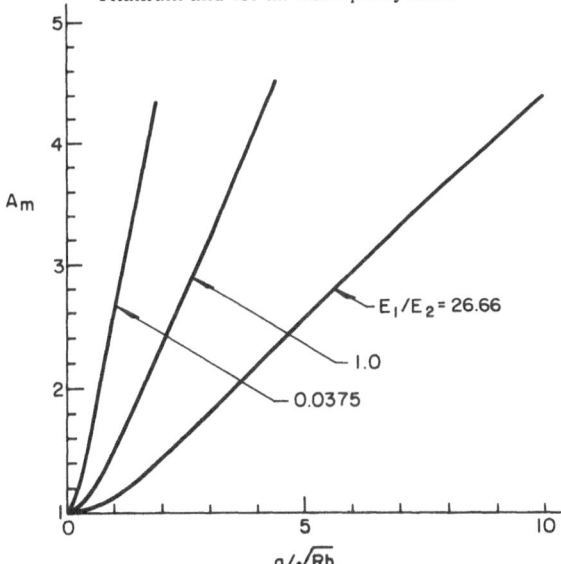

Figure 5.4. Membrane component of the stress intensity factor ratio A_m for a pressurized Graphite and for an isotropic cylinder

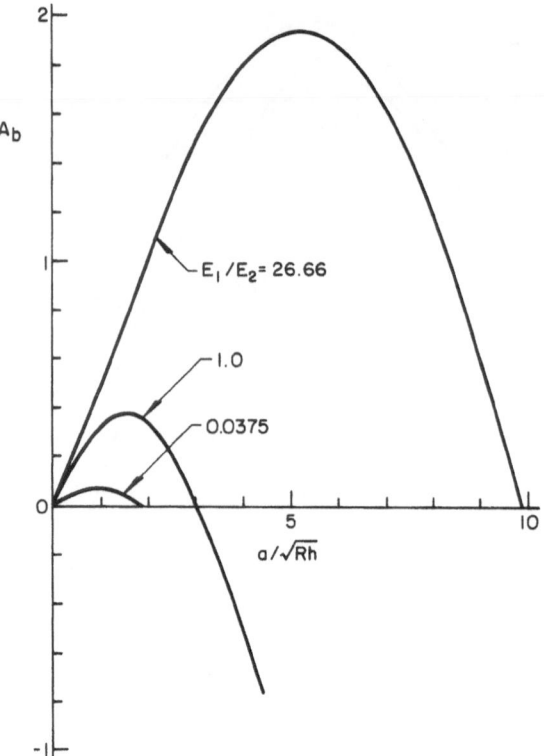

Figure 5.5. Bending component of the stress intensity factor ratio A_b for a pressurized Graphite and for an isotropic cylinder

obtained for one value of v only, namely $v = 1/3$. The effect of v on the stress intensity factors in isotropic shells is discussed in the following section. In each figure there are two sets of orthotropic results which correspond to the alignment of the stiff direction of the material in the axial or the circumferential direction of the cylinders.

The results indicate that in the specially orthotropic shells the stress intensity factors are strongly dependent on the modulus ratio E_1/E_2, and generally they increase with decreasing E_1/E_2, E_1 being the modulus in axial direction. This does not, of course, necessarily mean a reduction in the resistance of the shell to crack propagation as the shell becomes stiffer in circumferential direction. Any material, particularly a composite, which is not isotropic in elastic properties, would not be expected to be isotropic in

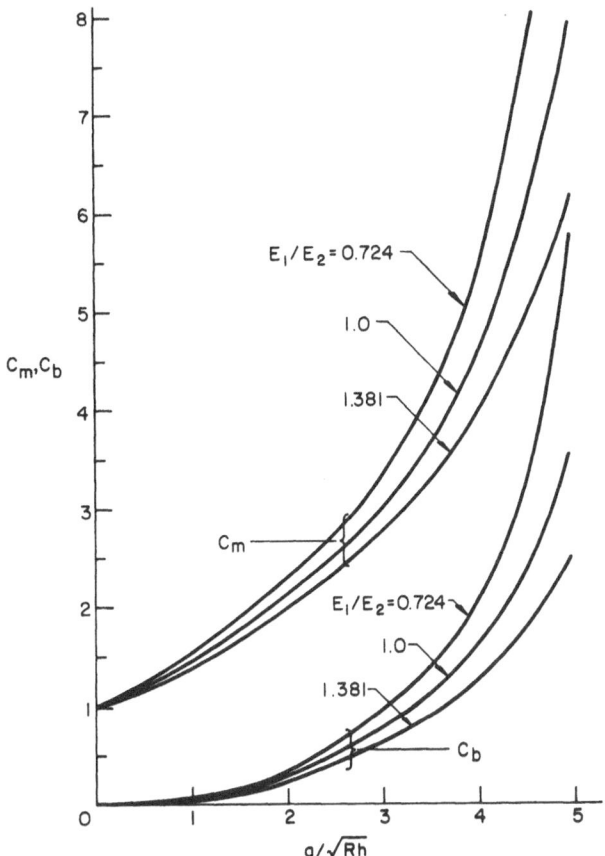

Figure 5.6. Membrane and bending components of the stress intensity factor ratio, C_m and C_b for a specially orthotropic (Titanium) and for an isotropic ($\nu = 1/3$) cylinder under torsion

its resistance to crack propagation. In each case the load-bearing strength of the structure would, of course, be decided by the ratio of the stress intensity factor or whatever the measure of the severity of the external loads and the crack geometry to the corresponding strength parameter of the material.

Figure 5.6 shows the results for a cylinder with an axial crack under skew symmetric loading. Here it is assumed that the cylinder is under torsion and away from the crack region the uniform shear $N_{xy} = N_0$ is the only nonzero stress component acting on the shell. Thus, the corresponding flat plate stress intensity factor is a mode II component given by $k_p = N_0\sqrt{(a)}/h$. The membrane and bending stress intensity factor ratios C_m and C_b shown in the figure are defined by equations (5.50), (5.51), and (5.54). In this example too $a/\sqrt{(Rh)}$ rather than the shell parameter λ is used as the independent variable and for the isotropic case (designated by $(E_1/E_2) = 1$) it is again assumed that $v = 1/3$. Figure 6 shows the same trend as Figures 5.2–5.5, namely, the stress intensity factors increase with decreasing E_1/E_2. This appears to be primarily due to the multiplicative factor $(E_1/E_2)^{\frac{1}{4}}$ in the expression of the shell parameter λ given by equation (5.85). In fact for a quick estimate of the stress intensity ratios in skew-symmetric as well as in symmetric problems for the specially orthotropic shells the isotropic results may be sufficient provided λ is calculated from equation (5.87).

Table 5.2, which shows the results for only one value of the variable $a/\sqrt{(Rh)} = 1.66$, gives some idea about the relative effect of material orthotropy. Here the results for graphite, titanium, and an isotropic material ($v = 1/3$) are compared. In this case too, the strong influence of material orthotropy is apparent.

TABLE 5.2

The effect of orthotropy on the stress intensity ratios ($a/\sqrt{(Rh)} = 1.66$)

	Isotropic Material	Titanium		Graphite	
E_1/E_2	1.0	1.381	0.724	26.667	0.0375
λ	3.0	2.811	3.304	1.359	7.018
C_m	1.942	1.880	2.044	1.340	4.045
C_b	0.199	0.158	0.239	0.019	1.241

5.6   The effect of Poisson's ratio

As indicated in the previous section, in the isotropic shells the Poisson's ratio v appears in the analysis explicitly as well as through λ defined by equation (5.86). This means that the stress intensity factors are functions of two independent variables, namely v and $a/\sqrt{(Rh)}$. However, since in most metallic structural materials v is in the neighborhood of 1/3 and since v affects the results partly if not mostly through λ, in practice the tendency has been to present the results by using only λ as the independent variable for a fixed Poisson's ratio, $v = 1/3$. To justify this or to throw some light on the approximation involved, the effect of v for some selected values of λ or $a/\sqrt{(Rh)}$ has to be studied.

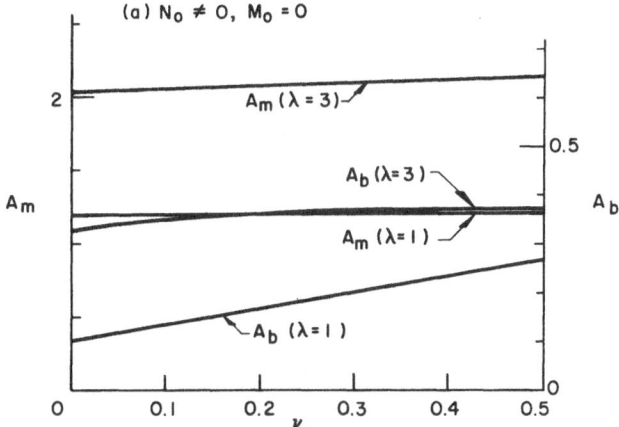

Figure 5.7. The effect of Poisson's ratio on the stress intensity factors in a pressurized isotropic cylinder with an axial crack

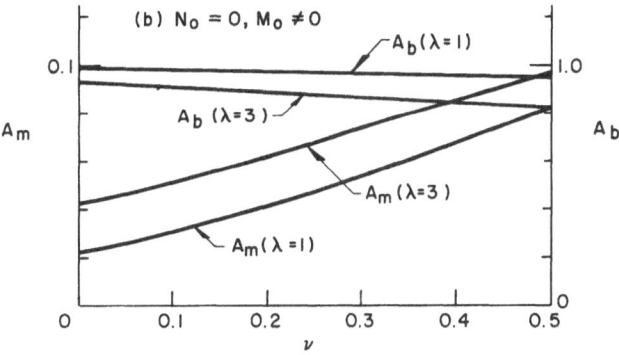

Figure 5.8. The effect of Poisson's ratio on the stress intensity factors in an axially cracked cylindrical shell under uniform bending, $M_{yy} = M_0$

Figures 5.7 to 5.9 show some results for a cylindrical shell with an axial crack. Figure 5.7 show the variation of the symmetric stress intensity factor ratios A_m and A_b for $\lambda = 1$ and $\lambda = 3$ in a pressurized cylinder where

$$A_m = \frac{k_s^m}{k_p}, \; A_b = \frac{k_s^b}{k_p}, \; k_p = \frac{N_0 \sqrt{a}}{h} = \frac{p_0 R \sqrt{a}}{h} . \tag{5.88}$$

In this case the effect of v on the main stress intensity component A_m appears to be negligible. Figure 5.8 shows some symmetric results for the shell under cylindrical bending only in which $M_{yy} = -M_0$ is the only nonzero crack surface loading. For this loading the corresponding flat plate stress intensity factor is defined by

$$k_p = \frac{6M_0}{h^2} \sqrt{a} \tag{5.89}$$

and A_m and A_b are again given by equations (5.88). In this case too the variation of the main stress intensity component A_b with v for the values of $\lambda = 1$ and $\lambda = 3$ does not appear to be significant. Even though there is a considerable relative change in A_m as v goes from zero to 0.5, it should be observed that the absolute value of A_m itself is rather small.

An example for the skew-symmetric problem is shown in Figure 5.9. Here it is assumed that a cylinder containing an axial crack is under torsion and $\lambda = 5$. The related stress intensity factors are defined by equations (5.50), (5.51), and (5.54). For this λ value, the effect of v again appears to be negligible.

It should be noted that in Figures 5.7 to 5.9 λ is used as a constant para-

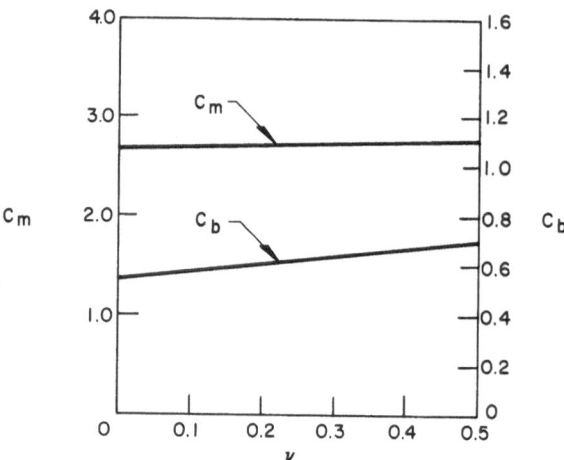

Figure 5.9. The effect of Poisson's ratio on the stress intensity factors in an axially cracked isotropic cylinder under torsion ($\lambda = 5$)

meter. Since λ is a decreasing function of v, this would compensate some of the increases in the stress intensity factor ratios observed for increasing v. A somewhat more meaningful result would be obtained by comparing the stress intensity factor ratios for different Poisson's ratios and a fixed geometric parameter $a/\sqrt{(Rh)}$. A very limited such comparison for the symmetric problem is shown in Table 5.3 which leads to the same general conclusion that the effect of v on the stress intensity factors is not very significant.

TABLE 5.3

The effect of Poisson's ratio

$\dfrac{a^2}{Rh}$	v	$N_0 \neq 0, M_0 = 0$		$N_0 = 0, M_0 \neq 0$	
		A_m	A_b	A_m	A_b
0.3	0.5	2.157	0.353	0.097	0.810
	1/3	2.163	0.364	0.078	0.865
2.63	0.15	2.066	0.352	0.057	0.912
	1/3	2.074	0.370	0.076	0.873
2.6	0	2.045	0.326	0.043	0.932
	1/3	2.059	0.372	0.076	0.875

5.7 Interaction of two cracks

In plane problems it is known that if the medium contains more than one crack, depending on the relative distance between the cracks, there could be a strong interaction between the respective stress fields and the stress intensity factors could be highly affected. In order to give some idea about the effect of interacting stress fields on the stress intensity factors in shells, in this section the results of a simple problem for a pressurized cylindrical shell containing two axial cracks are presented. From the formulation and the solution of the crack problem in shells given in Section 2–4 of this chapter it is clear that there is no major difficulty in formulating the problem and in deriving the governing system of singular integral equations if the shell contains, instead of a single crack, a set of collinear cracks. Therefore, there is no need to present further analytical details.

In the example under consideration the two cracks are assumed to be equal in length. The particular crack geometry and dimensions are shown by the

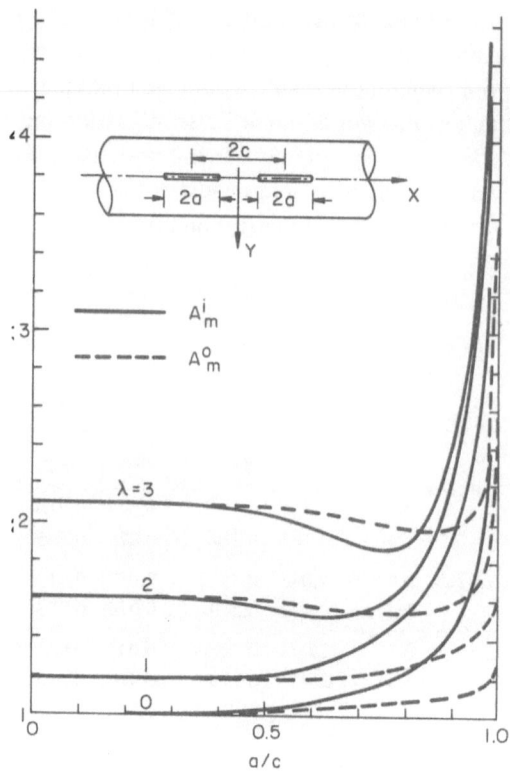

Figure 5.10. The membrane components of the stress intensity factor ratio in a pressurized isotropic cylinder ($v = 1/3$) with two collinear axial cracks. A_m^i for the inner crack tip, A_m^o for the outer crack tip

insert in Figure 5.10, and the results are shown in Figures 5.10 and 5.11. The stress intensity factor ratios A_m and A_b are again defined by equations (5.88). The superscripts i and o on A_m and A_b refer to the inner and outer crack tips, respectively. The figures show the results for $\lambda = 1, 2, 3$ where λ is defined by equation (5.86) and v again is assumed to be $1/3$. For the purpose of comparison, Figure 5.10 also shows the stress intensity factor ratios for the flat plate with the same crack geometry (i.e., for $\lambda = 0$) evaluated from [25]

$$A_m^i = \frac{k_p^i}{k_p} = \frac{b_1^2 E(m)/K(m) - a_1^2}{(b_1 - a_1)\left[a_1(b_1 + a_1)/2\right]^{\frac{1}{2}}},$$

(5.90a)

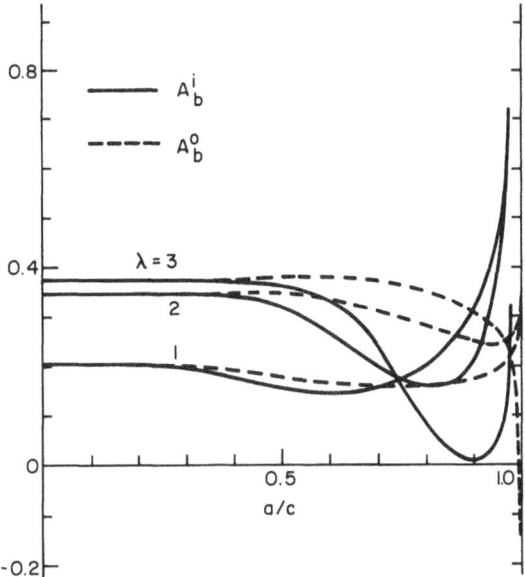

Figure 5.11. The bending components of the stress intensity factor ratio in a pressurized cylinder with two axial cracks. A_b^i for the inner crack tip, A_b^0 for the outer crack tip

$$A_m^0 = \frac{k_p^0}{k_p} = \frac{b_1^2[1 - E(m)/K(m)]}{(b_1 - a_1)\{b_1(b_1 + a_1)/2\}^{\frac{1}{2}}}, \qquad (5.90b)$$

where

$$k_p = \frac{p_0 R \sqrt{a}}{h}, \, a_1 = c - a, b_1 = c + a, m = 1 - \frac{a_1^2}{b_1^2}, \qquad (5.91)$$

and $K(m)$ and $E(m)$ are the complete elliptic integrals of the first and the second kind, respectively.

For $a/c = 0$ the two cracks are far apart, there is no interaction, and the results correspond to that of a single crack in a pressurized cylinder. On the other hand as $a/c \to 1$, i.e., as the length of the net ligament between the two cracks approaches zero, as expected, the stress intensity factor at the inner crack tip goes to infinity and that at the outer tip approaches the value obtained for a single crack of length $4a$. However, for $a/c > 0.4$ and $\lambda \geq 2$ the results show a somewhat unexpected behavior. In flat plates A_m^i is always greater than A_m^0 whereas in shells the results show that for certain ranges of a/c and λ it is possible to have $A_m^i < A_m^o$. This behavior seems to be even more pronounced for the bending stress intensity factors shown in Figure 5.11. A partial explanation of this phenomenon may be found in the distribution of the displacement component $w(x, y)$ normal to the shell surface. In a pressurized isotropic shell containing a single crack of length $2a$, evaluating w in the plane of the crack, i.e., for $x > 0$, $y = 0$, one obtains, for example for $\lambda = 2$, the result shown in Figure 5.12 (where, in the nota-

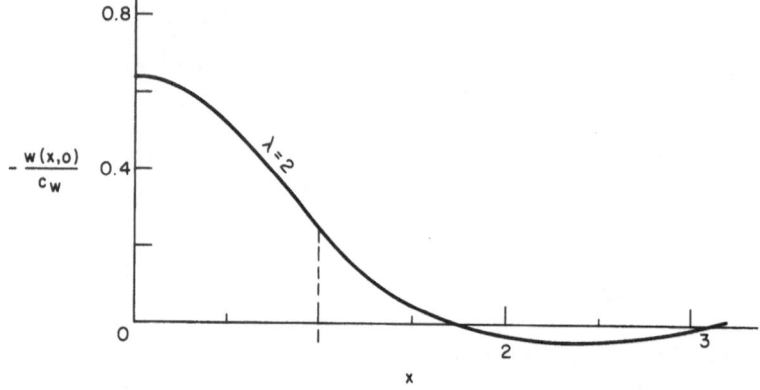

Figure 5.12. Displacement component w normal to the thell surface in the plane of the crack for a pressurized cylinder with an axial crack, $x = X/a$

tion of Figures 5.1 and 5.10, $w > 0$ inward). The normalization factor which appears in the figure is given by

$$c_w = \frac{p_0 R^2 \lambda^2}{2Eh},\qquad(5.92)$$

and the coordinate x is normalized with respect to a. The figure shows that, although around the crack there is an outward bulging in the shell, further along the x axis w changes sign and there is a zone of depression. When the distance c is small enough for the stress and displacement fields of the two cracks to interact, for a certain range of c this 'depression' may cause a reduction in the stress intensity factors.

5.8 Further results for isotropic shells

This section presents a summary of the calculated results for the three

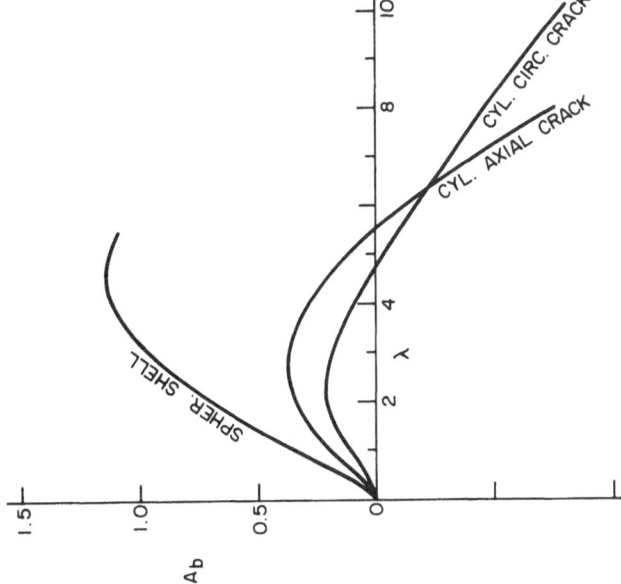

Figure 5.13. Membrane component of the stress intensity factor ratio A_m in symmetrically-loaded shells: $N_{yy} = N_0 \neq 0, M_{yy} = 0$

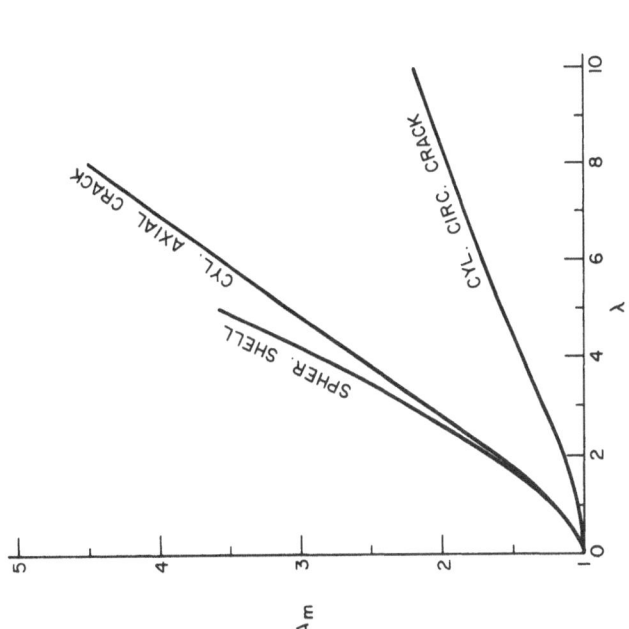

Figure 5.14. Bending component of the stress intensity factor ratio A_b in symmetrically-loaded shells: $N_{yy} = N_0 \neq 0, M_{yy} = 0$

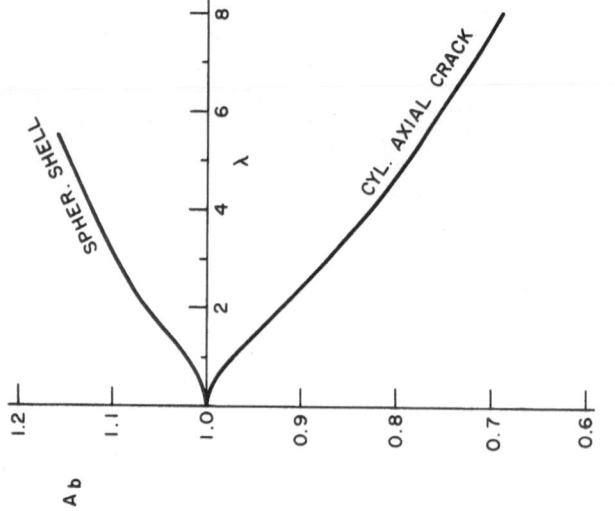

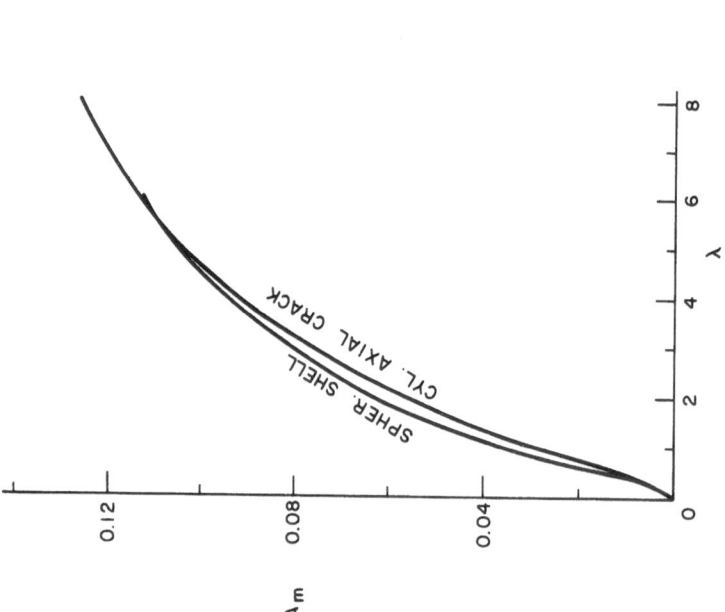

Figure 5.15. Membrane component of the stress intensity factor ratio A_m in shells symmetrically-loaded in bending: $M_{yy} = M_0 \neq 0$, $N_{yy} = 0$

Figure 5.16. Bending component of the stress intensity factor ratio A_b in shells symmetrically-loaded in bending: $M_{yy} = M_0 \neq 0$, $N_{yy} = 0$

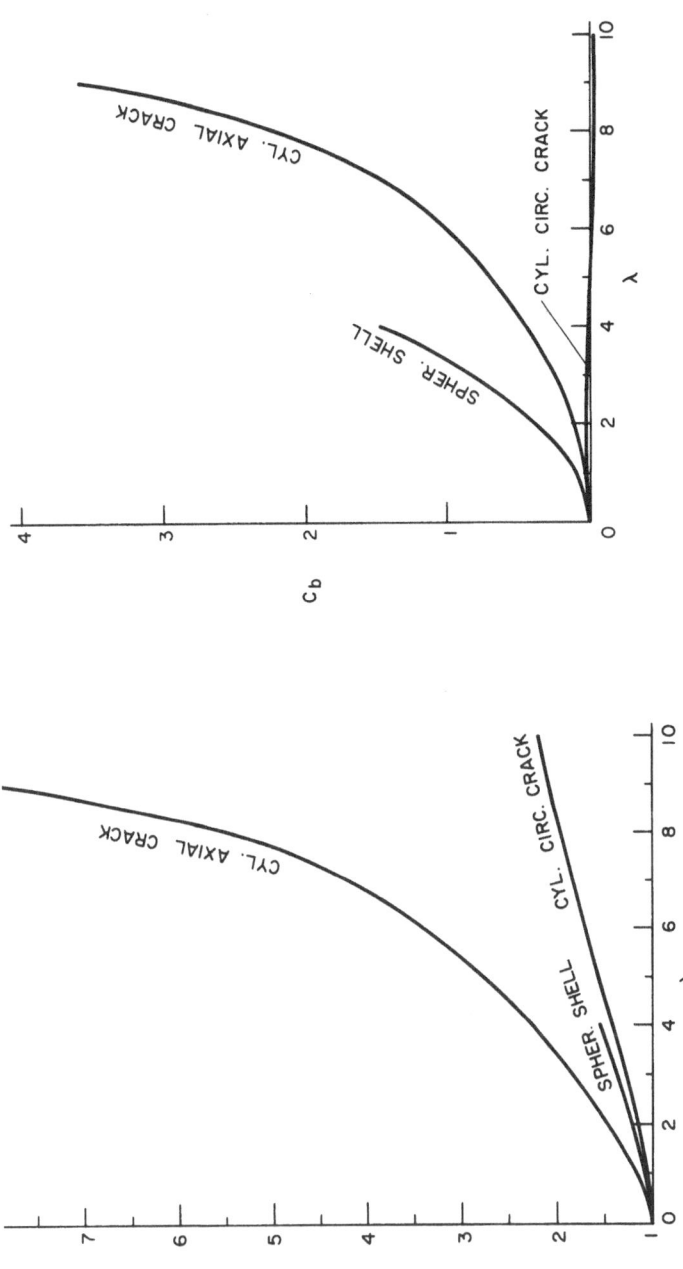

Figure 5.17. Membrane component of the stress intensity factor ratio C_m in shells under uniform skew-symmetric membrane loading: $N_{xy} = N_0$

Figure 5.18. Bending component of the stress intensity factor ratio C_b in shells under uniform skew-symmetric membrane loading: $N_{xy} = N_0$

idealized crack geometries, namely a cylindrical shell with an axial crack, that with a circumferential crack, and a spherical shell with a meridional crack. The loading condition is assumed to be homoengeous and either symmetric or skew-symmetric. The Poisson's ratio of $1/3$ is assumed in all calculations. The technique used to solve the related shell problems is similar to that described in Sections 5.3 and 5.4 of this chapter [6–9].

Figures 5.13 and 5.14 show, respectively, the membrane and bending, stress intensity factor ratios A_m and A_b for the three shell geometries. In this symmetric case the only nonzero crack surface traction is assumed to be $N_{yy} = -N_0 =$ constant, x being the coordinate along the crack.* The results for the symmetric problem in which $M_{yy} = -M_0 =$ constant is the only nonzero crack surface load are given in Figures 5.15 and 5.16. Finally, Figures 5.17 and 5.18 show the skew-symmetric results for the three crack geometries where $N_{xy} = -N_0 =$ constant is the nonzero crack surface traction. In presenting these results λ is defined by equation (5.86), A_m and A_b are defined by equations (5.88) (with k_p as given by equations (5.89) for Figures 5.15 and 5.16), and C_m and C_b are defined by equations (5.50), (5.51), and (5.54).

---

* It should be noted that there was a numerical error in A_b for the cylindrical shell with a circumferential crack given in [7].

References

[1] Copley, L. G. and Sanders, J. L., Jr., A longitudinal crack in a cylindrical shell under internal pressure, *International Journal of Fracture Mechanics*, 5, pp. 117–131 (1969).

[2] Duncan, M. E. and Sanders, J. L., Jr., The effect of circumferential stiffener on the stress in a pressurized cylindrical shell with a longitudinal crack, *International Journal of Fracture Mechanics*, 5, pp. 133–155 (1969).

[3] Folias, E. S., An axial crack in a pressurized cylindrical shell, *International Journal of Fracture Mechanics*, 1, pp. 104–113 (1965).

[4] Folias, E. S., A finite line crack in a pressurized spherical shell, *International Journal of Fracture Mechanics*, 1, pp. 20–46 (1965).

[5] Folias, E. S., A circumferential crack in a pressurized cylinder, *International Journal of Fracture Mechanics*, 3, pp. 1–12 (1967).

[6] Erdogan, F. and Kibler, J. J., Cylindrical and spherical shells with cracks, *International Journal of Fracture Mechanics*, 5, pp. 229–237 (1969).

[7] Erdogan, F. and Ratwani, M., Fatigue and fracture of cylindrical shells containing a circumferential crack, *International Journal of Fracture Mechanics*, 6, pp. 379–392 (1970).

[8] Erdogan, F. and Ratwani, M., Plasticity and crack opening displacement in shells, *International Journal of Fracture Mechanics*, 8, pp. 413–426 (1972).

[9] Erdogan, F. and Ratwani, M., A circumferential crack in a cylindrical shell under torsion, *International Journal of Fracture Mechanics*, 8, pp. 87–95 (1972).

[10] Murthy, M. V. V., Rao, K. P., and Rao, A. K., Stresses around an axial crack in a pressurized cylindrical shell, *International Journal of Fracture Mechanics*, 8, pp. 287–297 (1972).

[11] Murthy, M. V. V., Rao, K. P., and Rao, A. K., On the stress problem of large elliptical cutouts and cracks in circular cylindrical shells, *International Journal of Solids Structures*, 10, pp. 1243–1269 (1974).

[12] Erdogan, F. and Ratwani, M., A note on the interference of two collinear cracks in a cylindrical shell, *International Journal of Fracture*, 10, pp. 463–465 (1974).

[13] Erdogan, F. and Ratwani, M., Fracture of cylindrical and spherical shells containing a crack, *Nuclear Engineering and Design*, 20, pp. 265–286 (1972).

[14] Yuceoglu, U. and Erdogan, F., A cylindrical shell with an axial crack under skew-symmetric loading, *International Journal of Solids Structures*, 9, pp. 347–362 (1973).

[15] Erdogan, F., Ratwani, M., and Yuceoglu, U., On the effect of orthotropy in a cracked cylindrical shell, *International Journal of Fracture*, 10, pp. 369–374 (1974).

[16] Yuceoglu, U., *Ph. D. Dissertation*, Lehigh University (1971).

[17] Hartranft, R. J. and Sih, G. C., Effect of plate thickness on the bending stress distribution around through cracks, *Journal of Mathematics and Physics*, 47, pp. 276–291 (1969).

[18] Ambartsumyan, S. A., On the theory of anisotropic shallow shells, NACA Tech. Memo. 1424 (1956).

[19] Ambartsumyan, S. A., Theory of anisotropic shallow shells, NASA Tech. Transl. F-118 (1964).

[20] Flügge, W. and Conrad, D. A., *Singular solutions in the theory of shallow shells*, Tech. Report No. 101, Stanford University (1956).

[21] Apeland, K., Analysis of anisotropic shallow shells, *Acta Technica Scandinavica*, 22 (1963).

[22] Muskhelishvili, N. I., *Singular Integral Equations*, P. Noordhoff, Groningen (1953).

[23] Erdogan, F., Gupta, G. D., and Cook, T. S., Numerical solution of singular integral equations, in *Methods of Analysis and Solutions of Crack Problems*, G. C. Sih, ed., Noordhoff International Publishing, Leyden, The Netherlands (1973).

[24] Gradshteyn, I. S. and Ryzhic, I. M., *Tables of Integrals, Series and Products*, Academic Press (1965).

[25] Erdogan, F., On the stress distribution in plates with collinear cuts under arbitrary loads, *Proc. 4th U.S. National Congress Applied Mechanics*, 1, pp. 547–553 (1962).

G. C. Sih and H. C. Hagendorf

6 | On cracks in shells with shear deformation

6.1 Introduction

Although much work has been done on the stress analysis of cracks in initially flat plates [1], theoretical treatment of cracks in initially curved plates or shells has received attention only in recent years. The presence of curvature in a shell generates deviation from behavior of flat plates, in that stretching loads will induce both extensional and bending stresses, while bending loads will also lead to both types of stresses. In the neighborhood of geometric defects such as man-made flaws or cracks, the stresses are redistributed and can lead to a reduction of the load carrying capacity of shell structures.

One of the simplest shell geometries is that of a spherical shell whose curvature radius R is everywhere constant. The first investigation on the stresses in a spherical shell containing a crack was made by Ang, Folias and Williams [2], who associated the shell problem with that of an initially flat plate resting on an elastic foundation. The equivalence of the two problems was made by identifying the foundation modulus with Eh/R^2 where E is the Young's modulus and h the shell thickness. The general character of the crack tip stress field in a shell was studied by Sih and Setzer [3] who pointed out that the functional relationships of the local (extension-bending) stresses are identical with those obtained by superimposing the individual extensional and bending stresses of an initially flat plate. They further emphasized the coupling effects of extension and bending through the intensity of the local stress field. A separate treatment of the spherical shell problem was later given by Folias [4] who utilized singular integral equations with Cauchy type kernels, as devised by Knowles and Wang [5], for solving crack problems involving flat plates. The formulation in [4] relies on certain approximations in the kernels for small values of the curvature parameter. Other methods for solving shell problems with cracks have been presented by Erdogan and Kibler [6], and Sih and Dobreff [7]. Their results covered a significantly wider range of the curvature parameter. Problems of cracks in cylindrical shells have also been treated by Folias [8, 9], Copley and Sanders [10], and Erdogan and Ratwani [11].

One of the shortcomings inherent in all of the foregoing works, which are based on the classical shallow shell theory of Reissner [12], is that the boundary conditions on the crack surfaces are only satisfied approximately in the Kirchhoff sense. The drawback gives rise to a difference in the angular distribution of the extensional and bending stress fields near the crack tip*. This fundamental difficulty was overcome by Sih and Hagendorf [13]. They developed a tenth-order system of shell equations which accounted for the effect of transverse shear. The new results satisfy five individual boundary conditions on a free edge of the shell and acquired a three dimensional character in the crack front stresses resembling those found by Sih [1] for a through crack in a thick plate.

This chapter will be concerned with a tenth-order system of shell equations applied to solve shell problems with cracks. Fourier transform is employed reducing each problem to the solution of a system of coupled Fredholm integral equations of the second kind. Both symmetric and skew-symmetric loadings are considered and the resulting stress solutions can differ substantially from those based on the classical theory of Reissner [12].

6.2   Shell theory with shear deformation

The fundamental equations of the three-dimensional theory of elasticity involve overwhelming difficulties when applied to solve the shell problem unless simplifying assumptions are invoked. A reasonable description of the behavior of elastic shells can be made by assuming that the shell thickness is everywhere at least one order of magnitude smaller than the radius of curvature of the reference surface. On this basis, approximate theories of thin elastic shells may be developed and they generally can differ from one another essentially in the formulation of appropriate stress-strain relations.

A theory of thin elastic isotropic shells which accounts for transverse

---

* Since the extensional and bending stress intensity factors can no longer be conveniently combined, special caution should be exercised when the results are used to analyze failure.

shear deformation and normal stress was derived by Naghdi [14]. He obtained the stress-strain relations by applying Reissner's [15] variational theorem of elasticity. Kalnins [16] has also proposed a shear deformation theory of shells. In order to obtain a system of shell equations, which are amenable to effective solutions of cracks or cavities in shells, Sih and Hagendorf [13] developed a theory that satisfies five physical boundary conditions on an edge of the shell. This general theory of elastic shells is based on the following three assumptions:

(1) The displacement vector is linear in the normal coordinate curve to the middle surface,

(2) the distance between points on a normal to the undeformed middle surface do not change during deformation, and

(3) the stresses are replaced by a system of stress resultants and stress couples.

The stress-strain relations are deduced for an elastic orthotropic material satisfying a generalized Hooke's law that is consistent with the restriction imposed on the kinematics of deformation by assumption (2). The resulting system of equations are invariant with respect to rigid-body motion.

Shallow shell equations. A shell is said to be shallow (i.e., slightly curved) if the middle surface is sufficiently smooth and all points on this surface are sufficiently close to some plane. In a theory of shallow shells, the material particles on the middle surface are located by the Cartesian coordinates of their projection on the XY-plane, i.e.,

$$X = x, \ Y = y, \ Z = f(x, y) \tag{6.1}$$

in which x and y are the curvilinear coordinates and $f(x, y)$ is the equation of the middle surface of the shell. The points in the shell are described by (x, y, z) with reference to the undeformed middle surface, where z is the distance measured along the normal to the middle surface and is such that $-h/2 \leq z \leq h/2$. Here, h stands for the thickness of the shell as shown in Figure 6.1.

The theory of shallow shells is based on three simplifying assumptions [17]:

(1) Neglect the squares of the derivatives $\dfrac{\partial f}{\partial x}, \dfrac{\partial f}{\partial y}$ and their products in comparison to unity.

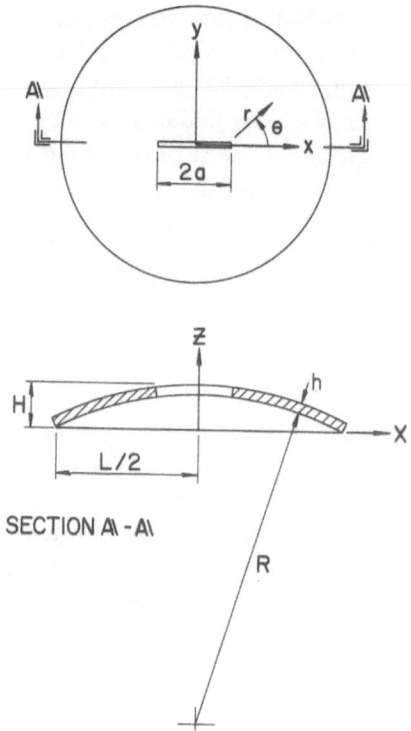

Figure 6.1. Spherical Cap with a Crack

(2) Neglect the transverse shear resultants Q_x, Q_y in the first two equations
of equilibrium.
(3) Neglect the tangential displacements u_x, u_y in the expressions for the
transverse shearing strains γ_{xz}, γ_{yz}.
In what follows, only the first assumption will be incorporated:

$$\left(\frac{\partial f}{\partial x}\right)^2 << 1, \left(\frac{\partial f}{\partial y}\right)^2 << 1, \frac{\partial f}{\partial x}\frac{\partial f}{\partial y} << 1 \tag{6.2}$$

Let the reference surface be the middle surface of a shallow spherical shell
of constant thickness, h. Then the principal radii of curvature are $R_x = R_y = = R$. The shell strain-displacement equations are

$$\varepsilon_{xx} = \frac{\partial u_x}{\partial x} + \frac{u_z}{R} \tag{6.3a}$$

$$\varepsilon_{yy} = \frac{\partial u_y}{\partial y} + \frac{u_z}{R} \tag{6.3b}$$

$$\gamma_{xy} = \frac{\partial u_x}{\partial y} + \frac{\partial u_y}{\partial x} \tag{6.3c}$$

$$\gamma_{xz} = \frac{\partial u_z}{\partial x} - \frac{u_x}{R} + \beta_x \tag{6.3d}$$

$$\gamma_{yz} = \frac{\partial u_z}{\partial y} - \frac{u_y}{R} + \beta_y \tag{6.3e}$$

in which u_x, u_y are the displacements tangential to the x and y coordinate lines and u_z is the displacement in the direction of the normal. The quantity β_x is the angle through which the normal **n** rotates in the direction of the x-coordinate line. A corresponding interpretation holds for β_y. The curvature relations can be written as

$$\kappa_{xx} = \frac{\partial \beta_x}{\partial x}, \ \kappa_{yy} = \frac{\partial \beta_y}{\partial y}, \ \kappa_{xy} = \frac{\partial \beta_x}{\partial y} + \frac{\partial \beta_y}{\partial x} \tag{6.4}$$

It follows that the equations of equilibrium which express the vanishing of the *force stress-resultants* at each point of the middle surface are

$$\frac{\partial N_{xx}}{\partial x} + \frac{\partial N_{yx}}{\partial y} + \frac{Q_x}{R} = 0 \tag{6.5a}$$

$$\frac{\partial N_{xy}}{\partial x} + \frac{\partial N_{yy}}{\partial y} + \frac{Q_y}{R} = 0 \tag{6.5b}$$

$$\frac{\partial Q_x}{\partial x} + \frac{\partial Q_y}{\partial y} - \frac{N_{xx} + N_{yy}}{R} = 0 \tag{6.5c}$$

Similarly, the equilibrium equations which express the vanishing of the *moment stress-resultants* at every point of the middle surface take the forms

$$\frac{\partial M_{xx}}{\partial x} + \frac{\partial M_{yx}}{\partial y} - Q_x = 0 \tag{6.6a}$$

$$\frac{\partial M_{xy}}{\partial x} + \frac{\partial M_{yy}}{\partial y} - Q_y = 0 \tag{6.6b}$$

$$N_{xy} - N_{yx} + \frac{1}{R} (M_{xy} - M_{yx}) = 0 \tag{6.6c}$$

For a homogeneous and isotropic elastic medium, the stress resultants are expressed in terms of the shell strains as

$$N_{xx} = \frac{2\mu h}{1 - \nu}(\varepsilon_{xx} + \nu\varepsilon_{yy}) \tag{6.7a}$$

$$N_{yy} = \frac{2\mu h}{1 - \nu}(\varepsilon_{yy} + \nu\varepsilon_{xx}) \tag{6.7b}$$

$$N_{xy} = N_{yx} = \mu h\gamma_{xy} \tag{6.7c}$$

and the shear resultants are given by

$$Q_x = \mu h\gamma_{xz}, \; Q_y = \mu h\gamma_{yz} \tag{6.8}$$

where ν is Poisson's ratio and μ the shear modulus of elasticity. The moment stress-resultants M_{xx}, M_{yy} are considered positive when they produce *positive stresses* on the part of the shell above the middle surface (Figures 6.2):

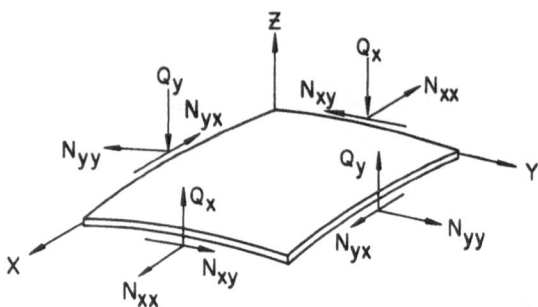

Figure 6.2a. Notations for Stress Resultants and Shear Forces

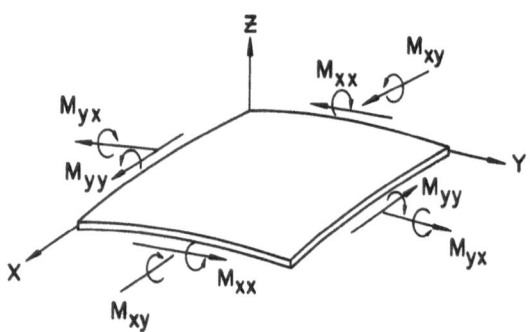

Figures 6.2b. Notations for Bending and Twisting Moments

$$M_{xx} = \frac{\mu h^3}{6(1 - v)} (\kappa_{xx} + v\kappa_{yy}) \tag{6.9a}$$

$$M_{yy} = \frac{\mu h^3}{6(1 - v)} (\kappa_{yy} + v\kappa_{xx}) \tag{6.9b}$$

$$M_{xy} = M_{yx} = \frac{\mu h^3}{12} \kappa_{xy} \tag{6.9c}$$

Equations (6.7) to (6.9) complete the system of equations which represent the mathematical model used for this shallow shell theory. It is obvious from the above relations that the last equation of equilibrium in equations (6.6) is satisfied identically and may be discarded from further consideration. A point to be made here is that since the present theory admits transverse shear strains γ_{xz} and γ_{yz}, the transverse shear forces Q_x and Q_y can be determined in the same way as the other stress resultants. The actual stresses are expressible in terms of the stress resultants as follows:

$$\left(1 + \frac{Z}{R}\right)\sigma_{xx} = \frac{1}{h} N_{xx} + \frac{12Z}{h^3} M_{xx} \tag{6.10a}$$

$$\left(1 + \frac{Z}{R}\right)\sigma_{yy} = \frac{1}{h} N_{yy} + \frac{12Z}{h^3} M_{yy} \tag{6.10b}$$

$$\left(1 + \frac{Z}{R}\right)\tau_{xy} = \frac{1}{h} N_{xy} + \frac{12Z}{h^3} M_{xy} \tag{6.10c}$$

$$\left(1 + \frac{Z}{R}\right)\tau_{xz} = \frac{1}{h} Q_x \tag{6.10d}$$

$$\left(1 + \frac{Z}{R}\right)\tau_{yz} = \frac{1}{h} Q_y \tag{6.10e}$$

The foregoing relations may be further rearranged to express the stress resultants N_{xx}, N_{yy}, etc., directly in terms of the displacements:

$$N_{xx} = \frac{2\mu h}{1 - v} \left\{ \frac{\partial u_x}{\partial x} + v \frac{\partial u_y}{\partial y} + \frac{1 + v}{R} u_z \right\} \tag{6.11a}$$

$$N_{yy} = \frac{2\mu h}{1 - v} \left\{ \frac{\partial u_y}{\partial y} + v \frac{\partial u_x}{\partial x} + \frac{1 + v}{R} u_z \right\} \tag{6.11b}$$

$$N_{xy} = N_{yx} = \mu h \left\{ \frac{\partial u_x}{\partial y} + \frac{\partial u_y}{\partial x} \right\} \tag{6.11c}$$

The moments M_{xx}, M_{yy}, etc., are

$$M_{xx} = \frac{\mu h^3}{6(1-v)} \left\{ \frac{\partial \beta_x}{\partial x} + v \frac{\partial \beta_y}{\partial y} \right\} \tag{6.12a}$$

$$M_{yy} = \frac{\mu h^3}{6(1-v)} \left\{ \frac{\partial \beta_y}{\partial y} + v \frac{\partial \beta_x}{\partial x} \right\} \tag{6.12b}$$

$$M_{xy} = M_{yx} = \frac{\mu h^3}{12} \left\{ \frac{\partial \beta_x}{\partial y} + \frac{\partial \beta_y}{\partial x} \right\} \tag{6.12c}$$

and the transverse shear resultants become*

$$Q_4 = \mu h \left\{ \frac{\partial u_z}{\partial x} - \frac{u_x}{R} + \beta_x \right\}, \quad Q_y = \mu h \left\{ \frac{\partial u_z}{\partial y} - \frac{u_y}{R} + \beta_y \right\} \tag{6.13}$$

Substituting the appropriate quantities in equations (6.11) to (6.13) into equations (6.5) and (6.6) render

$$\frac{\partial^2 u_x}{\partial x^2} + \frac{1-v}{2} \frac{\partial^2 u_x}{\partial y^2} + \frac{1+v}{2} \frac{\partial^2 u_y}{\partial x \partial y} + \frac{3+v}{2R} \frac{\partial u_z}{\partial x} + \frac{1-v}{2R}$$

$$\left\{ \beta_x - \frac{u_x}{R} \right\} = 0 \tag{6.14a}$$

$$\frac{\partial^2 u_y}{\partial y^2} + \frac{1-v}{2} \frac{\partial^2 u_y}{\partial x^2} + \frac{1+v}{2} \frac{\partial^2 u_x}{\partial x \partial y} + \frac{3+v}{2R} \frac{\partial u_z}{\partial y} + \frac{1-v}{2R}$$

$$\left\{ \beta_y - \frac{u_y}{R} \right\} = 0 \tag{6.14b}$$

$$\nabla^2 u_z + \frac{\partial \beta_x}{\partial x} + \frac{\partial \beta_y}{\partial y} - \frac{1}{R} \left(\frac{3+v}{1-v} \right) \left\{ \frac{\partial u_x}{\partial x} + \frac{\partial u_y}{\partial y} \right\} - \frac{4}{R^2}$$

$$\left(\frac{1+v}{1-v} \right) u_z = 0 \tag{6.14c}$$

$$\frac{\partial^2 \beta_x}{\partial x^2} + \frac{1-v}{2} \frac{\partial^2 \beta_x}{\partial y^2} + \frac{1+v}{2} \frac{\partial^2 \beta_y}{\partial x \partial y} - \frac{6(1-v)}{h^2} \left\{ \frac{\partial u_z}{\partial x} + \beta_x - \frac{u_x}{R} \right\}$$

$$= 0 \tag{6.14d}$$

* If Naghdi's [14] general shell equations were used, then the transverse shear resultants in equations (6.13) should be multiplied by the factor 5/6. The other expressions of N_{xx}, N_{yy}, etc., remain the same.

$$\frac{\partial^2 \beta_y}{\partial y^2} + \frac{1-v}{2} \frac{\partial^2 \beta_y}{\partial x^2} + \frac{1+v}{2} \frac{\partial^2 \beta_x}{\partial x \partial y} - \frac{6(1-v)}{h^2} \left\{ \frac{\partial u_z}{\partial y} + \beta_y - \frac{u_y}{R} \right\} = 0$$

$$(6.14e)$$

These coupled partial differential equations constitute the governing equations for the equilibrium of the middle surface of a shallow spherical shell in terms of the five displacement measures.

Introduction of stress and displacement functions. The system of equations (6.14) can be uncoupled by introducing the functions ϕ, ψ and Φ, Ψ such that

$$u_x = \frac{1}{R} \left(\frac{\partial \phi}{\partial x} - \psi \right), \ u_y = \frac{1}{R} \frac{\partial \phi}{\partial y} \qquad (6.15)$$

and

$$Q_x = \mu h \left(\frac{\partial \Phi}{\partial x} - \Psi \right), \ Q_y = \mu h \frac{\partial \Phi}{\partial y} \qquad (6.16)$$

After a considerable amount of algebra, the following system of equations* governing the non-symmetric deformation of a shallow spherical shell are obtained:

$$\nabla^6 u_z + \frac{1}{\lambda^4} \nabla^2 u_z = 0 \qquad (6.17a)$$

$$\nabla^4 \psi - \frac{1}{k^2} \nabla^2 \psi = 0 \qquad (6.17b)$$

$$\nabla^4 \phi_0 = 0 \qquad (6.17c)$$

---

* If all of the three simplifying assumptions stated earlier were used, then the application of equations (6.15) and (6.16) will lead to a different system of uncoupled equations given by

$$\nabla^6 u_z - [2(1+v)/R^2]/\nabla^4 u_z + [12(1-v^2)/h^2 R^2]\nabla^2 u_z = 0$$
$$\nabla^2 \psi = 0, \ \nabla^2 \phi_0 = 0$$
$$\nabla^2 \Psi - (12/h^2)\Psi = 0$$
with the provision that

$$\frac{\partial \psi}{\partial x} = 2 \left\{ u_z + \frac{hR}{2[3(1-v^2)]^{\frac{1}{2}}} \nabla^2 \left[\frac{hR}{2[3(1-v^2)]^{\frac{1}{2}}} \nabla^2 - \frac{h}{R} \left[\frac{1}{3} \left(\frac{1+v}{1-v} \right) \right]^{\frac{1}{2}} \right] u_z \right\}$$

provided that the normal displacement u_z and the functions ψ, ϕ_0 satisfy the conditions

$$\frac{\partial}{\partial x}(\psi - k^2\nabla^2\psi) = \frac{2(1-\varepsilon)}{1-2\varepsilon}(u_z + \lambda^4\nabla^4 u_z) \tag{6.18a}$$

$$\nabla^2\phi_0 = \frac{2\varepsilon}{1-2\varepsilon}(u_z + \lambda^4\nabla^4 u_z) \tag{6.18b}$$

in which k, λ and ε stand for

$$k^2 = \frac{h^2}{12l_1}, \quad \lambda^4 = \frac{h^2R^2}{12(1-v^2)l_2}, \quad \varepsilon = \frac{1+v}{1-v}\frac{\varepsilon_0}{l_3} \tag{6.19}$$

with ε_0 being $h^2/12R^2$. The quantities l_j ($j = 1, 2, 3$) are given by

$$l_1 = 1 + \varepsilon_0, \ l_2 = 1 + \frac{2}{1-v}\varepsilon_0, \ l_3 = 1 + \frac{3+v}{1-v}\varepsilon_0 \tag{6.20}$$

The remaining functions in equations (6.15) and (6.16) can be determined from

$$\Psi = -\nabla^2\psi \tag{6.21a}$$

$$\phi = [\varepsilon(1-\varepsilon)]^{\frac{1}{2}}(1-v)\lambda^2 u_z + \frac{(1+v)l_4}{l_3}\lambda^4\nabla^2 u_z + k^2\frac{\partial\psi}{\partial x} + \phi_0 \tag{6.21b}$$

$$\Phi = 2\varepsilon u_z - 2[\varepsilon(1-\varepsilon)]^{\frac{1}{2}}\lambda^2\nabla^2 u_z - \frac{\partial\psi}{\partial x} + \Phi_0 \tag{6.21c}$$

$$\Phi_0 = (1-2\varepsilon)\frac{\partial}{\partial x}(\psi - k^2\nabla^2\psi) \tag{6.21d}$$

in which

$$l_4 = 1 + \frac{4}{1-v}\varepsilon_0 \tag{6.22}$$

The appearance of the functions ϕ_0, Φ_0 in the expressions for ϕ, Φ respectively result from integrations after manipulating the equilibrium equations (6.14). A close examination of the equilibrium conditions leads to equations (6.18) and the expression for Φ_0 in equations (6.21). Upon defining a new function

$$\chi = u_z - \Phi - \frac{\phi}{R^2} \tag{6.23}$$

and making use of the transformations in equations (6.15) and (6.16), the stress resultants in terms of the normal displacement u_z and the auxiliary functions ϕ, ψ, Φ, Ψ and χ are found:

$$N_{xx} = \frac{2\mu}{1-v} \frac{h}{R} \left\{ \frac{\partial^2 \phi}{\partial x^2} + v \frac{\partial^2 \phi}{\partial y^2} + (1+v)u_z - \frac{\partial \psi}{\partial x} \right\} \tag{6.24a}$$

$$N_{yy} = \frac{2\mu}{1-v} \frac{h}{R} \left\{ \frac{\partial^2 \phi}{\partial y^2} + v \frac{\partial^2 \phi}{\partial x^2} + (1+v)u_z - v \frac{\partial \psi}{\partial x} \right\} \tag{6.24b}$$

$$N_{xy} = N_{yx} = \frac{2\mu h}{R} \left\{ \frac{\partial^2 \phi}{\partial x \partial y} - \frac{1}{2} \frac{\partial \psi}{\partial y} \right\} \tag{6.24c}$$

Similar expressions are obtained for the moments

$$M_{xx} = -\frac{\mu h^3}{6(1-v)} \left\{ \frac{\partial^2 \chi}{\partial x^2} + v \frac{\partial^2 \chi}{\partial y^2} + \frac{\partial}{\partial x} \left(\Psi + \frac{\psi}{R^2} \right) \right\} \tag{6.25a}$$

$$M_{yy} = -\frac{\mu h^3}{6(1-v)} \left\{ \frac{\partial^2 \chi}{\partial y^2} + v \frac{\partial^2 \chi}{\partial x^2} + v \frac{\partial}{\partial x} \left(\Psi + \frac{\psi}{R^2} \right) \right\} \tag{6.25b}$$

$$M_{xy} = M_{yx} = -\frac{\mu h^3}{6} \left\{ \frac{\partial^2 \chi}{\partial x \partial y} + \frac{1}{2} \frac{\partial}{\partial y} \left(\Psi + \frac{\psi}{R^2} \right) \right\} \tag{6.25c}$$

and shear resultants

$$Q_x = \mu h \left\{ \frac{\partial \Phi}{\partial x} - \Psi \right\}, \quad Q_y = \mu h \frac{\partial \Phi}{\partial y} \tag{6.26}$$

The set of equations (6.15) to (6.26) forms a complete system of equations for a theory of shallow spherical shells. The positive conventions for the stress resultants N_{xx}, N_{xy}, N_{yy}, shear forces Q_x, Q_y, and bending moments M_{xx}, M_{xy}, M_{yy} are shown in Figures 6.2a and 6.2b.

A quantitative estimate of the error inherent in this theory of shallow spherical shells resulting from the approximation introduced by equations (6.2) is given by the inequality

$$\frac{L}{R} \leq \left(\frac{8\delta}{1+2\delta} \right)^{\frac{1}{4}}, \quad \delta = \left[1 + \left(\frac{\partial f}{\partial x} \right)^2_{max} \right]^{\frac{1}{2}} - 1 \tag{6.27}$$

where L is the diameter of the plane outline of the shallow spherical cap shown in Figure 6.1, and $\delta \ll 1$ denotes the maximum deviation in approximating the Lamé parameters by unity [17]. For example, a deviation of 5 percent limits $R \geq 1.66\ L$.

6.3   Symmetric loading

Consider a portion of a spherical shell of constant thickness h containing a through crack at the apex of the shell. The shallowness condition requires that the crack of length $2a$ in Figure 6.1 is small in comparison with the radius of curvature R and hence the problem can be solved in the projected XZ-plane. The solution to this problem may be considered as the sum of two separate problems. The first concerns with loads applied along the edge of an uncracked shell. The membrane and moment stresses will be different from zero along the prospective crack site: $y = 0$, $|x| \leq a$. The second problem deals with a cracked shell with tractions specified on the crack surfaces. These tractions are equal and opposite to those found along $y = 0$, $|x| \leq a$ in the first problem. Since the solution for the uncracked shell is straightforward, attention will be focused only on the solution for a cracked shell subjected to surface tractions.

Let the shell in Figure 6.1 be loaded symmetrically with respect to the X-axis along which the line crack is located. From the symmetry conditions, the in-plane shear resultant, transverse shear and twisting moment vanish everywhere along the line $y = 0$, i.e.,

$$N_{xy}(x, 0) = Q_y(x, 0) = M_{xy}(x, 0) = 0 \text{ for all } x \tag{6.28}$$

The boundary conditions on the crack surfaces are

$$N_{yy}(x, 0) = N(x); \quad M_{yy}(x, 0) = M(x), \, |x| < a \tag{6.29}$$

and the continuity requirements are given by

$$\lim_{|y| \to 0} \beta_y = \lim_{|y| \to 0} u_y = 0, \quad |x| > a \tag{6.30}$$

where $u_y(x, y, z) = u_y(x, y) + z\beta_y(x, y)$. Imposed on the solution are also the regularity conditions that the stresses and displacements must be bounded as $(x^2 + y^2)^{\frac{1}{2}} \to \infty$. This implies that the functions u_z, ψ, ϕ_0 and their derivatives must be bounded at infinity. Although these conditions at infinity are not geometrically feasible in the shell problem, they are satisfied within the approximation introduced in the shallow shell theory by equations (6.2).

Integral representations. The integral solutions to the shallow shell equations (6.17) and (6.18) can be obtained by means of the Fourier sine and cosine transforms depending upon whether the transformed function is odd or even in x. Without going into the details, the integral representations for u_z, ψ amd ϕ_0 that possess the appropriate symmetrical behavior with respect to x and y and satisfy the regularity conditions at infinity are

$$u_z(x, y) = \int_0^\infty [(1 - 2\varepsilon) A_1 \exp(-s|y|) + A_5 \exp(-\alpha|y|)$$
$$+ A_3 \exp(-\bar{\alpha}|y|)] \cos(xs)\,ds \qquad (6.31a)$$

$$\psi(x, y) = \int_0^\infty [A_4 \exp(-s|y|) + A_2 \exp(-\beta|y|)] \sin(xs)\,ds \qquad (6.31b)$$

$$\phi_0(x, y) = \int_0^\infty [A_6 + |y| A_7] \exp(-s|y|) \cos(xs)\,ds \qquad (6.31c)$$

For abbreviation, the functions α and β stand for

$$\alpha = \alpha(s) = [s^2 + (i/\lambda^2)]^{\frac{1}{2}}, \beta = \beta(s) = [s^2 + (1/k^2)]^{\frac{1}{2}}$$

in which $i = \sqrt{-1}$ and $\bar{\alpha}$ is the complex conjugate of α. The functions $A_j = A_j(s)$ $(j = 1, 2, \ldots, 7)$ are not independent and are related to each other as

A_1, A_2 are real, $A_3 = \bar{A}_5$,

$$A_4 = 2(1 - \varepsilon)A_1/s, A_7 = -\varepsilon A_1/s \qquad (6.32)$$

It is convenient to define two additional functions A and B such that

$$A = \alpha A_5 + \bar{\alpha} A_3, iB = \bar{\alpha} A_3 - \alpha A_5 \qquad (6.33)$$

The satisfaction of the symmetry conditions in equations (6.24) to (6.26) leads to the following expressions for A and B in terms of $A_1(s)$ and $A_2(s)$

$$A = -(1 - 2\varepsilon)sA_1 - \frac{\beta}{2sk^2} A_2 \qquad (6.34a)$$

$$B = -2[\varepsilon(1 - \varepsilon)]^{\frac{1}{2}}sA_1 - \frac{s\beta}{2[\varepsilon(1 - \varepsilon)]^{\frac{1}{2}}}[1 + (\varepsilon/k^2 s^2)]A_2 \qquad (6.34b)$$

and A_6 is found in the same way:

$$s^2 A_6 = -\{1 - \varepsilon + 2[(1 - \varepsilon)(1 - 2\varepsilon) - (\varepsilon/\varepsilon_0)]k^2 s^2\}A_1$$

$$+ \frac{\beta(\varepsilon + k^2 s^2)}{2\varepsilon_0(1 - \varepsilon)} A_2 \qquad (6.35)$$

Making use of the above relations, the membrane stress resultants, transverse shears, and bending moments may be expresses in terms of only two unknowns $A_1(s)$ and $A_2(s)$. Since the resulting expressions are lengthy, only those quantities which are pertinent to the specification of symmetry and boundary conditions in equations (6.28) and (6.29) will be listed. The stress resultants N_{yy} and N_{xy} are

$$
N_{yy} = \frac{2\mu h}{R} \int_0^\infty \left[\left\{ \left[-\left(\frac{1+v}{1-v} + s\,|\,y\,| \right) \varepsilon + \frac{3+v}{1-v}(1-2\varepsilon)(ks)^2 \right] \right. \right.
$$

$$
\exp\left(-s\,|\,y\,|\right) - \left[\varepsilon + \frac{3+v}{1-v}(1-2\varepsilon)(ks)^2 \right] sF - \left[\varepsilon(1-\varepsilon) \right]^{\frac{1}{2}}
$$

$$
+ \frac{1-2\varepsilon}{1+\varepsilon_0}(1+v)(\lambda s)^2 \Big] sG \Big\} A_1(s) + \left\{ \frac{\varepsilon + (ks)^2}{2\varepsilon_0(1-\varepsilon)} \beta \exp\left(-s\,|\,y\,|\right) \right.
$$

$$
+ \left[1 + (ks)^2 \right] s \exp\left(-\beta\,|\,y\,|\right) - \frac{1}{2(1-\varepsilon)} \left[\frac{3-v-(5-v)\varepsilon}{1-v} \right. \tag{6.36a}
$$

$$
+ \frac{l_4(1-2\varepsilon)}{\varepsilon_0}(ks)^2 \Big] s\beta F - \frac{1}{2[\varepsilon(1-\varepsilon)]^{\frac{1}{2}}} \left[\frac{\varepsilon}{(ks)^2} + \frac{1+v-(1+3v)\varepsilon}{1-v} \right.
$$

$$
- (1-2\varepsilon)(ks)^2 \Big] s\beta G \Big\} A_2(s) \Big] \cos\left(xs\right) ds
$$

$$
N_{xy} = \frac{2\mu h}{R} \int_0^\infty \left[\left\{ \left[-\varepsilon s\,|\,y\,| + \frac{3+v}{1-v}(1-2\varepsilon)(ks)^2 \right] \exp\left(-s\,|\,y\,|\right) \right. \right.
$$

$$
- \frac{3+v}{1-v}(1-2\varepsilon)(ks)^2 F^* - \frac{1-2\varepsilon}{1+\varepsilon_0}(1+v)(\lambda s)^2 G^* \Big\} A_1(s)
$$

$$
+ \left\{ \frac{\varepsilon + (ks)^2}{2\varepsilon_0(1-\varepsilon)} \exp\left(-s\,|\,y\,|\right) + \left[\frac{1}{2} + (ks)^2 \right] \exp\left(-\beta\,|\,y\,|\right) \right.
$$

$$
- \frac{1-2\varepsilon}{2(1-\varepsilon)} \left[\frac{2}{1-v} + \frac{l_4}{\varepsilon_0}(ks)^2 \right] F^*
$$

$$
- \frac{1-2\varepsilon}{2[\varepsilon(1-\varepsilon)]^{\frac{1}{2}}} \left[\frac{1+v}{1-v} - (ks)^2 \right] G^* \Big\} \beta A_2(s) \Big] \sin\left(xs\right) ds \tag{6.36b}
$$

and the moment expressions M_{yy} and M_{xy} take the forms

$$
M_{yy} = 2\mu h \int_0^\infty \left[\left\{ -\left[\left(1 + \frac{2\varepsilon_0}{1-v} \right) \varepsilon + \varepsilon \varepsilon_0\, s\,|\,y\,| \right. \right. \right.
$$

$$
+ (1-2\varepsilon)(ks)^2 \Big] \exp(-s\,|\,y\,|) + \left[\varepsilon + (1-2\varepsilon)(ks)^2 \right] sF
$$

$$+\left[\left[\varepsilon(1-\varepsilon)\right]^{\frac{1}{2}}-(1-v)\varepsilon(\lambda s)^2\right]sG\Big\}A_1(s)$$

$$+\left\{\frac{\varepsilon+(ks)^2}{2(1-\varepsilon)}\beta\exp(-s\,|\,y\,|)-[1+(ks)^2]s\exp(-\beta\,|\,y\,|)\right.$$

$$+\frac{1}{2(1-\varepsilon)}[2-3\varepsilon+(1-2\varepsilon)(ks)^2]s\beta F$$

$$+\frac{1}{2[\varepsilon(1-\varepsilon)]^{\frac{1}{2}}}\left[\frac{\varepsilon}{(ks)^2}+\varepsilon-(1-2\varepsilon)(ks)^2\right]s\beta G\Big\}A_2(s)\bigg]\cos(xs)ds$$

$$\tag{6.37a}$$

$$M_{xy}=2\mu h\int_0^\infty\bigg[\{-[\varepsilon\varepsilon_0\,s\,|\,y\,|+(1-2\varepsilon)(ks)^2]\exp(-s\,|\,y\,|)$$

$$+(1-2\varepsilon)(ks)^2F^*-(1-v)\varepsilon(\lambda s)^2G^*\}A_1(s)$$

$$+\left\{\frac{\varepsilon+(ks)^2}{2(1-\varepsilon)}\exp(-s\,|\,y\,|)-[\tfrac{1}{2}+(ks)^2]\exp(-\beta\,|\,y\,|)\right.$$

$$+\frac{1-2\varepsilon}{2(1-\varepsilon)}[1+(ks)^2]F^*-\tfrac{1}{2}(1-v)(\lambda s)^2G^*\Big\}\beta A_2(s)\bigg]\sin(xs)ds$$

$$\tag{6.37b}$$

The transverse shear resultant is

$$Q_y=2\mu h\int_0^\infty\bigg\{\left[\varepsilon\exp(-s\,|\,y\,|)-\varepsilon F^*-[\varepsilon(1-\varepsilon)]^{\frac{1}{2}}G^*\right]sA_1(s)$$

$$+\frac{1}{2}\left[\exp(-\beta\,|\,y\,|)-F^*-\left(\frac{\varepsilon}{1-\varepsilon}\right)^{\frac{1}{2}}\left(\frac{1}{k^2s^2}+1\right)G^*\right]s\beta A_2(s)\bigg\}$$

$$\cos(xs)ds \tag{6.38}$$

In equations (6.36) to (6.38), the functions F and G stand for

$$F(|\,y\,|;s)=\frac{1}{2}\left[\frac{1}{\bar\alpha}\exp(-\bar\alpha\,|\,y\,|)+\frac{1}{\alpha}\exp(-\alpha\,|\,y\,|)\right] \tag{6.39a}$$

$$G(|\,y\,|;s)=\frac{1}{2i}\left[\frac{1}{\bar\alpha}\exp(-\bar\alpha\,|\,y\,|)-\frac{1}{\alpha}\exp(-\alpha\,|\,y\,|)\right] \tag{6.39b}$$

while F^* and G^* are given by

$$F^*(|\,y\,|;s)=\frac{1}{2}[\exp(-\bar\alpha\,|\,y\,|)+\exp(-\alpha\,|\,y\,|)] \tag{6.40a}$$

$$G^*(|y|; s) = \frac{1}{2i} \left[\exp(-\bar{\alpha}|y|) - \exp(-\alpha|y|) \right] \tag{6.40b}$$

Similar expressions for the remaining stress resultants, bending moments and transverse shear may also be obtained.

Fredholm integral equations.   The crack problem will now be reduced to a system of dual integral equations which can be solved numerically. By substituting the appropriate stress resultants, shear resultant and moments into equations (6.28) and (6.29), it can be easily shown that the unknowns $A_1(s)$ and $A_2(s)$ are governed by

$$\left. \begin{array}{l} \displaystyle\int_0^\infty \frac{1}{s} A_1(s) \cos(xs) ds = 0 \\[2ex] \displaystyle\int_0^\infty \frac{\beta}{s} A_2(s) \cos(xs) ds = 0 \end{array} \right\} |x| > a \tag{6.41a}$$

and

$$\left. \begin{array}{l} \displaystyle\int_0^\infty [a_{11}(s) A_1(s) + a_{12}(s) \beta A_2(s)] \cos(xs) ds = \frac{M(x)}{2\mu h} \\[2ex] \displaystyle\int_0^\infty [a_{21}(s) A_1(s) + a_{22}(s) \beta A_2(s)] \cos(xs) ds = \frac{RN(x)}{2\mu h} \end{array} \right\} |x| < a \tag{6.41b}$$

The functions $a_{ij}(s)$ $(i, j = 1, 2)$ depend on the various physical parameters are given by equations (A 6.1) in the Appendix. Equations (6.41) can be satisfied by the following representations of $A_j(s)$ $(j = 1, 2)$:

$$A_1(s) = \frac{2\sqrt{as}}{Ehl_s} \int_0^a \sqrt{t}\, \phi_1(t) J_0(st) dt \tag{6.42a}$$

$$A_2(s) = \frac{4\sqrt{as}}{Eh\beta} \int_0^a \sqrt{t}\, \phi_2(t) J_0(st) dt \tag{6.42b}$$

where J_0 is the zero order Bessel function of the first kind and

$$l_s = 1 + \frac{1+\nu}{1-\nu} \varepsilon \tag{6.43}$$

The newly introduced functions $\phi_j(t)$ $(j = 1, 2)$ are assumed continuous over the interval $[o, a]$ and required to satisfy the condition

$$\lim_{t \to 0} \left[t^{-\frac{1}{2}} \phi_j(t) \right] = 0, \quad j = 1, 2$$

A further change of the unknown functions by expressing $\phi_j(\xi)$ in terms of $\Phi_j(\xi)$ for $j = 1, 2$ according to

$$- \phi_1(\xi) = \frac{R\Phi_1(\xi) + (h/6)\Phi_2(\xi)}{1 + \varepsilon_0} \tag{6.44a}$$

$$- \phi_2(\xi) = \frac{\varepsilon_0 R\Phi_1(\xi) - (h/6)\Phi_2(\xi)}{1 + \varepsilon_0} \tag{6.44b}$$

it is found that the functions $\Phi_j(\xi)$ $(j = 1, 2)$ are solutions to the following system of coupled Fredholm integral equations

$$\Phi_1(\xi) + \int_0^1 \left[K_{11}(\xi, \eta) \, \Phi_1(\eta) + \frac{1}{3} \frac{h}{R} \, K_{12}(\xi, \eta)\Phi_2(\eta) \right] d\eta$$

$$= \frac{2\sqrt{\xi}}{\pi} \int_0^\xi \frac{N(ax)}{(\xi^2 - x^2)^{\frac{1}{2}}} \, dx \tag{6.45a}$$

$$\Phi_2(\xi) + \int_0^1 \left[\frac{h}{R} \, K_{21}(\xi, \eta)\Phi_1(\eta) + K_{22}(\xi,\eta)\Phi_2(\eta) \right] d\eta$$

$$= \frac{2\sqrt{\xi}}{\pi} \int_0^\xi \frac{(6/h) \, M \, (ax)}{(\xi^2 - x^2)^{\frac{1}{2}}} \, dx \tag{6.45b}$$

where $0 < \xi \le 1$ and the symmetric kernels are given by

$$K_{ij}(\xi, \eta) = (\xi\eta)^{\frac{1}{2}} \int_0^\infty sc_{ij}(s/a)J_0(\xi s)J_0(\eta s)ds, \, 0 < \xi \le 1; 0 < \eta \le 1 \tag{6.46}$$

The quantities c_{ij} $(i, j = 1, 2)$ are given in equations (A 6.3) of the Appendix. Equations (6.45) can be solved numerically for $\Phi_j(\xi)$ $(j = 1, 2)$ once the stress resultant $N(x)$ and bending moment $M(x)$ on the crack surfaces are specified.

Stress intensity factor. The stress intensity factor may be extracted from the singular contributions of the stress resultants and moments given by the improper integrals in equations (6.36) and (6.37). These integrals are convergent everywhere except at the crack tips $x = \pm a$ and $y = 0$ at which points they become unbounded. To obtain the singular terms, equations (6.42) will first be integrated by parts giving

$$A_1(s) = \frac{2a}{Ehl_s} \left\{ \phi_1(1)J_1(as) - \int_0^1 \xi J_1(as\xi) \frac{d}{d\xi} \left[\frac{\phi_1(\xi)}{\sqrt{\xi}} \right] d\xi \right\} \tag{6.47a}$$

$$\beta(s)A_2(s) = \frac{4a}{Eh} \left\{ \phi_2(1)J_1(as) - \int_0^1 \xi J_1(as\xi) \frac{d}{d\xi} \left[\frac{\phi_2(\xi)}{\sqrt{\xi}} \right] d\xi \right\}$$

(6.47b)

and then the results can be inserted into equations (6.36) and (6.37). A solution near the crack may be obtained by retaining only the first term in the above expressions for A_1 and A_2, and expanding the integrands for large values of the argument s.

The near field solution is obtained by retaining only the first term in the above expressions for $A_1(s)$ and $A_2(s)$. Let r and θ denote the polar coordinates measured from the line of expected crack extension, the positive X-axis in Figure 6.1. For symmetric loading case, the asymptotic expansions* for the stress resultants in the neighborhood of the crack tip are found to be (Sih and Hagendorf [13])

$$N_{xx} = -\Phi_1(1)(a/2r)^{\frac{1}{2}} \cos(\theta/2)[1 - \sin(\theta/2)\sin(3\theta/2)] + \dots \quad (6.48a)$$

$$N_{yy} = -\Phi_1(1)(a/2r)^{\frac{1}{2}} \cos(\theta/2)[1 + \sin(\theta/2)\sin(3\theta/2)] + \dots \quad (6.48b)$$

$$N_{xy} = -\Phi_1(1)(a/2r)^{\frac{1}{2}} \cos(\theta/2)\sin(\theta/2)\cos(3\theta/2) + \dots \quad (6.48c)$$

and the moments are

$$M_{xx} = -\frac{h}{6} \Phi_2(1)(a/2r)^{\frac{1}{2}} \cos(\theta/2)[1 - \sin(\theta/2)\sin(3\theta/2)] + \dots$$

(6.49a)

$$M_{yy} = -\frac{h}{6} \Phi_2(1)(a/2r)^{\frac{1}{2}} \cos(\theta/2)[1 + \sin(\theta/2)\sin(3\theta/2)] + \dots$$

(6.49b)

$$M_{xy} = -\frac{h}{6} \Phi_2(1)(a/2r)^{\frac{1}{2}} \cos(\theta/2)\sin(\theta/2)\cos(3\theta/2) + \dots \quad (6.49c)$$

in which the nonsingular terms in r have been neglected.

An important point to be made here is that the angular distribution of the stress resultants and moments around the crack tip are identical. Equations (6.48) and (6.49) differ only in the coefficients or amplitude of the singular stress field determined by $\Phi_1(1)$ and $\Phi_2(1)$. This feature of the solution permits the membrane and moment stress fields to be combined in a natural fashion by equations (6.10) whereas in the classical shell theory

---

* At this point, the quantities ε_0 and ε are neglected in comparison to unity.

[4, 6, 7] the corresponding expressions for M_{xx}, M_{yy} and M_{xy} are incompatible with those for N_{xx}, N_{yy} and N_{xy} in that the θ-dependence for the stress resultants and moments are entirely different. It is obviously more meaningful to consider the combined stress field:

$$\sigma_{xx} = \frac{k_1(z)}{(2r)^{\frac{1}{2}}} \cos(\theta/2) \left[1 - \sin(\theta/2) \sin(3\theta/2)\right] + O(1) \tag{6.50a}$$

$$\sigma_{yy} = \frac{k_1(z)}{(2r)^{\frac{1}{2}}} \cos(\theta/2) \left[1 + \sin(\theta/2) \sin(3\theta/2)\right] + O(1) \tag{6.50b}$$

$$\tau_{xy} = \frac{k_1(z)}{(2r)^{\frac{1}{2}}} \cos(\theta/2) \sin(\theta/2) \cos(3\theta/2) + O(1) \tag{6.50c}$$

$$\tau_{xz} = \tau_{yz} = O(1) \tag{6.50d}$$

The stress intensity factor $k_1(z)$ in equations (6.50) depends on the thickness coordinate z and is defined as

$$k_1(z) = -\left[\frac{\Phi_1(1) + (2z/h)\Phi_2(1)}{1 + (z/R)}\right] \frac{\sqrt{a}}{h} \tag{6.51}$$

It is seen that the effects of shell curvature enter into the local stress field only through the stress intensity factor $k_1(z)$. Furthermore, it is clear that in the shell problem, both the extensional and bending effects are always coupled regardless of the nature of the loading. The numerical results for three specific cases will be discussed subsequently.

Extensional load: $N(x) = N_0$. Let the crack surfaces be opened out by uniform extensional loads of constant magnitude N_0. The loading situation is illustrated in Figure 6.3. Of interest are the stress intensity factors at the top and bottom surface layers $z = \pm h/2$ of the shell. Equation (6.51) will be rewritten for the extensional case as

$$k_1^{(e)}(\pm h/2) = \left[\frac{\psi_1^{(e)}(1) \pm \psi_2^{(e)}(1)}{1 \pm (h/2R)}\right] \frac{N_0 \sqrt{a}}{h} \tag{6.52}$$

where $\psi_j^{(e)}(1)$ are related to $\Phi_j(1)$ ($j = 1, 2$) by

$$\Phi_j(\xi) = -N_0 \psi_j^{(e)}(\xi) \quad \text{for} \quad \xi = 1$$

The numerical values of $k_1^{(e)}$ at the top layer of the shell are shown graphically in Figure 6.3 as a function of $h/2a$ for a Poisson's ratio of $v = 0.3$ and

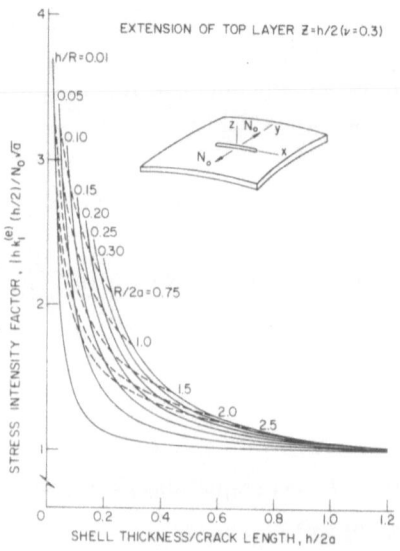

Figure 6.3. Variations of Extensionally-Loaded Stress-Intensity Factor at Outer Layer with Shell Thickness to Crack Length Ratio

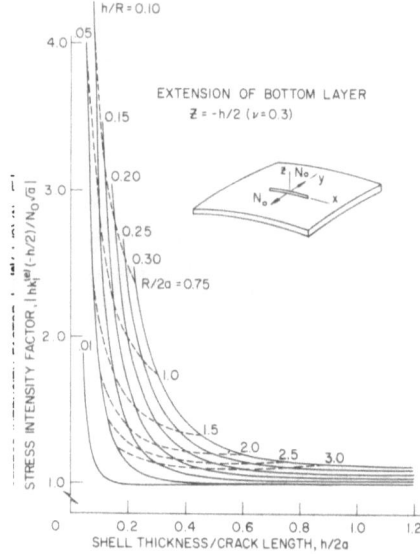

Figure 6.4. Variations of Extensionally-Loaded Stress-Intensity Factor at Inner Layer with Shell Thickness to Crack Length Ratio

h/R ranging from 0.01 to 0.30 inclusive. These curves depict the effect of increasing radius of curvature, R. Note that for a given value of $h/2a$, the normalized stress intensity factor $hk_1^{(e)}(h/2)/N_0\sqrt{a}$ decreases with increasing radius of curvature. This effect is more pronounced as R is increased. The dotted curves in Figure 6.3 represent lines of constant $R/2a$, and hence show the thickness effect. A similar set of curves for $k_1^{(e)}$ at the bottom layer of the shell $z = -h/2$ are given in Figure 6.4. For large values of $R/2a$, say 3, $k_1^{(e)}(-h/2)$ does not change appreciably with the shell thickness to crack length ratio, $h/2a$, i.e., with increasing thickness. A comparison of the stress intensity factors $k_1^{(e)}$ for $z = \pm h/2$ is made in Figure 6.5 for $v = 1/3$ and a

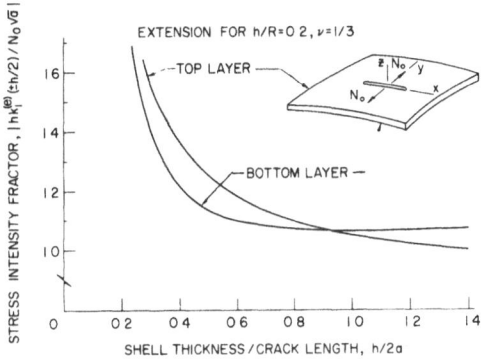

Figure 6.5. Comparison of Extensionally-Loaded Results at Outer and Inner Layer of Shell

particular value of $h/R = 0.2$. For the given loading condition, the $k_1^{(e)}$ value at the top layer is slightly higher than that at the bottom layer. The point of intersection corresponds to the case of a short crack which falls outside the validity of the present theory of thin shells. Figure 6.6 illustrates the quantitative difference in the values of $k_1^{(e)}$ at the inner layer of the shell $z = -h/2$ between the proposed theory and the classical shell theory[6, 7]. The proposed theory predicts larger values of $k_1^{(e)}$ and the difference becomes more pronounced for longer cracks, i.e., as $h/2a \to 0$.

Bending load: $M(x) = M_0$. When the loading on the crack surfaces are of the bending type, say equal and opposite uniform moments with magnitude

---

* This comparison can only be made in the region ahead of the crack, i.e., for $\theta = 0$, as the extensional and bending effects in the classical theory cannot be combined for other values of θ.

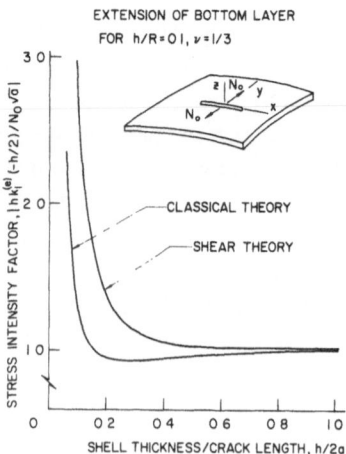

Figure 6.6. Classical Theory versus Shear Theory for Stretching Load

M_0, the results are quite different. The functions $\Phi_j(1)$ in equation (6.51) are equal to

$$\Phi_j(\xi) = \frac{6M_0}{h} \psi_j^{(b)}(\xi) \quad \text{for} \quad \xi = 1$$

while the stress intensity factor for $z = \pm h/2$ becomes

$$k_1^{(b)}(\pm h/2) = \mp \left[\frac{\psi_2^{(b)}(1) \pm \psi_1^{(b)}(1)}{1 \pm (h/2R)}\right] \frac{6M_0}{h} \sqrt{a} \tag{6.53}$$

Under bending, one side of the crack tends to close as it is subjected to compression. The maximum value of $k_1^{(b)}$ will always occur on the tension side. Hence, $k_1^{(b)}(h/2)$ and $k_1^{(b)}(-h/2)$ at the outer and inner surfaces of the shell will correspond to couples $-M_0$ and M_0, respectively. With the tension side at $z = h/2$, Figure 6.7 gives a plot of $h^2 k_1^{(b)}(h/2)/6M_0\sqrt{a}$ versus $h/2a$ for $v = 0.3$ and $h/R = 0.01, 0.05$, etc. Unlike the results for extensional loading, the magnitude of $k_1^{(b)}$ increases monotonically with $h/2a$ for a fixed value of h/R. This behavior was also observed in the flat plate theory by Hartranft and Sih [18]. The dotted curves in Figure 6.7 shows the variations of $k_1^{(b)}$ along the lines of $R/2a = $ constant as the thickness of the shell is changed. Figure 6.8 gives the results for the situation where the tension side is reversed to the inner surface of the shell, $z = -h/2$. By fixing $h/2a < 0.8$, $k_1^{(b)}$ increases with decreasing values of h/R. The opposite effect is seen when

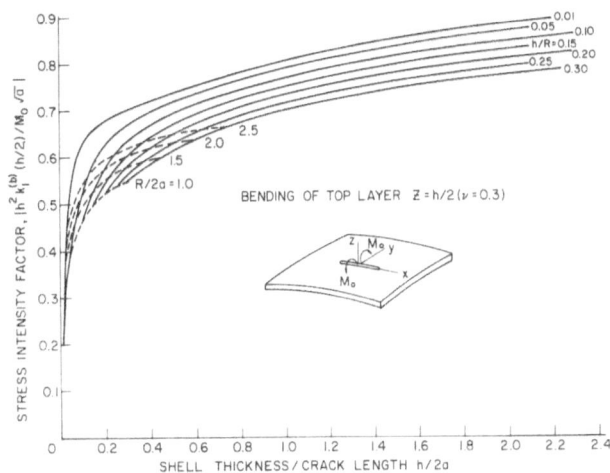

Figure 6.7. Stress-Intensity Factor at Outer Layer Induced by Bending Load

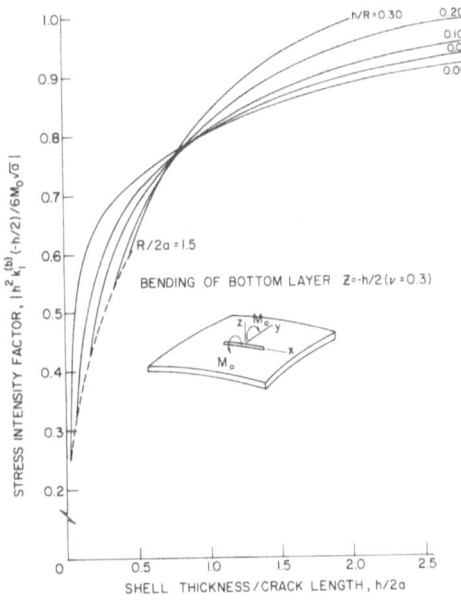

Figure 6.8. Stress-Intensity Factor at Inner Layer Induced by Bending Load

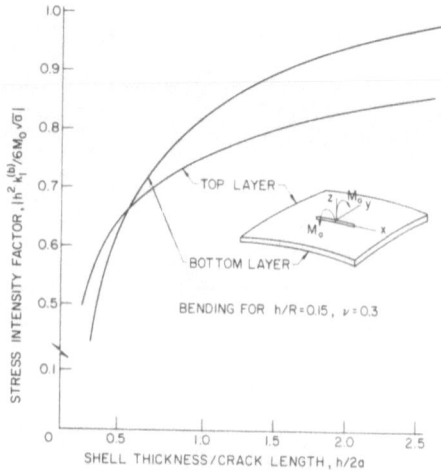

Figure 6.9. Results at Outer and Inner Layer of Shell Subjected to Bending Load

$h/2a$ is approximately greater than 0.8. At this ratio of $h/2a$ the shell is relatively thick in comparison with the crack length $2a$ and the assumption of
thin shell is violated.

Figure 6.9 makes a comparison of $k_1^{(b)}$ for $z = \pm h/2$ with $h/R = 0.15$ and
$v = 0.3$. The two curves intersect at $h/2a \simeq 0.48$ meaning that
$k_1^{(b)}(h/2) > k_1^{(b)}(-h/2)$ for $h/2a < 0.48$ and $k_1^{(b)}(h/2) < k_1^{(b)}(-h/2)$ for

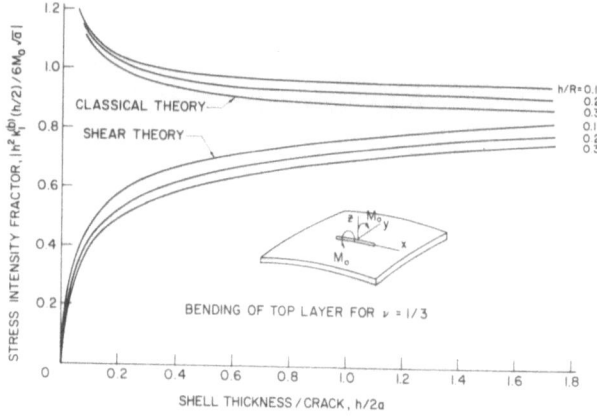

Figure 6.10. Comparison of Classical and Shear Deformation Theory for Bending Load

$h/2a > 0.48$. It should be borne in mind that these results correspond to two different loading conditions, namely applied moments $\pm M_0$, i.e., opposite in sign.

A dramatic departure of the present results from those obtained by the classical theory is shown in Figure 6.10 where the trend of the variations of $k_1^{(b)}$ with $h/2a$ is, in fact, opposite. The difference becomes more and more significant as the ratio $h/2a$ decreases. The results based on the classical theory is known to be approximate since the boundary conditions on the crack surfaces are satisfied only in the Kirchhoff sense. A more detailed discussion on the approximate nature of the classical shell theory has been given by Sih and Dobreff [7].

Combined loading: $6M_0/h^2 = N_0/h = \sigma$. In general, both extensional and bending loads may be present and the total normal tractions applied to the crack surfaces consists of $N_{yy}(x, o)$ and that due to the couple $M_{yy}(x, o)$. Consider the special case where the crack surfaces are not permitted to come into contact. This condition can be satisfied by taking

$$|-N_0/h| \geq |\pm 6M_0/h^2|$$

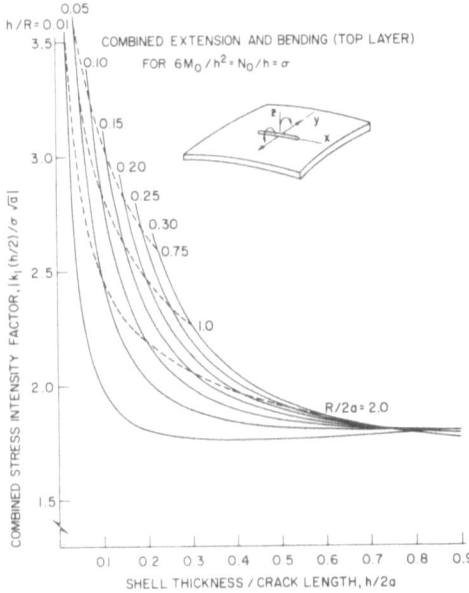

Figure 6.11. Results at Outer Layer for Combined Extensional and Bending Loads

where both N_0 and M_0 are constants. According to the above equation, there arises two possible conditions of combined loading which are shown in Figures 6.11 and 6.12.

In Figure 6.11, the maximum tension occurs at the outer layer of the shell, $z = h/2$ with the combined stress intensity factor $k_1(h/2)$. For convenience, let the crack surfaces be subjected to the combined load such that

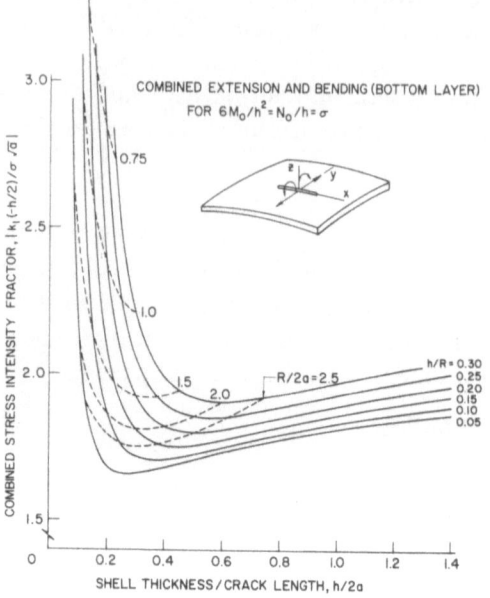

Figure 6.12. Results at Inner Layer for Combined Extensional and Bending Loads

$|-N_0/h| = |\pm 6M_0 h^2|$. A plot of $k_1(h/2)/\sigma\sqrt{a}$ versus $h/2a$ is shown in Figure 6.11 and shows that $k_1(h/2)$ is reduced sharply with decreasing crack length. Curves for $h/R = 0.01$ to 0.30 inclusive are given. By following the dotted curve for $R/2a = 2$, it can be seen that an increase in the shell thickness from $h = 0.04a$ to $h = 0.2a$ corresponds to a reduction in the stress intensity factor of approximately 45 percent. If the direction of the applied couple is reversed, then the side of maximum tension switches to the inner layer of the shell at $z = -h/2$. The numerical results for $k_1(-h/2)$ are shown graphically in Figure 6.12. In the range $1.5 \leq R/2a \leq 2.5$, $k_1(-h/2)$ first decreases reaching a minimum at $h \simeq 0.6a$ and then increases as $h/2a$ is raised.

Appendix: Integrand and Kernel Functions

The quantities $a_{ij}(s)$ in the dual integral equations (6.41) are complicated functions of the geometric parameters of the cracked spherical shell and their expressions are listed below. In addition, $c_{ij}(s)$ ($j = 1, 2$) in equations (6.46) are given separately in the Appendix so that unnecessary complications in the presentation can be avoided.

A6.1 *Integrand functions*
The integrand functions $a_{ij}(s)$ ($i, j = 1, 2$) in the dual integral equations (6.41) are given by

$$a_{11}(s) = -\left(1 + \frac{2}{1 - v}\, \varepsilon_0\right)\varepsilon - (1 - 2\varepsilon)k^2 s^2$$
$$+ \{\varepsilon + (1 - 2\varepsilon)k^2 s^2\}F_0(\lambda s)$$
$$+ \{[\varepsilon(1 - \varepsilon)]^{\frac{1}{2}} - (1 - v)\varepsilon\lambda^2 s^2\}G_0(\lambda s) \qquad (A6.1a)$$

$$a_{12}(s) = \frac{1}{2}\frac{\varepsilon + k^2 s^2}{1 - \varepsilon} - ks(1 + k^2 s^2)^{\frac{1}{2}}$$
$$+ \frac{1}{2(1 - \varepsilon)}\{2 - 3\varepsilon + (1 - 2\varepsilon)k^2 s^2\}F_0(\lambda s)$$
$$+ \frac{1}{2[\varepsilon(1 - \varepsilon)]^{\frac{1}{2}}}\left[\frac{\varepsilon}{k^2 s^2} + \varepsilon - (1 - 2\varepsilon)k^2 s^2\right]G_0(\lambda s) \qquad (A6.1b)$$

$$a_{21}(s) = -\frac{1 + v}{1 - v}\varepsilon - \varepsilon F_0(\lambda s) + \frac{3 + v}{1 - v}(1 - 2\varepsilon)k^2 s^2\{1 - F_0(\lambda s)\}$$
$$- \left\{[\varepsilon(1 - \varepsilon)]^{\frac{1}{2}} + \frac{1 - 2\varepsilon}{1 + \varepsilon_0}(1 + v)\lambda^2 s^2\right\}G_0(\lambda s) \qquad (A6.1c)$$

$$a_{22}(s) = \frac{1}{2}\frac{\varepsilon + k^2 s^2}{\varepsilon_0(1 - \varepsilon)} + ks(1 + k^2 s^2)^{\frac{1}{2}}$$
$$- \frac{1}{2(1 - \varepsilon)}\left\{\frac{3 - v - (5 - v)\varepsilon}{1 - v} + \frac{l_4(1 - 2\varepsilon)}{\varepsilon_0}k^2 s^2\right\}F_0(\lambda s)$$
$$- \frac{1}{2[\varepsilon(1 - \varepsilon)]^{\frac{1}{2}}}\left\{\frac{\varepsilon}{k^2 s^2} + \frac{1 + v - (1 + 3v)\varepsilon}{1 - v}\right.$$
$$\left. - (1 - 2\varepsilon)k^2 s^2\right\}G_0(\lambda s) \qquad (A6.1d)$$

in which F_0, G_0 are defined

$$F_0(\lambda s) = sF(0; s), \quad G_0(\lambda s) = sG(0; s)$$

leading to

$$F_0(s) = \{[s^2(1 + s^4)^{\frac{1}{2}} + s^4]/2(1 + s^4)\}^{\frac{1}{2}} \tag{A6.2a}$$

$$G_0(s) = s/\{[2(1 + s^4)][(1 + s^4)^{\frac{1}{2}} + s^2]\}^{\frac{1}{2}} \tag{A6.2b}$$

and the parameters k, λ, ε, etc., can be found in equations (6.19) and (6.20).

A6.2 *Integrand Kernel function*

The functions $c_{ij}(s)$ ($i, j = 1, 2$) in the kernel K_{ij} of equation (6.46) stand for

$$c_{11}(s) = \frac{2\varepsilon_0}{1 + \nu}[\{(1 + k^2s^2)^{\frac{1}{2}} + ks\}]^{-2} - \frac{8}{1 - \nu^2}[\varepsilon_0 + k^2s^2][1 - F_0(\lambda s)]$$

$$- \{1 - 2\lambda^2s^2G_0(\lambda s)\} + \frac{4k^2}{(1 - \nu^2)\lambda^2}G_0(\lambda s)$$

$$+ \frac{2\varepsilon_0}{(1 - \nu^2)\lambda^2s^2}G_0(\lambda s) \tag{A6.3a}$$

$$c_{12}(s) = k_{21}(s) = -\frac{1}{1 + \nu}[\{(1 + k^2s^2)^{\frac{1}{2}} + ks\}]^{-2}$$

$$+ \frac{3 - \nu}{1 - \nu^2}[1 - F_0(\lambda s)] - \frac{1}{(1 - \nu^2)\lambda^2s^2}G_0(\lambda s)$$

$$+ \frac{(1 - \nu)\lambda^4s^2}{k^2}[1 - F_0(\lambda s)] - \frac{1}{1 + \nu}\{1 - 2\lambda^2s^2G_0(\lambda s)\}$$

$$- \frac{\lambda^2}{k^2}G_0(\lambda s) \tag{A6.3b}$$

$$c_{22}(s) = \frac{2}{1 + \nu}[\{(1 + k^2s^2)^{\frac{1}{2}} + ks\}]^{-2}$$

$$- \left[\frac{4}{1 + \nu} + \frac{4}{1 - \nu}\varepsilon_0k^2s^2\right][1 - F_0(\lambda s)]$$

$$+ \frac{1 - \nu}{1 + \nu}\{1 - 2\lambda^2s^2G_0(\lambda s)\} + \frac{2}{1 - \nu^2}\frac{1}{\lambda^2s^2}G_0(\lambda s) \tag{A6.3c}$$

References

[1] Sih, G. C., A review of the three-dimensional stress problem for a cracked plate, *International Journal of Fracture Mechanics*, 7, pp. 39–61 (1971).

[2] Ang, D. D., Folias, E. S. and Williams, M. L., The bending stress in a cracked plate on an elastic foundation, *Journal of Applied Mechanics*, 30, pp. 245–251 (1964).

[3] Sih, G. C. and Setzer, D. E., Discussion of The bending stress in a cracked plate on an elastic foundation, *Journal of Applied Mechanics*, 31, pp. 365–367 (1964).

[4] Folias, E. S., The stresses in a cracked spherical shell, *Journal of Mathematics and Physics*, 44, pp. 164–176 (1965).

[5] Knowles, J. K. and Wang, N. M., On the bending of an elastic plate containing a crack, *Journal of Mathematics and Physics*, 39, pp. 223–236 (1960).

[6] Erdogan, F. and Kibler, J. J., Cylindrical and spherical shells with cracks, *International Journal of Fracture Mechanics*, 5, pp. 229–237 (1969).

[7] Sih, G. C. and Dobreff, P. S., Crack-like imperfections in a spherical shell, *Glasgow Mathematical Journal*, 12, pp. 65–88 (1971).

[8] Folias, E. S., An axial crack in a pressurized cylindrical shell, *International Journal of Fracture Mechanics*, 1, pp. 104–113 (1965).

[9] Folias, E. S., A circumferential crack in a pressurized cylindrical shell, *International Journal of Fracture Mechanics*, 3, pp. 1–12 (1967).

[10] Copley, L. G. and Sanders, J. L., A longitudinal crack in a cylindrical shell under internal pressure, *International Journal of Fracture Mechanics*, 5, pp. 113–131 (1969).

[11] Erdogan, F. and Ratwani, M., A circumferential crack in a cylindrical shell under torsion, *International Journal of Fracture Mechanics*, 8, pp. 87–95 (1972).

[12] Reissner, E., On some problems in shell theory, *Structural Mechanics*, Proceedings First Symposium on Naval Structural Machenics, pp. 74–113 (1958).

[13] Sih, G. C. and Hagendorf, H. C., A new theory of spherical shells with cracks, *Thin-Shell Structures: Theory, Experiment, and Design*, Edited by Fung, Y. C. and Sechler, E. E., Prentice-Hall, New Jersey, pp. 519–545 (1974).

[14] Naghdi, P. M., On the theory of thin elastic shells, *Quarterly Applied Mathematics*, 54, pp. 369–380 (1957).

[15] Reissner, E., On a variational theorem in elasticity, *Journal of Mathematics and Physics*, 29, pp. 90–95 (1950).

[16] Kalnins, A., On the derivation of a general theory of elastic shells, *Indian Journal of Mathematics*, 9, pp. 381–425 (1967).

[17] Novozhilov, V. V., *Thin Shell Theory*, Second edition, Noordhoff, Groningen, Netherlands (1964).

[18] Hartranft, R. J. and Sih, G. C., Effect of plate thickness on the bending stress distribution around through cracks, *Journal of Mathematics and Physics*, 47, pp. 276–291 (1968).

G. C. Sih and E. P. Chen

7 | *Dynamic analysis of cracked plates in bending and extension*

7.1 Introduction

The present chapter deals with dynamic loading of flat plates containing a through crack where inertia effects can no longer be neglected. Two types of dynamic-load sources are considered: namely, vibratory and impact. When the applied load is time dependent, waves travel through the plate and result in a complex stress pattern upon striking the crack modeled as a plane of discontinuity in the material. Because of the complexities encountered in the treatment of elastodynamic crack problems in three dimensions, simplifying assumptions are made to relax the system of equations in the theory of elasticity. The classical approach in the development of plate theories [1–3] is to assume that the stress distribution in the thickness direction of the plate is known as a priori. For the elastostatic crack problem, Hartranft and Sih [4] suggested to determine the stress variations through the plate thickness from the plane strain condition. Although their method also applies to elastodynamic crack problems, it will not be discussed here.

One of the most commonly used plate bending theories is identified with the Kirchhoff boundary conditions [1]. In this theory, only two boundary conditions are applied for obtaining the plate deflection that satisfies a fourth order differential equation whereas physical reasoning requires three conditions. This inconsistency has been removed in the Mindlin's [2] theory of plate bending which accounts for the rotatory inertia and shear effects and hence involves the satisfaction of three boundary conditions on the crack surface. In plane extension, Kane and Mindlin [3] have developed a higher order plate theory that couples the extensional motion and the thickness mode vibration. The dynamic stress or moment intensity factor solutions are obtained for these higher order plate theories and are compared with those found from the classical thin plate theory of Kirchhoff and the plane theory of elasticity.

7.2 Classical plate bending theory

In the application of the classical plate bending theory, the crack length and input wave length are assumed to be large in comparison with the plate thickness h. The input waves are generated by a combination of bending moments applied to the plate edge causing the plate to vibrate in the transverse direction. As mentioned earlier, since the classical plate bending theory fails to account for all the physical boundary conditions on the crack surface, the stress distribution in the immediate neighborhood of the crack will naturally be affected. A discussion of this will follow.

Lagrange equation.   Let the coordinate axes x and y be chosen such that they are in the middle plane of the plate and the z-axis is perpendicular to this plane as illustrated in Figure 7.1. It is assumed that the normals of the xy-plane before bending remain as normals of the same plane after bending while the middle or xy-plane is unstrained during loading. Under these considerations, the rectangular displacement components u_x, u_y and u_z become

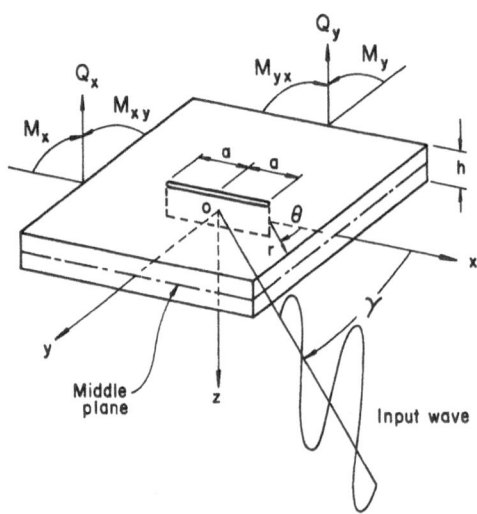

Figure 7.1. Flexural waves impinging on a cracked plate

$$u_x = z \, \frac{\partial w}{\partial x}, \, u_y = z \, \frac{\partial w}{\partial y}, \, u_z = w \tag{7.1}$$

in which w represents the deflection of the middle plane of the plate and it depends on the variables x, y and t. Referring to Figure 7.1, the bending and twisting moments per unit length, M_x, M_y and M_{xy} may be expressed in terms of $w(x, y, t)$ by making use of the stress-strain relations and collecting the moments produced by σ_x, σ_y and τ_{xy} acting on any section of the plate parallel to the xz and yz planes. The results are

$$M_x = D\left(\frac{\partial^2 w}{\partial x^2} + v \, \frac{\partial^2 w}{\partial y^2} \right) \tag{7.2a}$$

$$M_y = D\left(\frac{\partial^2 w}{\partial y^2} + v \, \frac{\partial^2 w}{\partial x^2} \right) \tag{7.2b}$$

$$M_{xy} = (1 - v) D \, \frac{\partial^2 w}{\partial x \partial y} \tag{7.2c}$$

in which

$$D = \frac{Eh^3}{12(1 - v^2)} \tag{7.3}$$

is called the flexural rigidity of the plate. The Young's modulus and Poisson's ratio are denoted by E and v. Similarly, the vertical shearing forces Q_x and Q_y acting on the plate edge due to the shearing stresses τ_{zx} and τ_{zy} may also be obtained:

$$Q_x = D \, \frac{\partial}{\partial x} \left(\frac{\partial^2 w}{\partial x^2} + \frac{\partial^2 w}{\partial y^2} \right) \tag{7.4a}$$

$$Q_y = D \, \frac{\partial}{\partial y} \left(\frac{\partial^2 w}{\partial x^2} + \frac{\partial^2 w}{\partial y^2} \right) \tag{7.4b}$$

By neglecting the rotatory inertia which is of order h^3, the equation of motion yields

$$D\nabla^4 w + \rho h \, \frac{\partial^2 w}{\partial t^2} = 0 \tag{7.5}$$

where ∇^4 is the biharmonic operator in two dimensions and ρ is the mass density of the material.

Incident and scattered waves. Let a plate with a crack of length $2a$ as shown in Figure 7.1 be excited by propagating flexural waves which when encountered by the crack are reflected and refracted, that is, scattered. The source that emits the incident waves corresponds to moments applied symmetrically about the crack plane $y = 0$. The expression for the incident wave may be written as

$$w^{(i)}(x, y, t) = w_0 \cos (\alpha y \sin \gamma) \exp \left[- i(\alpha x \cos \gamma + \omega t) \right] \qquad (7.6)$$

in which w_0 is the amplitude of the input wave and ω is the circular frequency. The angle of incidence γ lies between the limits $-\pi$ and π and is measured from the positive x-axis. In equation (7.6), the parameter

$$\alpha^4 = \frac{\rho h}{D} \omega^2 \qquad (7.7)$$

is determined such that $w^{(i)}$ satisfies equation (7.5). The frequency ω in equation (7.7) corresponds to the Rayleigh-Lamb frequency [5] for long waves, a limitation of the classical plate bending theory.

The total wave field in the plate can be separated into two parts, one associated with incident wave and the other with scattered wave:

$$w(x, y, t) = w^{(i)} (x, y, t) + w^{(s)} (x, y, t) \qquad (7.8)$$

In the same way, the plate deflection, moments and shear forces may also be divided into two parts. For example,

$$M_y(x, y, t) = M_y^{(i)} (x, y, t) + M_y^{(s)} (x, y, t), \text{ etc.} \qquad (7.9)$$

From equation (7.6), all the quantities associated with the incident waves can be easily derived. Hence, the flexure problem is primarily concerned with finding the scattered field.

For a stress-free crack in plate bending, the three quantities M_y, M_{xy} and Q_y should vanish individually for $|x| < a$ and $y = 0$. The classical theory, however, involves only two conditions. This is done by combining M_{xy} and Q_y into a single one, V_y, referred to as the equivalent shear force [1], i.e.,

$$M_y(x, o, t) = V_y(x, o, t) = 0, |x| < a \qquad (7.10)$$

The conditions to be specified on the crack for the scattered field become

$$M_y^{(s)} = - M_y^{(i)}; V_y^{(s)} = - V_y^{(i)} = 0, \quad |x| < a; \quad y = 0 \qquad (7.11)$$

Outside the crack, the problem is symmetric about the x-axis and hence

$$u_y^{(s)} = V_y^{(s)} = 0, \quad |x| \geq a; \quad y = 0 \tag{7.12}$$

The quantities $M_y^{(i)}$ and $V_y^{(i)}$ in equations (7.11) are given by

$$M_y^{(i)}(x, o, t) = -Dw_0 \alpha^2 (\sin^2 \gamma + v \cos^2 \gamma) \exp[-i(\alpha x \cos \gamma + \omega t)] \tag{7.13a}$$

$$V_y^{(i)}(x, o, t) = \left[\frac{\partial M_{xy}^{(i)}}{\partial x} + Q_y^{(i)} \right]_{y=0} = 0 \tag{7.13b}$$

in which $M_{xy}^{(i)}$ and $Q_y^{(i)}$ may be computed from equations (7.2c), (7.4b) and (7.6).

Fredholm integral equations. Application of Fourier transform to the variable x in equation (7.5) leads to an ordinary differential equation in the variable y whose time dependent solution for the scattered wave field takes the form

$$w^{(s)}(x, y, t) = \frac{1}{2\pi} \int_{-\infty}^{\infty} \{A_1(s) \exp[-y(s^2 + \alpha^2)^{\frac{1}{2}}]$$

$$+ A_2(s) \exp[-y(s^2 - \alpha^2)^{\frac{1}{2}}]\} \exp[-i(sx + \omega t)] ds, \quad y \geq 0 \tag{7.14}$$

which satisfies the regularity condition that $w^{(s)} \to 0$ as $y \to \infty$. It follows from equation (7.14) that the conditions in equations (7.11) and (7.12) are satisfied if $A_1(s)$ and $A_2(s)$ are expressed in terms of a single function $A(s)$:

$$\begin{bmatrix} A_1(s) \\ A_2(s) \end{bmatrix} = \frac{A(s)}{2\alpha_2^2} \begin{bmatrix} [(1 - v)s^2 + \alpha^2]/(s^2 + \alpha^2)^{\frac{1}{2}} \\ -[(1 - v)s^2 - \alpha^2]/(s^2 - \alpha^2)^{\frac{1}{2}} \end{bmatrix} \tag{7.15}$$

such that $A(s)$ satisfies the dual integral equations

$$\frac{1}{2\pi} \int_{-\infty}^{\infty} A(s) \exp(-isx) ds = 0, \, |x| \geq a \tag{7.16a}$$

$$\frac{1}{2\pi} \int_{-\infty}^{\infty} f_1(s) A(s) \exp(-isx) ds = \frac{M_0}{D} (\sin^2 \gamma + v \cos^2 \gamma) \exp(-i\alpha x \cos \gamma),$$

$$|x| < a \tag{7.16b}$$

in which

$$f_1(s) = \frac{(s^2 - \alpha^2)^{\frac{1}{2}} [(1 - v) s^2 + \alpha^2]^2 - (s^2 + \alpha^2)^{\frac{1}{2}} [(1 - v) s^2 - \alpha^2]^2}{2\alpha^2 (s^4 - \alpha^4)^{\frac{1}{2}}}$$

The magnitude of the maximum moment generated at the incident wave front is given by $M_0 = Dw_0 \alpha^2$ which is assumed to remain finite as $\omega \to 0$.

Equations (7.16) may be solved by a method described in [6]. Without going into details, a solution is obtained:

$$A(s) = \frac{2\pi M_0 a^2}{(1-v)(3+v)D} (\sin^2 \gamma + v \cos^2 \gamma) \left[\int_0^1 \sqrt{\xi}\, \Phi_1(\xi)\, J_0(sa\xi) d\xi \right.$$

$$\left. + \int_0^1 \sqrt{\xi}\, \Psi_1(\xi)\, J_1(sa\xi) d\xi \right] \tag{7.17}$$

with J_0 and J_1 being the first kind Bessel functions or order zero and one. The functions $\Phi_1(\xi)$ and $\Psi_1(\xi)$ may be calculated numerically from the Fredholm integral equations of the second kind:

$$\Phi_1(\xi) + \int_0^1 L_1(\xi, \eta)\, \Phi_1(\eta) d\eta = \sqrt{\xi}\, J_0(\alpha a\xi \cos \gamma) \tag{7.18a}$$

$$\Psi_1(\xi) + \int_0^1 L_2(\xi, \eta)\, \Psi_1(\eta) d\eta = \sqrt{\xi}\, J_1(\alpha a\xi \cos \gamma) \tag{7.18b}$$

whose kernels L_1 and L_2 are given by the expressions

$$L_1(\xi, \eta) = (\xi\eta)^{\frac{1}{2}} \int_0^\infty s[F_1(s/a) - 1]\, J_0(\xi s)\, J_0(\eta s) ds \tag{7.19a}$$

$$L_2(\xi, \eta) = (\xi\eta)^{\frac{1}{2}} \int_0^\infty s[F_1(s/a) - 1]\, J_1(\xi s)\, J_1(\eta s) ds \tag{7.19b}$$

Note that equations (7.19) is symmetric in the variables ξ and η and $F_1(s)$ stands for

$$F_1(s) = \frac{2}{(1-v)(3+v)} \frac{f_1(s)}{s} \tag{7.20}$$

Once $\Phi_1(\xi)$ and $\Psi_1(\xi)$ are known, the moments and shear forces in the cracked plate can be determined.

Moment intensity factor. The moment and shear force distribution around the crack can be determined from the asymptotic solution expressed in terms of a set of polar coordinates r and θ referred to the crack tip as shown in Figure 7.1 with

$$r^2 = (x - a)^2 + y^2, \theta = \tan^{-1}\left(\frac{y}{x-a}\right) \tag{7.21}$$

Near the crack tip $x = a$ and $y = 0$, the scattered field solution being singular dominates while the incident field contributes at distances away from the crack. Inserting $A(s)$ in equation (7.17) into the appropriate moment expressions, it can be shown that as $r \to 0$

$$M_x = M_x^{(s)} = \frac{K_1}{(2r)^{\frac{1}{2}}} \left(\frac{1-v}{3+v}\right) \left[\frac{1}{2} \sin\theta \sin\frac{3\theta}{2} - \cos\frac{\theta}{2}\right] + O(1) \qquad (7.22a)$$

$$M_y = M_y^{(s)} = \frac{K_1}{(2r)^{\frac{1}{2}}} \left[\frac{1}{2}\left(\frac{1-v}{3+v}\right) \sin\theta \sin\frac{3\theta}{2} - \cos\frac{\theta}{2}\right] + O(1) \qquad (7.22b)$$

$$M_{xy} = M_{xy}^{(s)} = \frac{K_1}{(2r)^{\frac{1}{2}}} \left(\frac{1}{3+v}\right) \left[2 \sin\frac{\theta}{2} + \frac{1-v}{2} \sin\theta \cos\frac{3\theta}{2}\right] + O(1)$$

$$(7.22c)$$

The coefficient K_1 in equations (7.22) is defined as the moment intensity factor

$$K_1 = M_0 \sqrt{a} \, (\sin^2\gamma + v\cos^2\gamma) \left[\Phi_1(1) - i\Psi_1(1)\right] \exp(-i\omega t) \qquad (7.23)$$

Since the stresses σ_x, σ_y and τ_{xy} are related to the moments M_x, M_y and M_{xy} by the relations

$$\sigma_x = \frac{12z}{h^3} M_x, \sigma_y = \frac{12z}{h^3} M_y, \tau_{xy} = \frac{12z}{h^3} M_{xy} \qquad (7.24)$$

the stress intensity factor k_1 that varies linearly with z in the thickness direction of the plate may be obtained by multiplying equation (7.23) by the factor $12z/h^3$. Note that equations (7.22) for M_x, M_y and M_{xy} possess the same inverse square-root of r singularity as in plane elasticity. The angular distribution of the moments, however, is a function of the Poisson's ratio of the material. A physically unrealistic feature of the classical plate bending theory is that the shear forces

$$Q_x = Q_x^{(s)} = \frac{K_1}{(2r)^{3/2}} \left(\frac{2}{3+v}\right) \cos\left(\frac{3\theta}{2}\right) + \dots \qquad (7.25a)$$

$$Q_y = Q_y^{(s)} = \frac{K_1}{(2r)^{3/2}} \left(\frac{2}{3+v}\right) \sin\left(\frac{3\theta}{2}\right) + \dots \qquad (7.25b)$$

have a singularity of order $r^{-3/2}$ as $r \to 0$. This higher order singularity arises from the failure of satisfying all three boundary conditions on the crack surface and can be removed by a theory in which shear deformation and rotatory inertia are included. In subsequent work, it will be shown that Q_x and Q_y actually remain finite as the crack tip $r = 0$ is approached for pure bending loads.

The numerical results can be best displayed by expressing equation (7.23) as

$$K_1 = M_0 \sqrt{a} \, |K_1^{(1)}| \exp[-i\omega(t - \delta_1)] \tag{7.26}$$

in which $K_1^{(1)}$ represents the amplitude of the moment intensity factor and is given by

$$K_1^{(1)} = (\sin^2 \gamma + v \cos^2 \gamma) [\Phi_1(1) - i\Psi_1(1)] \tag{7.27}$$

The phase angle δ_1 between K_1 and incident wave in equation (7.26) is

$$\delta_1 = \frac{\tau}{2\pi} \tan^{-1} \left\{ \frac{\text{Im}[\Phi_1(1)] - \text{Re}[\Psi_1(1)]}{\text{Re}[\Phi_1(1)] + \text{Im}[\Psi_1(1)]} \right\} \tag{7.28}$$

with $\tau = 2\pi/\omega$ being the period of oscillation. In the limit as $\omega \to 0$, the static value of $K_1 = M_0\sqrt{a}$ is recovered [7]. The numerical values of Φ_1 and Ψ_1 in equation (7.27) are obtained from the Fredholm integral equations (7.18) for $v = 0.25$ and flexural waves at normal incidence, i.e., $\gamma = 90°$. Figure 7.2 shows a plot of $|K_1^{(1)}|$ versus frequency ω normalized

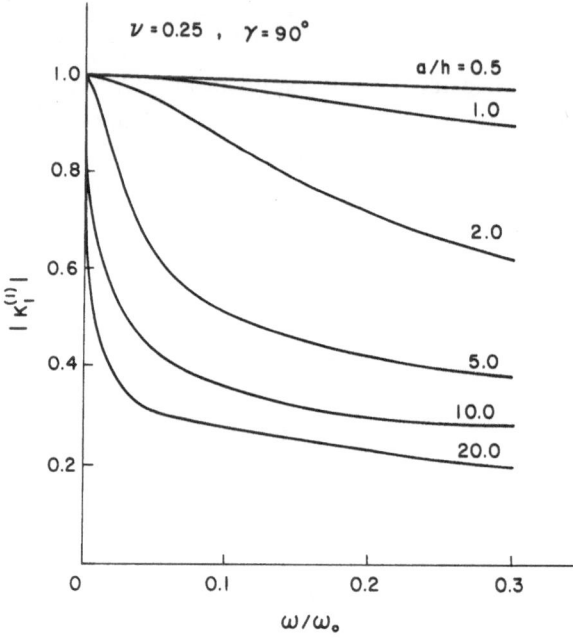

Figure 7.2. Normalized moment intensity factor versus frequency for different a/h values

against the cut-off frequency of the plate, $\omega_0 = \pi c_2/h$, where c_2 is the shear wave velocity of the material. The dynamic moment intensity factor is found to be smaller than the static case and decrease in magnitude as the frequency is increased. This effect becomes more and more pronounced as the crack length to plate thickness ratio increases.

7.3 Mindlin's theory of plate bending

Since the classical plate theory is not expected to be valid for sharp transients or for the frequencies of modes of higher order, Mindlin [2] advanced a higher order theory which includes shear deformation as well as rotatory inertia. The same difference prevails between the Bernoulli-Eulear beam and the Timoshenko beam. For crack problems, it is essential that the three physical boundary conditions of vanishing bending moment, twisting moment, and transverse shear force are satisfied individually on the crack surface.

Governing equations of flexural motions. In the Mindlin's theory [2] of flexural motions of plates, three types of flexural waves are considered: slow flexural, fast flexural and thickness shear waves. In general, any one of these waves propagating toward a crack gives rise to reflected waves of all three types. If the cracked plate is set into steady-state motion by the propagating flexural waves, the rectangular components of the displacement vector may assume the forms

$$u_x(x, y, z, t) = z\psi_x(x, y) \exp(-i\omega t) \tag{7.29a}$$

$$u_y(x, y, z, t) = z\psi_y(x, y) \exp(-i\omega t) \tag{7.29b}$$

$$u_z(x, y, z, t) = \psi_z(x, y) \exp(-i\omega t) \tag{7.29c}$$

In equations (7.29), ω is the circular frequency of the harmonic wave and z is the thickness coordinate. The normal displacement of the plate is ψ_z and the rotations of the normals about the x- and y-axes are denoted by ψ_x and ψ_y as shown in Figure 7.3. From equation (7.29c), it is clear that the normal strain in the thickness direction has been neglected.

With reference to the rectangular coordinate system x, y and z of Figure 7.3, the bending and twisting moments can be expressed in terms of ψ_x, ψ_y and ψ_z as

$$M_x = D\left(\frac{\partial \psi_x}{\partial x} + v \, \frac{\partial \psi_y}{\partial y}\right) \exp(-i\omega t) \tag{7.30a}$$

$$M_y = D\left(\frac{\partial\psi_y}{\partial y} + v\,\frac{\partial\psi_x}{\partial x}\right)\exp(-i\omega t) \tag{7.30b}$$

$$M_{xy} = \left(\frac{1-v}{2}\right) D\left(\frac{\partial\psi_x}{\partial y} + \frac{\partial\psi_y}{\partial x}\right)\exp(-i\omega t) \tag{7.30c}$$

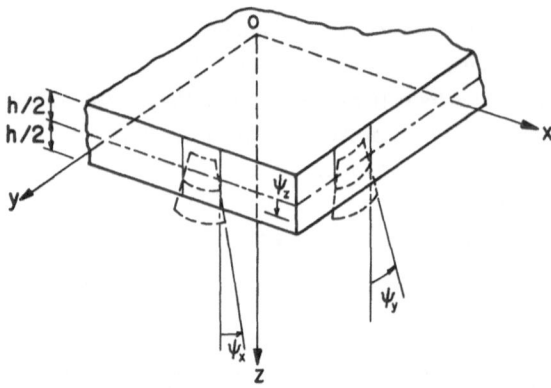

Figure 7.3. Displacement functions in the shear deformation theory of Mindlin

where $D = Eh^3/12(1-v^2)$ is the flexural rigidity of the plate. The shear forces Q_x and Q_y per unit length of the plate are given by

$$Q_x = \kappa^2\mu h\left(\frac{\partial\psi_z}{\partial x} + \psi_x\right)\exp(-i\omega t) \tag{7.31a}$$

$$Q_y = \kappa^2\mu h\left(\frac{\partial\psi_z}{\partial y} + \psi_y\right)\exp(-i\omega t) \tag{7.31b}$$

The shear coefficient κ^2 assumes the value $\pi^2/12$ so that the frequency of the first anti-symmetric mode of thickness-shear vibration matches that of the exact three-dimensional theory. Making use of equations (7.30) and (7.31), the three equations of motion become

$$6D\left[(1-v)\nabla^2\psi_x + (1+v)\frac{\partial}{\partial x}\left(\frac{\partial\psi_x}{\partial x} + \frac{\partial\psi_y}{\partial y}\right)\right]$$

$$- \pi^2\mu h\left(\psi_x + \frac{\partial\psi_z}{\partial x}\right) = -\rho h^3\omega^2\psi_x \tag{7.32a}$$

$$6D\left[(1-v)\nabla^2\psi_y + (1+v)\frac{\partial}{\partial y}\left(\frac{\partial\psi_x}{\partial x} + \frac{\partial\psi_y}{\partial y}\right)\right]$$

$$- \pi^2 \mu h \left(\psi_y + \frac{\partial \psi_z}{\partial y} \right) = - \rho h^3 \omega^2 \psi_y \tag{7.32b}$$

$$\pi^2 \mu h \left(\nabla^2 \psi_z + \frac{\partial \psi_x}{\partial x} + \frac{\partial \psi_y}{\partial y} \right) = - 12 \rho h \omega^2 \psi_z \tag{7.32c}$$

The Laplacian operator ∇^2 in two dimensions is $\partial^2/\partial x^2 + \partial^2/\partial y^2$.

It is convenient to introduce three displacement potentials ϕ_1, ϕ_2 and ϕ_3 such that they are related to ψ_x, ψ_y and ψ_z as follows:

$$\psi_x = (a_2 - 1) \frac{\partial \phi_1}{\partial x} + (a_1 - 1) \frac{\partial \phi_2}{\partial x} + \frac{\partial \phi_3}{\partial y} \tag{7.33a}$$

$$\psi_y = (a_2 - 1) \frac{\partial \phi_1}{\partial y} + (a_1 - 1) \frac{\partial \phi_2}{\partial y} - \frac{\partial \phi_3}{\partial x} \tag{7.33b}$$

$$\psi_z = \phi_1 + \phi_2 \tag{7.33c}$$

in which

$$a_j = \frac{2}{1 - v} \left(\frac{\alpha_j}{\alpha_3} \right)^2, j = 1, 2 \tag{7.34}$$

The wave numbers α_j ($j = 1, 2, 3$) stand for

$$\alpha_1^2 = \frac{1}{2} \left(\frac{\omega}{\omega_0} \right)^2 \left\{ \frac{1}{S} + \frac{1}{R} + \left[\left(\frac{1}{S} - \frac{1}{R} \right)^2 + \frac{4}{RS} \left(\frac{\omega_0}{\omega} \right)^2 \right]^{\frac{1}{2}} \right\} \tag{7.35a}$$

$$\alpha_2^2 = \frac{1}{2} \left(\frac{\omega}{\omega_0} \right)^2 \left\{ \frac{1}{S} + \frac{1}{R} - \left[\left(\frac{1}{S} - \frac{1}{R} \right)^2 + \frac{4}{RS} \left(\frac{\omega_0}{\omega} \right)^2 \right]^{\frac{1}{2}} \right\} \tag{7.35b}$$

$$\alpha_3^2 = \left(\frac{\pi}{h} \right)^2 \left[\left(\frac{\omega}{\omega_0} \right)^2 - 1 \right] \tag{7.35c}$$

where the cut-off frequency is $\omega_0 = \pi c_2/h$ and $c_2 = (\mu/\rho)^{\frac{1}{2}}$ is the shear wave velocity. The rotatory inertia and transverse shear effects are associated with R and S as given by

$$R = \frac{h^2}{12}, S = \frac{12D}{\pi^2 \mu h} \tag{7.36}$$

Substituting equations (7.33) into (7.32), the potentials ϕ_1, ϕ_2 and ϕ_3 that generate slow flexural, fast flexural and thickness shear waves satisfy differential equations of the Helmholtz type:

$$(\nabla^2 + \alpha_j^2) \phi_j(x, y) = 0, j = 1, 2, 3 \tag{7.37}$$

Equations (7.35) show that the three wave numbers α_j ($j = 1, 2, 3$) are dependent on the frequencies; thus, all three flexural waves are dispersive. In addition, α_1 is real for all frequencies, whereas α_2 and α_3 may be real or imaginary depending upon whether ω is greater or smaller than ω_0.

Incident flexural waves. Suppose that the plate with a crack of length $2a$ in Figure 7.1 is set into motion by a slow flexural wave ($\omega < \omega_0$)* arriving from distances far away from the crack. At low frequencies, it is the slow flexural waves that are of interest since the fast flexural and thickness shear waves attenuate gradually from the source of loading. The form of the input flexural wave that results in symmetric bending with reference to the line crack can be written as

$$\phi_1^{(i)} = \phi_0\{\exp\left[-i\alpha_1(x + y)/\sqrt{2}\right] + \exp\left[-i\alpha_1(x - y)/\sqrt{2}\right]\} \exp\left(-i\omega t\right)$$
(7.38a)

$$\phi_2^{(i)} = \phi_3^{(i)} = 0$$
(7.38b)

with ϕ_0 being the amplitude of the input wave. Along the axis of symmetry $y = 0$, the moments of the incident field are

$$M_x^{(i)}(x, 0, t) = (a_2 - 1)\, M_0(1 + v) \exp\left[-i\left(\frac{\alpha_1 x}{\sqrt{2}} + \omega t\right)\right]$$
(7.39a)

$$M_y^{(i)}(x, 0, t) = (a_2 - 1)\, M_0(1 + v) \exp\left[-i\left(\frac{\alpha_1 x}{\sqrt{2}} + \omega t\right)\right]$$
(7.39b)

$$M_{xy}^{(i)}(x, 0, t) = 0$$
(7.39c)

The incident moment $M_0 = \phi_0\alpha_1^2 D$ is assumed to remain finite as $\omega \to 0$. Similarly, the incident shear forces are

$$Q_x^{(i)}(x, 0, t) = -\frac{i\pi^2}{6\sqrt{2}}\, \phi_0\alpha_1 a_2\mu h \exp\left[-i\left(\frac{\alpha_1 x}{\sqrt{2}} + \omega t\right)\right]$$
(7.40a)

$$Q_y^{(i)}(x, 0, t) = 0$$
(7.40b)

The complete solution of the waves as diffracted by the crack is obtained by adding the incident and scattered waves, i.e.,

$$\phi_j(x, y, t) = \phi_j^{(i)}(x, y, t) + \phi_j^{(s)}(x, y, t), \quad j = 1, 2, 3$$
(7.41)

* The case of fast flexural wave $\omega > \omega_0$ may be treated by the same method.

Likewise, the plate displacements, moments and shears can also be found by superposing the incident and scattered parts and the results are obvious. For a traction-free crack, the quantities M_y, M_{yx} and Q_y must each vanish for $|x| < a$ and $y = 0$. This implies that

$$M_y^{(s)} = -M_y^{(i)}, \; M_{yx}^{(s)} = -M_{yx}^{(i)} = 0, \; Q_y^{(s)} = -Q_y^{(i)} = 0 \qquad (7.42)$$

for $|x| < a$ and $y = 0$ while

$$\psi_y^{(s)} = Q_y^{(s)} = M_{yx}^{(s)} = 0, \; |x| > a; \quad y = 0 \qquad (7.43)$$

must hold outside the crack. There remains the determination of the scattered wave field.

Solution of dual integral equation. With the aid of Fourier transform, a solution of equation (7.37) is

$$\phi_j^{(s)}(x, y, t) = \frac{1}{2\pi} \int_{-\infty}^{\infty} B_j(s) \exp\{-[\beta_j y + i(sx + \omega t)]\}ds, \; y \ge 0 \qquad (7.44)$$

where the functions

$$\beta_j = (s^2 - \alpha_j^2)^{\frac{1}{2}} = -i(\alpha_j^2 - s^2)^{\frac{1}{2}}, \; j = 1, 2, 3 \qquad (7.45)$$

represent the branch cuts. Equation (7.44) satisfies the conditions of vanishing displacements, moments and shear forces for the scattered waves sufficiently far away from the crack.

Let the unknowns $B_j(s)$ ($j = 1, 2, 3$) in equations (7.44) be related to a single function $B(s)$ as

$$B_j(s) = (-1)^j \frac{(1 - v)s^2 - \alpha_j^2}{(\alpha_1^2 - \alpha_2^2)\beta_j} B(s), \; j = 1, 2 \qquad (7.46a)$$

$$B_3(s) = -\frac{2is}{\alpha_3^2} B(s) \qquad (7.46b)$$

The conditions in equations (7.42) and (7.43) can then be satisfied if $B(s)$ is governed by the following system of dual integral equations

$$\frac{1}{2\pi} \int_{-\infty}^{\infty} B(s) \exp(-isx)ds = 0, \; |x| \ge a \qquad (7.47a)$$

$$\frac{1}{2\pi} \int_{-\infty}^{\infty} sf_2(s) B(s) \exp(-isx)ds = \frac{2(a_2 - 1)M_0}{(1 - v)D} \exp\left(-\frac{i\alpha_1 x}{\sqrt{2}}\right), \; |x| < a \qquad (7.47b)$$

in which $f_2(s)$ is known:

$$f_2(s) = \frac{2}{(1-v^2)s}\left\{ -\frac{(a_2-1)\left[(1-v)s^2-\alpha_1^2\right]^2}{(\alpha_1^2-\alpha_2^2)\beta_1} \right.$$
$$\left. +\frac{(a_1-1)\left[(1-v)s^2-\alpha_2^2\right]^2}{(\alpha_1^2-\alpha_2^2)\beta_2} -\frac{2(1-v)s^2\beta_3}{\alpha_3^2} \right\} \tag{7.48}$$

By following a procedure outlined in [6], a solution of equations (7.47) is obtained:

$$B(s) = -\frac{2\pi(a_2-1)M_0a^2}{(1-v)D}\left[\int_0^1 \sqrt{\xi}\,\Phi_2(\xi)\,J_0(sa\xi)d\xi \right.$$
$$\left. +\int_0^1 \sqrt{\xi}\,\Psi_2(\xi)\,J_1(sa\xi)d\xi \right] \tag{7.49}$$

The functions Φ_2 and Ψ_2 can be calculated from the Fredholm integral equations

$$\Phi_2(\xi) - \int_0^1 L_3(\xi,\eta)\,\Phi_2(\eta)d\eta = \sqrt{\xi}\,J_0\left(\frac{\alpha_1 a\xi}{\sqrt{2}}\right) \tag{7.50}$$

$$\Psi_2(\xi) - \int_0^1 L_4(\xi,\eta)\,\Psi_2(\eta)d\eta = \sqrt{\xi}\,J_1\left(\frac{\alpha_1 a\xi}{\sqrt{2}}\right) \tag{7.50b}$$

whose symmetric kernels are given by

$$L_3(\xi,\eta) = (\xi\eta)^{\frac{1}{2}}\int_0^\infty s[f_2(s/a)+1]J_0(s\xi)\,J_0(s\eta)ds \tag{7.51a}$$

$$L_4(\xi,\eta) = (\xi\eta)^{\frac{1}{2}}\int_0^\infty s[f_2(s/a)+1]J_1(s\xi)\,J_1(s\eta)ds \tag{7.51b}$$

These equations can be solved numerically for the intensity of the moment distribution around the crack tip.

Crack front moment distribution. The asymptotic behavior of the moment distribution near the crack can be found by introducing the polar coordinates r and θ centered at $x = a$ and $y = 0$ such that

$$x = a + r\cos\theta,\ y = r\sin\theta \tag{7.52}$$

Keeping in mind that the moments of the incident waves are nonsingular, it is only necessary to consider the scattered field for small r:

$$M_x = M_x^{(s)} = \frac{K_1}{(2r)^{\frac{1}{2}}}\cos\frac{\theta}{2}\left(1 - \sin\frac{\theta}{2}\sin\frac{3\theta}{2}\right) + O(1) \tag{7.53a}$$

$$M_y = M_y^{(s)} = \frac{K_1}{(2r)^{\frac{1}{2}}} \cos\frac{\theta}{2}\left(1 + \sin\frac{\theta}{2}\sin\frac{3\theta}{2}\right) + O(1) \tag{7.53b}$$

$$M_{xy} = M_{xy}^{(s)} = \frac{K_1}{(2r)^{\frac{1}{2}}} \cos\frac{\theta}{2}\sin\frac{\theta}{2}\cos\frac{3\theta}{2} + O(1) \tag{7.53c}$$

The dynamic moment intensity factor is

$$K_1 = (1 - a_2)(1 + v) M_0 \sqrt{a}\, [\Phi_2(1) + i\Psi_2(1)] \exp(-i\omega t) \tag{7.54}$$

It is worthwhile noting that the functional dependence of M_x, M_y and M_{xy} on θ is identical with that found by Sih and Loeber [8] for plane extensional waves impinging on the crack. In this connection, the plate may be viewed as an assembly of thin sheets bonded together and each of the sheets is behaving locally at the crack as if it were being stretched and compressed by oscillating membrane forces. Note also that the bending loads yield finite values of Q_x and Q_y at $r = 0$ which is in contrast to the singular solution in the classical plate bending theory.

An alternative way of expressing K_1 in equation (7.54) is

$$K_1 = M_0 \sqrt{a}\, |K_1^{(2)}| \exp[-i\omega(t - \delta_2)] \tag{7.55}$$

in which $K_1^{(2)}$ is given by

$$K_1^{(2)} = (1 - a_2)(1 + v)\, [\Phi_2(1) + i\Psi_2(1)] \tag{7.56}$$

There is a phase angle between K_1 and the input wave which can be computed from

$$\delta_2 = \frac{\tau}{2\pi} \tan^{-1}\left\{\frac{\mathrm{Im}[\Phi_2(1)] + \mathrm{Re}[\Psi_2(1)]}{\mathrm{Re}[\Phi_2(1)] - \mathrm{Im}[\Psi_2(1)]}\right\} \tag{7.57}$$

Recall that the input waves in equation (7.38a) correspond to

$$M_y^{(i)} = -(1 + v)M_0 \cos\left(\frac{\alpha_1 x}{2}\right), \quad M_{yx}^{(i)} = Q_y^{(i)} = 0 \tag{7.58}$$

with $M_y^{(i)}$ being the only non-vanishing moment. Hence, in the limit as $\alpha_1 \to 0$, a state of pure bending with an applied moment of magnitude $(1 + v)M_0$ prevails in the region where the stresses are undisturbed by the crack. Moreover, it can be shown that equation (7.55) reduces to the static result of Hartranft and Sih [9] based on the Reissner theory of plate bending, if $h/\sqrt{10}$ is replaced by h/π.

Displayed in Figure 7.4 is the variations of the moment intensity factor

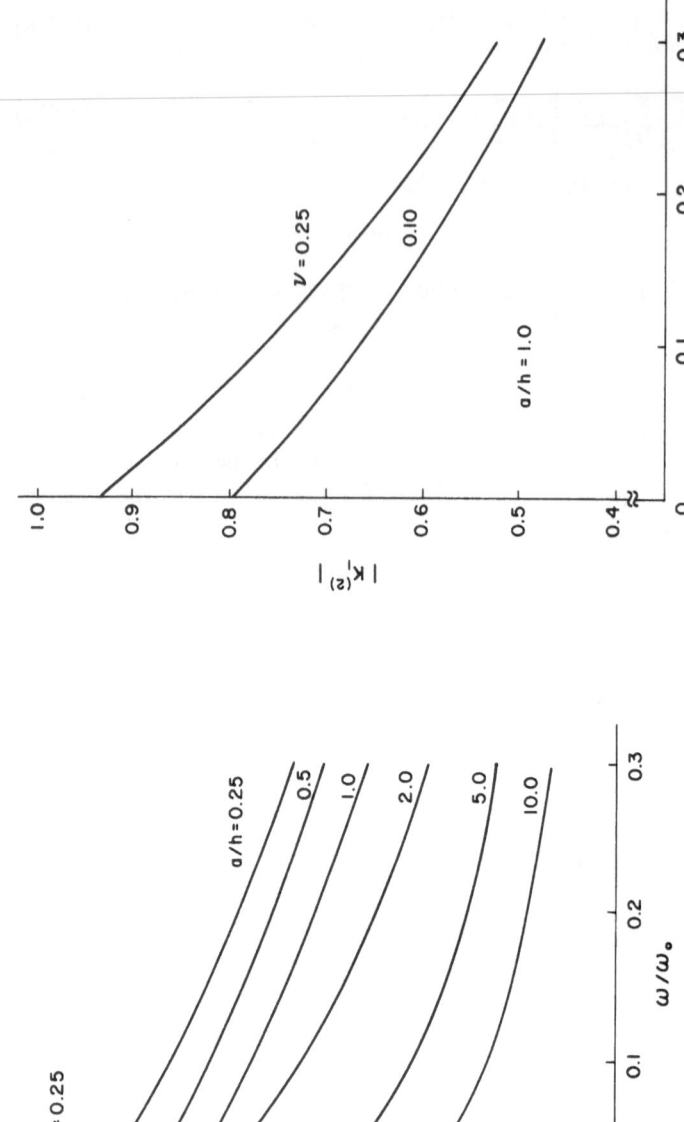

Figure 7.4. Moment inteosity factor as a function of frequency for various ratios of *a/h* in Mindlin's theory

Figure 7.5. Effect of Poisson's ratio on moment intensity factor for *a/h* = 1

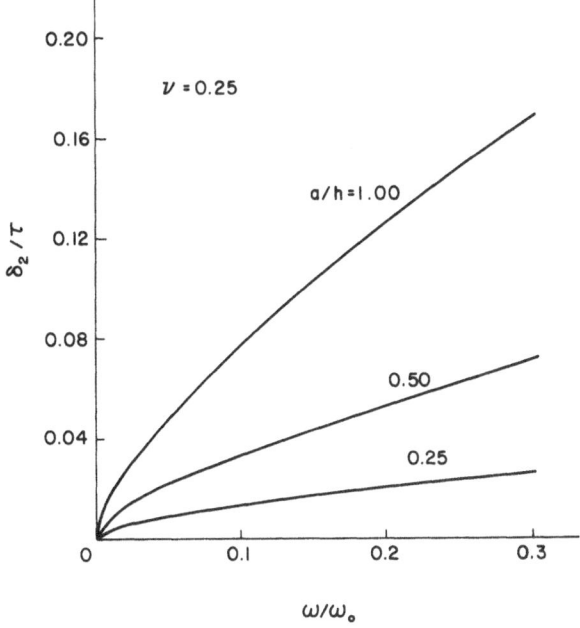

Figure 7.6. Phase angle versus frequency for $a/h = 1.0, 0.5$ and 0.25

$|K_1^{(2)}|$ with ω/ω_0 for different ratio of a/h. All the curves are below the static value of 1.25 and $|K_1^{(2)}|$ decreases more rapidly with frequency when a/h is increased. This general trend is similar to that of the classical plate theory and the corresponding problem of moment concentration for a circular hole [10] in a Mindlin plate. The effect of Poisson's ratio on the moment intensity factor is shown in Figure 7.5 for $a/h = 1.0$. A larger value of v tends to increase $|K_1^{(2)}|$ for a given frequency. Finally, the phase angle δ_2 is plotted against ω for $v = 0.25$ and $a/h = 1.0, 0.5$ and 0.25. For fixed values of ω and a, the phase angle between K_1 and the incident waves is seen to increase as the plate thickness is decreased as shown in Figure 7.6.

7.4 Kane-Mindlin's equation in plate extension

Sih [11] has discussed the two-dimensional problem of plane waves impinging on a through crack. The dynamic equations of plane elasticity are applicable to the extensional vibration of plates provided the wave lengths are large in comparison with the thickness of the plate. In terms of frequencies, the plane elasticity equations yield good results for frequencies which are lower than the frequency of the first mode of thickness vibration of the plate.

In order to overcome some of the foregoing shortcomings of the plane elasticity equations, the higher order extensional vibration theory of Kane and Mindlin [3] will be employed. This theory is also two-dimensional being reduced from the three-dimensional theory of elasticity. However, the reduction is not carried out as far as to arrive at the elastodynamic equations in plane elasticity and does account for the coupling between extensional motions and the first mode of thickness vibration.

Governing equations. Consider a plate bounded by planes $z = \pm h/2$ and a right handed cartesian coordinate system (x, y, z) with the xy-plane lying in the middle plane of the plate as shown in Figure 7.7. The components of displacement are assumed to be approximated by

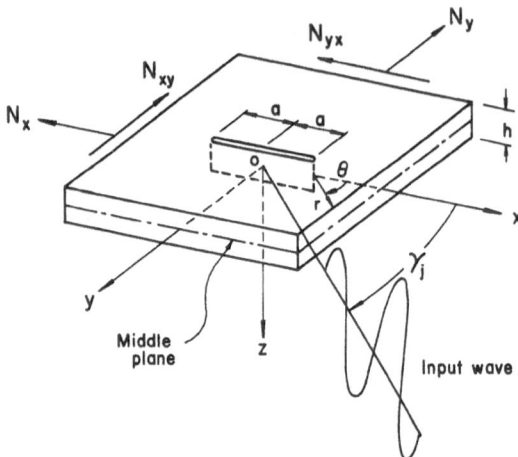

Figure 7.7. Extensional waves striking a through crack

$$u_x(x, y, t) = v_x(x, y) \exp(-i\omega t) \tag{7.59a}$$

$$u_y(x, y, t) = v_y(x, y) \exp(-i\omega t) \tag{7.59b}$$

$$u_z(x, y, t) = \frac{2z}{h} v_z(x, y) \exp(-i\omega t) \tag{7.59c}$$

for incident waves that vary sinusoidally in time with circular frequency ω. In terms of the potentials of the dilatation and rotation in the plane of the plate, the displacement functions v_x, v_y and v_z in equations (7.59) may be expressed as

$$v_x = \frac{\partial \phi}{\partial x} + \frac{\partial \psi}{\partial y} \tag{7.60a}$$

$$v_y = \frac{\partial \phi}{\partial y} - \frac{\partial \psi}{\partial x} \tag{7.60b}$$

$$v_z = e_1 \phi_1 + e_2 \phi_2 \tag{7.60c}$$

where $\phi = \phi_1 + \phi_2$ and

$$e_j = \frac{6h}{\lambda \pi^2} (\lambda + 2\mu) \left[\delta_j^2 - \left(\frac{\omega}{c_1} \right)^2 \right], j = 1, 2 \tag{7.61}$$

The dilatational wave velocity is $c_1 = [(\lambda + 2\mu)/\rho]^{\frac{1}{2}}$ with λ and μ being the Lamé constant and ρ the mass density of the material. The three independent displacement potentials ϕ_1, ϕ_2 and ψ satisfy the equations

$$(\nabla^2 + \delta_j^2)\phi_j = 0, j = 1, 2 \tag{7.62a}$$

$$(\nabla^2 + \delta_3^2)\psi = 0 \tag{7.62b}$$

as derived from the equations of motion. The numbers δ_j ($j = 1, 2, 3$) stand for

$$\delta_j^2 = \frac{1}{2\beta} \left(\frac{\pi}{h} \right)^2 \left[(\alpha + \beta) \left(\frac{\omega}{\bar{\omega}} \right)^2 - 1 - (-1)^j H \right], j = 1, 2 \tag{7.63a}$$

$$\delta_3^2 = (\omega/c_2)^2 \tag{7.63b}$$

The quantities α, β and H are given by

$$\alpha = \frac{(\lambda + 2\mu)^2}{4\mu(\lambda + \mu)}, \beta = \frac{\lambda + 2\mu}{4(\lambda + \mu)} \tag{7.64a}$$

$$H = \left\{ \left[(\alpha + \beta) \left(\frac{\omega}{\bar{\omega}} \right)^2 - 1 \right]^2 + 4\alpha\beta \left(\frac{\omega}{\bar{\omega}} \right)^2 \left[1 - \left(\frac{\omega}{\bar{\omega}} \right)^2 \right] \right\}^{\frac{1}{2}} \tag{7.64b}$$

The frequency

$$\bar{\omega} = \frac{\pi c_1}{h} \tag{7.65}$$

makes the Kane-Mindlin equations applicable to much higher frequencies than those accommodated by the dynamic equations of plane elasticity.

Examination of δ_j ($j = 1, 2, 3$) in equations (7.63) shows that δ_1^2 and δ_3^2 are always larger than zero but δ_2^2 is greater than zero for $\omega/\bar{\omega} > 1$ and less than zero for $\omega/\bar{\omega} < 1$. At the cut-off frequency $\omega = \bar{\omega}$, δ_2^2 is identically zero.

Referring to Figure 7.7, N_x, N_y and N_{xy} are the in-plane forces per unit of length defined as

$$(N_x, N_y, N_{xy}) = \int_{-h/2}^{h/2} (\sigma_x, \sigma_y, \tau_{xy})dz \qquad (7.66)$$

which in terms of the displacement potentials ϕ_1, ϕ_2 and ψ may be written as

$$N_x = 2\mu h \left\{ \sum_{j=1}^{2} \left[\left(\delta_j^2 - \frac{1}{2}\delta_3^2 \right)\phi_j + \frac{\partial^2 \phi_j}{\partial x^2} \right] + \frac{\partial^2 \psi}{\partial x \partial y} \right\} \qquad (7.67a)$$

$$N_y = 2\mu h \left\{ \sum_{j=1}^{2} \left[\left(\delta_j^2 - \frac{1}{2}\delta_3^2 \right)\phi_j + \frac{\partial^2 \phi_j}{\partial y^2} \right] - \frac{\partial^2 \psi}{\partial x \partial y} \right\} \qquad (7.67b)$$

$$N_{xy} = \mu h \left[2 \sum_{j=1}^{2} \frac{\partial^2 \phi_j}{\partial x \partial y} - \left(\frac{\partial^2 \psi}{\partial x^2} - \frac{\partial^2 \psi}{\partial y^2} \right) \right] \qquad (7.67c)$$

The force N_z is h times the average transverse stress σ_z and is given by

$$N_z = \frac{\pi \mu h}{2\sqrt{3}v} \sum_{j=1}^{2} \phi_j [2\delta_j^2 - (1 - v)\delta_3^2] \qquad (7.68)$$

The shear forces R_x and R_y are obtained from

$$(R_x, R_y) = \int_{-h/2}^{h/2} (\tau_{zx}, \tau_{zy})z \, dz \qquad (7.69)$$

which in turn yields

$$R_x = \frac{\mu h^2}{6} \sum_{j=1}^{2} e_j \frac{\partial \phi_j}{\partial x} \qquad (7.70a)$$

$$R_y = \frac{\mu h^2}{6} \sum_{j=1}^{2} e_j \frac{\partial \phi_j}{\partial y} \qquad (7.70b)$$

These quantities play a role analogous to that of transverse shear forces Q_x and Q_y in the corresponding plate bending theory.

Incident wave field. Let the crack of length $2a$ in Figure 7.7 be engulfed by a plane incident wave. The resulting wave field can be separated into two parts: an incident wave field and a scattered wave field, i.e.,

$$\phi_j = \phi_j^{(i)} + \phi_j^{(s)}, \ j = 1, 2 \qquad (7.71a)$$

$$\psi = \psi^{(i)} + \psi^{(s)} \tag{7.71b}$$

The potentials $\phi_j^{(i)}$ ($j = 1, 2$) and $\psi^{(i)}$ are specified while the potentials $\phi_j^{(s)}$ ($j = 1, 2$) and $\psi^{(s)}$ for the scattered waves must be found by solving equations (7.62). Three types of input waves will be considered.

Suppose that the incident wave is a plane harmonic compressional wave making an angle γ_1 with the x-axis along which the crack lies, i.e.,

$$\phi_1^{(i)} = \Phi_1 \exp\{-i[\delta_1(x \cos \gamma_1 + y \sin \gamma_1) + \omega t]\} \tag{7.72a}$$

$$\phi_2^{(i)} = \psi^{(i)} = 0 \tag{7.72b}$$

in which Φ_1 represents the amplitude of the incident P-wave. From equations (7.67) and (7.68) the in-plane and transverse normal forces per unit length are obtained:

$$N_x^{(i)} = \mu h \phi_1^{(i)} (2\delta_1^2 \sin^2 \gamma_1 - \delta_3^2) \tag{7.73a}$$

$$N_y^{(i)} = \mu h \phi_1^{(i)} (2\delta_1^2 \cos^2 \gamma_1 - \delta_3^2) \tag{7.73b}$$

$$N_{xy}^{(i)} = -2\mu h \phi_1^{(i)} \delta_1^2 \sin \gamma_1 \cos \gamma_1 \tag{7.73c}$$

$$N_z^{(i)} = \frac{\pi \mu h}{2\sqrt{3}v} \phi_1^{(i)} [2\delta_1^2 - (1-v)\delta_3^2] \tag{7.73d}$$

while the incident shear forces are given by

$$R_x^{(i)} = -\frac{i\mu h^2}{6} \phi_1^{(i)} e_1 \delta_1 \cos \gamma_1 \tag{7.74a}$$

$$R_y^{(i)} = -\frac{i\mu h^2}{6} \phi_1^{(i)} e_1 \delta_1 \sin \gamma_1 \tag{7.74b}$$

In the case of an incident wave with amplitude Φ_2 impinging on the crack at an angle γ_2 with the x-axis, the potentials are

$$\phi_2^{(i)} = \Phi_2 \exp\{-i[\delta_2(x \cos \gamma_2 + y \sin \gamma_2) + \omega t]\} \tag{7.75a}$$

$$\phi_1^{(i)} = \psi^{(i)} = 0 \tag{7.75b}$$

The corresponding in-plane and transverse normal forces become

$$N_x^{(i)} = \mu h \phi_2^{(i)} (2\delta_2^2 \sin^2 \gamma_2 - \delta_3^2) \tag{7.76a}$$

$$N_y^{(i)} = \mu h \phi_2^{(i)} (2\delta_2^2 \cos^2 \gamma_2 - \delta_3^2) \tag{7.76b}$$

$$N_{xy}^{(i)} = -2\mu h \phi_2^{(i)} \delta_2^2 \sin \gamma_2 \cos \gamma_2 \tag{7.76c}$$

$$N_z^{(i)} = \frac{\pi\mu h}{2\sqrt{3}v} \, \phi_2^{(i)} \left[2\delta_2^2 - (1-v)\delta_3^2\right] \tag{7.76d}$$

Similarly, the transverse shear forces are

$$R_x^{(i)} = -\frac{i\mu h^2}{6} \, \phi_2^{(i)} \, e_2\delta_2 \cos\gamma_2 \tag{7.77a}$$

$$R_y^{(i)} = -\frac{i\mu h^2}{6} \, \phi_2^{(i)} \, e_2\delta_2 \sin\gamma_2 \tag{7.77b}$$

When $\omega/\bar\omega < 1$, $\phi_2^{(i)}$ goes to zero since in this case δ_2 is imaginary. This implies that the incident wave attenuates near infinity.

If the incident wave is associated with δ_3 and makes angle γ_3 with the x-axis, then the following representation prevails:

$$\psi^{(i)} = \psi_3 \exp\left\{-i\left[\delta_3(x\cos\gamma_3 + y\sin\gamma_3) + \omega t\right]\right\} \tag{7.78a}$$

$$\phi_1^{(i)} = \phi_2^{(i)} = 0 \tag{7.78b}$$

with ψ_3 being the wave amplitude. Equations (7.78) give rise only to the in-plane forces

$$N_x^{(i)} = -2\mu h\psi^{(i)} \, \delta_3^2 \sin\gamma_3 \cos\gamma_3 \tag{7.79a}$$

$$N_y^{(i)} = 2\mu h\psi^{(i)} \, \delta_3^2 \sin\gamma_3 \cos\gamma_3 \tag{7.79b}$$

$$N_{xy}^{(i)} = \mu h\psi^{(i)} \, \delta_3^2 \cos 2\gamma_3 \tag{7.79c}$$

while $N_x^{(i)}$, $R_x^{(i)}$ and $R_y^{(i)}$are all zero.

Boundary conditions on the crack. The crack in Figure 7.7 lying along the x-axis from $x = -a$ to $x = a$ is assumed to be free of surface tractions. This means that the forces per unit length N_y, N_{xy} and R_y must vanish on $|x| < a$ and $y = 0$, i.e.,

$$N_y^{(i)} + N_y^{(s)} = 0 \tag{7.80a}$$

$$N_{xy}^{(i)} + N_{xy}^{(s)} = 0 \tag{7.80b}$$

$$R_y^{(i)} + R_y^{(s)} = 0 \tag{7.80c}$$

The problem can thus be formulated in three parts in terms of the symmetry conditions of equations (7.80).

The symmetric portion of the scattered wave problem refers to equation

(7.80a) such that

$$N_y^{(s)}(x, 0) = - N_y^{(i)}(x, 0), \ |x| < a \tag{7.81a}$$

$$v_y^{(s)}(x, 0) = 0, \ |x| \geq a \tag{7.81b}$$

$$N_{xy}^{(s)}(x, 0) = R_y^{(s)}(x, 0) = 0, \text{ for all } x \tag{7.81c}$$

The skew-symmetric scattered wave field is determined by the conditions

$$N_{xy}^{(s)}(x, 0) = - N_{xy}^{(i)}(x, 0), \ |x| < a \tag{7.82a}$$

$$v_x^{(s)}(x, 0) = 0, \ |x| \geq a \tag{7.82b}$$

$$N_y^{(s)}(x, 0) = R_y^{(s)}(x, 0) = 0, \text{ for all } x \tag{7.82c}$$

Equation (7.80c) represents the specification of the negative transverse shear force $R_y^{(i)}$ on the crack for the scattered wave problem i.e.,

$$R_y^{(s)}(x, 0) = - R_y^{(i)}(x, 0), \ |x| < a \tag{7.83a}$$

$$v_z^{(s)}(x, 0) = 0, \ |x| \geq a \tag{7.83b}$$

$$N_y^{(s)}(x, 0) = N_{xy}^{(s)}(x, 0) = 0, \text{ for all } x \tag{7.83c}$$

This is analogous to the problem of anti-plane shear in the theory of elasticity.

The quantities $N_y^{(i)}$, $N_{xy}^{(i)}$ and $R_y^{(i)}$ in the above equations are those given in equations (7.73), (7.74), (7.76), (7.77) and (7.79). Three different formulations referred individually to equations (7.81), (7.82) and (7.83) will now be treated.

In-plane normal force specified. With the aid of Fourier transforms and inversion theorem, equations (7.62) can be solved:

$$\phi_j^{(s)}(x, y) = \frac{1}{2\pi} \int_{-\infty}^{\infty} A_j(s) \exp[-(\beta_j y + isx)]ds. \ (j = 1, 2); \ y \geq 0 \tag{7.84a}$$

$$\psi^{(s)}(x, y) = \frac{1}{2\pi} \int_{-\infty}^{\infty} A_3(s) \exp[-(\beta_3 y + isx)]ds, \ y \geq 0 \tag{7.84b}$$

The branch cuts in the problem are described by

$$\beta_j = (s^2 - \delta_j^2)^{\frac{1}{2}} = - i(\delta_j^2 - s^2)^{\frac{1}{2}}, \ j = 1, 2, 3 \tag{7.85}$$

The three unknowns $A_j(s) \ (j = 1, 2, 3)$ may be expressed in terms of a single function $A(s)$ as given by

$$A_1(s) = \frac{e_2(2s^2 - \delta_3^2)}{(e_2 - e_1)\beta_1\delta_3^2} A(s), \quad A_2(s) = -\frac{e_1(2s^2 - \delta_3^2)}{(e_2 - e_1)\beta_2\delta_3^2} A(s),$$

$$A_3(s) = -\frac{2is}{\delta_3^2} A(s) \tag{7.86}$$

The satisfaction of equations (7.81a) and (7.81b) requires that $A(s)$ be determined from the dual integral equations

$$\int_{-\infty}^{\infty} A(s) \exp(-isx)ds = 0, \ |x| \ge a \tag{7.87a}$$

$$\int_{-\infty}^{\infty} g_1(s) A(s) \exp(-isx)ds = \frac{\pi P_j}{\mu[1 - (c_2/c_1)^2]} \exp(-ix\delta_j \cos \gamma_j), \ |x| < a \tag{7.87b}$$

where $j = 1, 2, 3$ and $g_1(s)$ is a known function:

$$g_1(s) = \left[\frac{(2s^2 - \delta_3^2)^2}{2(e_2 - e_1)\delta_3^2}\left(\frac{e_2}{\beta_1} - \frac{e_1}{\beta_2}\right) - 2\beta_3\left(\frac{s}{\delta_3}\right)^2\right]\left[1 - \left(\frac{c_2}{c_1}\right)^2\right]^{-1} \tag{7.88}$$

In equation (7.87b), P_j ($j = 1, 2, 3$) are defined as

$$P_j = \mu\Phi_j[\delta_3^2 - 2\delta_j^2 \cos^2 \gamma_j], j = 1, 2 \tag{7.89a}$$

$$P_3 = -2\mu\psi_3\delta_3^2 \sin \gamma_3 \cos \gamma_3 \tag{7.89b}$$

It can be shown that

$$A(s) = \frac{\pi P_j a^2}{2\mu}\left[1 - \left(\frac{c_2}{c_1}\right)^2\right]^{-1} \int_0^1 \sqrt{\xi} \left[\Psi_1(\xi) J_0(sa\xi) + \Psi_2(\xi)\right.$$

$$J_1(sa\xi)]d\xi \tag{7.90}$$

satisfies the dual integral equations (7.87) provided that Ψ_1 and Ψ_2 are determined from the following system of Fredholm integral equations

$$\Psi_1(\xi) - \int_0^1 L_1(\xi, \eta) \Psi_1(\eta)d\eta = -\sqrt{\xi} J_0(\delta_j a\xi \cos \gamma_j) \tag{7.91a}$$

$$\Psi_2(\xi) - \int_0^1 L_2(\xi, \eta) \Psi_2(\eta)d\eta = -\sqrt{\xi} J_1(\delta_j a\xi \cos \gamma_j) \tag{7.91b}$$

The kernels L_1 and L_2 which are symmetric in ξ and η take the forms

$$L_1(\xi, \eta) = (\xi\eta)^{\frac{1}{2}}\int_0^\infty s[G_1(s/a) + 1]J_0(s\xi) J_0(s\eta)ds \tag{7.92a}$$

$$L_2(\xi, \eta) = (\xi\eta)^{\frac{1}{2}}\int_0^\infty s[G_1(s/a) + 1]J_1(s\xi) J_1(s\eta)ds \tag{7.92b}$$

in which G_1 is given by

$$G_1(s) = \frac{1}{s} g_1(s) \tag{7.93}$$

In-plane shear force specified. A solution of equations (7.62) that pertains to the conditions specified in equations (7.82) is

$$\phi_j^{(s)}(x, y) = \frac{1}{2\pi} \int_{-\infty}^{\infty} B_j(s) \exp\left[-(\beta_j y + isx)\right]ds, \ (j = 1, 2); \ y \geq 0 \tag{7.94a}$$

$$\psi^{(s)}(x, y) = \frac{1}{2\pi} \int_{-\infty}^{\infty} B_3(s) \exp\left[-(\beta_3 y + isx)\right]ds, \ y \geq 0 \tag{7.94b}$$

From equation (7.82c), the three unknown $B_j(s)$ $(j = 1, 2, 3)$ can be related to $B(s)$ by the following relationships

$$B_1(s) = \frac{2i\beta_2 e_2 s}{(\beta_2 e_2 - \beta_1 e_1)\delta_3^2} B(s), \ B_2(s) = -\frac{2i\beta_1 e_1 s}{(\beta_2 e_2 - \beta_1 e_1)\delta_3^2} B(s) \tag{7.95a}$$

$$B_3(s) = \frac{2s^2 - \delta_3^2}{\beta_3 \delta_3^2} B(s) \tag{7.95b}$$

Equations (7.82a) and (7.82b) are then satisfied by requiring that $B(s)$ be the solution of the dual integral equations

$$\int_{-\infty}^{\infty} B(s) \exp(-isx)ds = 0, \ |x| \geq a \tag{7.96a}$$

$$\int_{-\infty}^{\infty} g_2(s) B(s) \exp(-isx)ds = \frac{2\pi Q_j}{\mu[(1-(c_2/c_1)^2]} \exp(-ix\delta_j \cos \gamma_j),$$
$$|x| < a \tag{7.96b}$$

and $g_2(s)$ is given by

$$g_2(s) = \left[\frac{2\beta_1 \beta_2 (e_1 - e_2)s^2}{(\beta_2 e_2 - \beta_1 e_1)\delta_3^2} + \frac{(2s^2 - \delta_3^2)^2}{2\beta_3 \delta_3^2}\right]\left[1 - \left(\frac{c_2}{c_1}\right)^2\right]^{-1} \tag{7.97}$$

The amplitude of the input wave is embedded in Q_j $(j = 1, 2, 3)$:

$$Q_j = \mu \Phi_j \delta_j^2 \sin 2\gamma_j, \ j = 1, 2 \tag{7.98a}$$

$$Q_3 = -\mu \psi_3 \delta_3^2 \cos 2\gamma_3 \tag{7.98b}$$

Solving equations (7.96), an integral representation of $B(s)$ is found:

$$B(s) = \frac{\pi Q_j a^2}{2\mu} \left[1 - \left(\frac{c_2}{c_1}\right)^2\right]^{-1} \int_0^1 \sqrt{\xi} \left[\Psi_3(\xi) J_0(sa\xi) + \Psi_4(\xi)\right]$$

$$J_1(sa\xi)]d\xi \tag{7.99}$$

where Ψ_3 and Ψ_4 can be obtained by solving the Fredholm integral equations

$$\Psi_3(\xi) - \int_0^1 L_3(\xi,\eta)\ \Psi_3(\eta)d\eta = -\sqrt{\xi}\ J_0(\delta_j a\xi \cos \gamma_j) \tag{7.100a}$$

$$\Psi_4(\xi) - \int_0^1 L_4(\xi,\eta)\ \Psi_4(\eta)d\eta = -\sqrt{\xi}\ J_1(\delta_j a\xi \cos \gamma_j) \tag{7.100b}$$

The kernel functions L_3 and L_4 stand for

$$L_3(\xi,\eta) = (\xi\eta)^{\frac{1}{2}} \int_0^\infty s[G_2(s/a) + 1]\ J_0(s\xi)\ J_0(s\eta)ds \tag{7.101a}$$

$$L_4(\xi,\eta) = (\xi\eta)^{\frac{1}{2}} \int_0^\infty s[G_2(s/a) + 1]\ J_1(s\xi)\ J_1(s\eta)ds \tag{7.101b}$$

with $G_2(s) = g_2(s)/s$.

Transverse shear force specified. Referring to the conditions specified in equations (7.83), the following potentials are taken as solutions to equations (7.62):

$$\phi_j^{(s)}(x,y) = \frac{1}{2\pi}\int_{-\infty}^\infty C_j(s) \exp\left[-(\beta_j y + isx)\right]ds, (j = 1, 2); \ y \geq 0 \tag{7.102a}$$

$$\psi^{(s)}(x,y) = \frac{1}{2\pi}\int_{-\infty}^\infty C_3(s) \exp\left[-(\beta_3 y + isx)\right]ds, \ y \geq 0 \tag{7.102b}$$

As before, the symmetry conditions given by equation (7.83c) reduce the number of unknowns from $C_j(s)$ $(j = 1, 2, 3)$ to $C(s)$:

$$C_1(s) = \frac{(2s^2 - \delta_3^2)^2 - 4\beta_2\beta_3 s^2}{(e_1 - e_2)(2s^2 - \delta e)^2 + 4\beta_3(e_2\beta_1 - e_1\beta_2)s^2} C(s) \tag{7.103a}$$

$$C_2(s) = -\frac{(2s^2 - \delta_3^2)^2 - 4\beta_1\beta_3 s^2}{(e_1 - e_2)(2s^2 - \delta_3^2)^2 + 4\beta_3(e_2\beta_1 - e_1\beta_2)s^2} C(s) \tag{7.103b}$$

$$C_3(s) = -\frac{i(2s^2 - \delta_3^2)}{2s\beta_3}[C_1(s) + C_2(s)] \tag{7.103c}$$

The remaining conditions in equations (7.83) yield

$$\int_{-\infty}^\infty C(s) \exp(-isx)ds = 0, \ |x| \geq a \tag{7.104a}$$

$$\int_{-\infty}^\infty g_3(s)\ C(s) \exp(-isx)ds = \frac{2\pi i R_j}{\mu} \exp(-ix\delta_j \cos \gamma_j), \ |x| < a \tag{7.104b}$$

in which $g_3(s)$ is given:

$$g_3(s) = -\frac{(\beta_1 e_1 - \beta_2 e_2)(2s^2 - \delta_3^2)^2 + 4\beta_1\beta_2\beta_3(e_2 - e_1)s^2}{(e_1 - e_2)(2s^2 - \delta_3^2)^2 + 4\beta_3(\beta_1 e_2 - \beta_2 e_1)s^2} \qquad (7.105)$$

In equation (7.104b), R_j ($j = 1, 2$) takes the form

$$R_j = \mu\Phi_j\, e_j\delta_j \sin \gamma_j, \, j = 1, 2 \qquad (7.106)$$

The system of dual integral equations (7.104) may be solved to give

$$C(s) = \frac{i\pi R_j a^2}{\mu}\int_0^1 \sqrt{\xi}\,[\Psi_5(\xi)\, J_0(sa\xi) + \Psi_6(\xi)\, J_1(sa\xi)]\mathrm{d}\xi \qquad (7.107)$$

such that Ψ_5 and Ψ_6 are to be found from

$$\Psi_5(\xi) - \int_0^1 L_5(\xi, \eta)\,\Psi_5(\eta)\mathrm{d}\eta = -\sqrt{\xi}\, J_0(\delta_j a\xi \cos \gamma_j) \qquad (7.108a)$$

$$\Psi_6(\xi) - \int_0^1 L_6(\xi,\eta)\,\Psi_6(\eta)\mathrm{d}\eta = -\sqrt{\xi}\, J_1(\delta_j a\xi \cos \gamma_j) \qquad (7.108b)$$

with L_5 and L_6 being the kernels given by

$$L_5(\xi, \eta) = (\xi\eta)^{\frac{1}{2}}\int_0^\infty s[G_3(s/a) + 1]\, J_0(s\xi)\, J_0(s\eta)\mathrm{d}s \qquad (7.109a)$$

$$L_6(\xi, \eta) = (\xi\eta)^{\frac{1}{2}}\int_0^\infty s[G_3(s/a) + 1]\, J_1(s\xi)\, J_1(s\eta)\mathrm{d}s \qquad (7.109b)$$

The function $G_3(s) = g_3(s)/s$ is given since $g_3(s)$ has already been defined by equation (7.105).

Singular stresses. In terms of the local polar coordinates r and θ measured from the crack edge as shown in Figure 7.7, the normal forces N_x, N_y, etc., and shear forces R_x, R_y, etc., are unbounded as $r \to 0$. The nature of this singularity can be found from the appropriate integral solutions that represent N_x, N_y, ..., and R_x, R_y, ... By carrying the integration for large values of the variable of integration s, it is found that near the crack edge the following results prevail:

$$h^{-1}\, N_x = \frac{k_1}{(2r)^{\frac{1}{2}}}\cos\frac{\theta}{2}\left(1 - \sin\frac{\theta}{2}\sin\frac{3\theta}{2}\right)$$

$$- \frac{k_2}{(2r)^{\frac{1}{2}}}\sin\frac{\theta}{2}\left(2 + \cos\frac{\theta}{2}\cos\frac{3\theta}{2}\right) + O(1) \qquad (7.110a)$$

$$h^{-1}\, N_y = \frac{k_1}{(2r)^{\frac{1}{2}}}\cos\frac{\theta}{2}\left(1 + \sin\frac{\theta}{2}\sin\frac{3\theta}{2}\right)$$

$$+ \frac{k_2}{(2r)^{\frac{1}{2}}} \sin \frac{\theta}{2} \cos \frac{\theta}{2} \cos \frac{3\theta}{2} + O(1) \qquad (7.110b)$$

$$h^{-1} N_{xy} = \frac{k_1}{(2r)^{\frac{1}{2}}} \cos \frac{\theta}{2} \sin \frac{\theta}{2} \sin \frac{3\theta}{2}$$

$$+ \frac{k_2}{(2r)^{\frac{1}{2}}} \cos \frac{\theta}{2} \left(1 - \sin \frac{\theta}{2} \sin \frac{3\theta}{2} \right) + O(1) \qquad (7.110c)$$

$$h^{-1} N_z = 2v \left(\frac{\pi}{2\sqrt{3}} \right) \left[\frac{k_1}{(2r)^{\frac{1}{2}}} \cos \frac{\theta}{2} - \frac{k_2}{(2r)^{\frac{1}{2}}} \sin \frac{\theta}{2} \right] + O(1) \qquad (7.110d)$$

The transverse shear forces become

$$6h^{-2} R_x = \frac{k_3}{(2r)^{\frac{1}{2}}} \sin \frac{\theta}{2} + O(1) \qquad (7.111a)$$

$$6h^{-2} R_y = \frac{k_3}{(2r)^{\frac{1}{2}}} \cos \frac{\theta}{2} + O(1) \qquad (7.111b)$$

Equations (7.110) and (7.111) show that the crack front stress field in the Kane-Mindlin theory is three-dimensional in character. In fact, the functional dependence of the six stress components on r and θ is the same as that found by Sih [12] from the three-dimensional theory of elasticity. The only exception is that equations (7.102) do not yield the exact plane strain condition since

$$N_z = \frac{\pi}{2\sqrt{3}} v(N_x + N_y) \qquad (7.112)$$

However, the discrepancy is not serious as the factor $\pi/2\sqrt{3}$ does not deviate appreciable from unity. The interesting aspect of this theory is that there is a coupling between extensional motions and thickness shear vibration as evidenced by the existence of all three stress intensity factors

$$k_1 = [\Psi_1(1) - i\Psi_2(1)] P_j \sqrt{a}, \, j = 1k, 2, 3 \qquad (7.113a)$$

$$k_2 = [\Psi_3(1) - i\Psi_4(1)] Q_j \sqrt{a}, \, j = 1, 2, 3 \qquad (7.113b)$$

$$k_3 = [\Psi_6(1) + i\Psi_5(1)] R_j \sqrt{a}, \, j = 1, 2 \qquad (7.113c)$$

The functions $\Psi_1, \Psi_2, \ldots, \Psi_6$ in equations (7.113) are calculated numerically from equations (7.91), (7.100) and (7.108) and evaluated at the crack tip where the dimensionless variable $\xi = 1$.

The numerical values of the stress inetnsity factors in equations (7.113) have been obtained [13] for $v = 0.25$ and various values of a/h and $\omega a/c_2$.

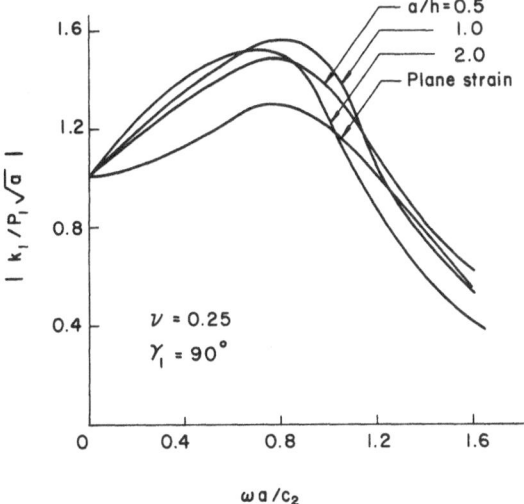

Figure 7.8. Extensional stress intensity factor versus wave number for input *P*-wave at normal incidence

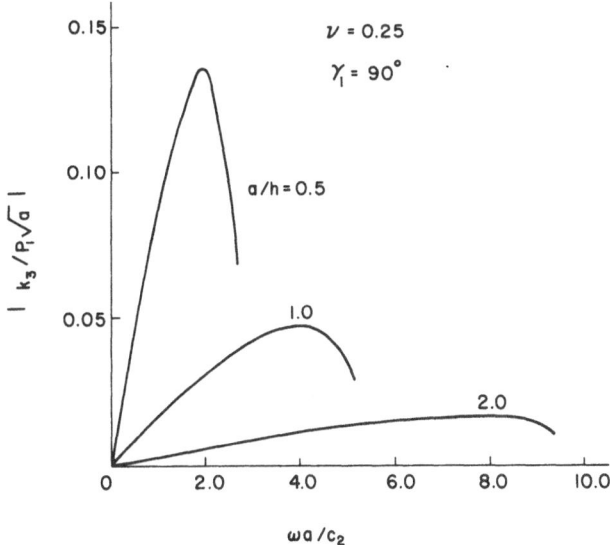

Figure 7.9. Thickness mode stress intensity factor versus wave number for input *P*-wave at normal incidence

Figure 7.8 shows a plot of the normalized stress intensity factor $| k_1/P_1\sqrt{a} |$ in equation (7.113a) versus the normalized wave number $\omega a/c_2$ which varies from 0.0 to 1.6. The magnitude of the stress at the front of the input P-wave is $P_1 = \mu\Phi_1\delta_3^2$ for normal incidence, i.e., $\gamma_1 = 90°$. It is seen that the k_1 values increase at first reaching a maximum and then decrease. The peak value of k_1 increases as a/h is raised from 0.5 to 1.0 and then it starts to decline with the plane strain solution as a lower limit. The coupling between the extensional and thickness motion gives rise to the stress intensity factor k_3 in equation (7.113c) which when normalized with respect to P_1, i.e., $| k_3/P_1\sqrt{a} |$ is plotted in Figure 7.9 as a function of $\omega a/c_2$. Note that the amplitude of k_3 is much lower than that of k_1 shown in Figure 7.8. The curves have the same general trend of increasing to a peak and then decays in amplitude. The peak decreases in magnitude as a/h is increased. This shows that the thickness coupling effect becomes less and less important as the plate thickness is decreased. The occurrence of k_3 alone corresponds to anti-plane shear motion [14] excited by input SH-waves.

If ψ_3 is specified, while $\Phi_1 = \Phi_2 = 0$, then k_2 in equation (7.113b) is the only nonzero stress intensity factor. At normal incidence, $\gamma_3 = 90°$, equation (7.98b) gives $Q_3 = \mu\psi_3\delta_3^2$ for the magnitude of the stress at the input

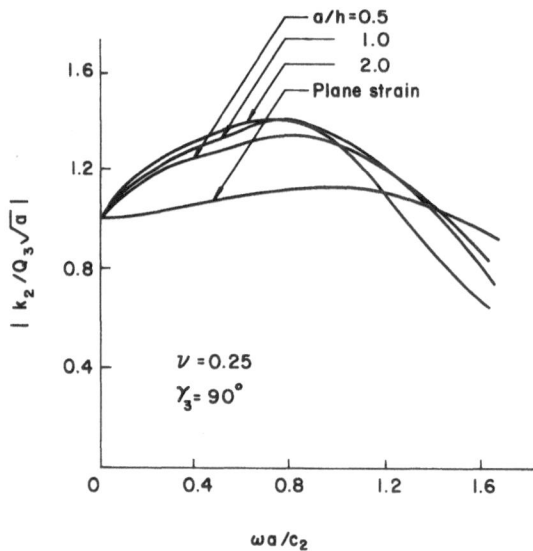

Figure 7.10. In-plane shear stress intensity factor versus wave number for input shear wave at normal incidence

wave front. Plotting $|k_2/Q_3\sqrt{a}|$ as a function of $\omega a/c_2$, the results are displayed graphically in Figure 7.10. Again, the curves increase in amplitude for small values of a/h and then decrease as a/h is increased with the plane strain solution as the limit of an infinitely thick plate.

7.5 Plates subjected to sudden loading

Sudden loading is also referred to as shock or impact where the interval of time involved in applying the load is very short. Common examples are collision, an explosion, or some other mechanical means. Since impact is unavoidably accompanied by vibration, the duration of impact should be compared with the fundamental period of natural vibration of the system being considered. Generally speaking, if the time required to apply load, i.e., increase the load from zero to its maximum value, is less than half the fundamental natural period of the system, then the loading is regarded as impact. In other words, the dynamic effects are negligible only if the time of loading* is many times greater than the fundamental natural period.

The results obtained in this section on the transient response of cracked plates apply to waves impinging on the crack as caused by the sudden application of load and the case of sudden appearance of crack in a plate under impact.

Classical plate under impact. The classical plate bending theory is used to solve the problem of a cracked plate under impact. The source of impact is the sudden appearance of uniform bending moment on the surface of a crack of length $2a$ lying on the x-axis. Because of symmetry with reference to the x-axis, it suffices to consider only the region of the plate where $y > 0$. Hence the following mixed boundary conditions

$$M_y(x, 0, t) = -M_0 H(t), \ V_y(x, 0, t) = 0, \ |x| < a \tag{7.114a}$$

$$u_y(x, 0, t) = V_y(x, 0, t) = 0, \ |x| \geq a \tag{7.114b}$$

are imposed on the solution of the differential equation (7.5). In equation (7.114a), $H(t)$ is the Heaviside unit step function. The initial conditions of this problem is assumed to be zero while the displacements at distances far away from the crack are required to vanish, i.e.,

---

* Periodically repeated loading has been discussed earlier.

$$\lim_{(x^2 + y^2) \to \infty} [u_x(x, y, t), u_y(x, y, t), u_z(x, y, t)] = 0 \tag{7.115}$$

The formulation is similar to that of section 7.2 except that the problem will now be solved in the Laplace transform domain. Let the Laplace transform pair be written as

$$g^*(p) = \int_0^\infty g(t) \exp(-pt) dt \tag{7.116a}$$

$$g(t) = \frac{1}{2\pi i} \int_{Br} g^*(p) \exp(pt) dt \tag{7.116b}$$

where the integral in equation (7.116b) is taken over the Bromwich path. Applying equation (7.116a) to equation (7.5), the deflection w^* in the transform domain depends only on the space variables x and y:

$$D\nabla^4 w^* + \rho h p^2 w^* = 0 \tag{7.117}$$

In addition, if the Fourier cosine transform is employed on the variable x, equation (7.117) reduces to an ordinary differential equation which can be readily solved. With the aid of the Fourier inversion theorem, a solution in terms of p is found:

$$w^*(x, y, p) = \frac{2}{\pi} \int_0^\infty \{ A_1^*(s, p) \exp[-(s^2 + \bar{\alpha}^2)^{\frac{1}{2}}]$$

$$+ A_2^*(s, p) \exp[-(s^2 - \bar{\alpha}^2)^{\frac{1}{2}}] \} \cos(sx) ds, \ y \geq 0 \tag{7.118}$$

The parameter $\bar{\alpha}$ is given by

$$\bar{\alpha}^4 = -\frac{\rho h}{D} p^2 \tag{7.119}$$

Solving for the appropriate moment and equivalent shear expressions in the transform domain and making use of equation (7.118), the conditions in equations (7.114) lead to a pair of dual integral equations

$$\frac{2}{\pi} \int_0^\infty A^*(s, p) \cos(sx) ds = 0, \ x \geq a \tag{7.120a}$$

$$\frac{2}{\pi} \int_0^\infty f_1^*(s, p) A^*(s, p) \cos(sx) ds = \frac{M_0}{pD}, \ x < a \tag{7.120b}$$

in which f_1^* is known, i.e.,

$$f_1^*(s, p) = \frac{(s^2 - \bar{\alpha}^2)^{\frac{1}{2}} [(1 - v)s^2 + \bar{\alpha}^2]^2 - (s^2 + \bar{\alpha}^2)^{\frac{1}{2}} [(1 - v)s^2 - \bar{\alpha}^2]^2}{2\bar{\alpha}^2 (s^4 - \bar{\alpha}^4)^{\frac{1}{2}}}$$

$$\tag{7.121}$$

Once $A^*(s, p)$ in equations (7.120) is found, A_1^* and A_2^* in equation (7.118) may be obtained from

$$\begin{bmatrix} A_1^*(s, p) \\ \\ A_2^*(s, p) \end{bmatrix} = \frac{1}{2\bar{\alpha}^2} A^*(s, p) \begin{bmatrix} \dfrac{(1 - v)s^2 + \bar{\alpha}^2}{(s^2 + \bar{\alpha}^2)^{\frac{1}{2}}} \\ \\ -\dfrac{(1 - v)s^2 + \bar{\alpha}^2}{(s^2 - \bar{\alpha}^2)^{\frac{1}{2}}} \end{bmatrix} \tag{7.122}$$

Following the procedure for solving dual integral equations as outlined in [6], it is found that

$$A^*(s, p) = \frac{\pi M_0 a^2}{(1 - v)(3 + v)Dp} \int_0^1 \sqrt{\xi}\, \Phi_1^*(\xi, p)\, J_0(sa\xi)d\xi \tag{7.123}$$

such that Φ_1^* satisfies the Fredholm integral equation

$$\Phi_1^*(\xi, p) + \int_0^1 L_1^*(\xi, \eta, p)\, \Phi_1^*(\xi, p)d\eta = \sqrt{\xi} \tag{7.124}$$

whose kernel being symmetric in ξ and η is

$$L_1^*(\xi, \eta, p) = (\xi\eta)^{\frac{1}{2}} \int_0^\infty s[F_1^*(s/a, p) - 1]\, J_0(s\xi)\, J_0(s\eta)ds \tag{7.125}$$

The function F_1^* takes the form

$$F_1^*(s, p) = \frac{2f_1^*(s, p)}{(1 - v)(3 + v)s} \tag{7.126}$$

It is apparent from equations (7.123) and (7.118) that Φ_1^* determines the expressions of moments and shear forces in the Laplace transform domain. Referred to a system of local polar coordinates r and θ measured from the crack tip, the moments are

$$M_x^* = \frac{K_1^*(p)}{(2r)^{\frac{1}{2}}} \left(\frac{1 - v}{3 + v}\right) \left[\frac{1}{2} \sin\theta \sin\frac{3\theta}{2} - \cos\frac{\theta}{2}\right] + O(1) \tag{7.127a}$$

$$M_y^* = \frac{K_1^*(p)}{(2r)^{\frac{1}{2}}} \left[\frac{1}{2}\left(\frac{1 - v}{3 + v}\right) \sin\theta \sin\frac{3\theta}{2} - \cos\frac{\theta}{2}\right] + O(1) \tag{7.127b}$$

$$M_{xy}^* = \frac{K_1^*(p)}{(2r)^{\frac{1}{2}}} \left(\frac{1}{3 + v}\right) \left[2 \sin\frac{\theta}{2} + \frac{1 - v}{2} \sin\theta \cos\frac{3\theta}{2}\right] + O(1) \tag{7.127c}$$

The Laplace transform of the moment intensity factor is

$$K_1^*(p) = M_0 \sqrt{a} \; \frac{\Phi_1^*(1, p)}{p} \tag{7.128}$$

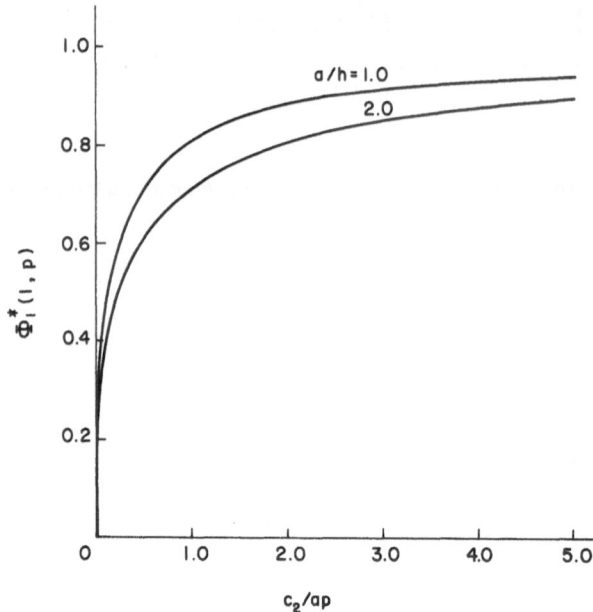

Figure 7.11. Solution of Fredholm integral equation in the Laplace transform domain for a Classical plate

Numerical values of Φ_1^* versus c_2/ap is shown in Figure 7.11 for $a/h = 1$ and 2. Note that Φ_1^* rises sharply at first and then levels off. The r- and θ-dependence of the moments for transient response in equations (7.127) is the same as that in equations (7.22) for the steady-state case. This conclusion was first made by Sih [15] who discussed the character of the transient stresses around a crack under anti-plane strain. The transverse shear forces

$$Q_x^* = \frac{K_1^*(p)}{(2r)^{\frac{3}{2}}} \left(\frac{2}{3 + v} \right) \cos \frac{3\theta}{2} + \dots \tag{7.129a}$$

$$Q_y^* = \frac{K_1^*(p)}{(2r)^{\frac{3}{2}}} \left(\frac{2}{3 + v} \right) \sin \frac{3\theta}{2} + \dots \tag{7.129b}$$

are again found to be singular of the order $r^{-3/2}$ as in equations (7.25). Hence, the nature of dynamic loading affects only the intensity of the moment field in the neighborhood of the crack tip and does not influence the functional dependence of the moments on r and θ.

Now, let the inverse Laplace transform of the function $\Phi_1^*(1, p)/p$ be denoted by $N_1(t)$. Then, the moment intensity factor as a function of time becomes

$$K_1(t) = M_0 \sqrt{a}\, N_1(t) \tag{7.130}$$

A plot of the normalized moment intensity factor $K_1(t)/M_0\sqrt{a}$ against $c_2 t/a$ is given in Figure 7.12 for $a/h = 1$ and 2. The dynamic moment

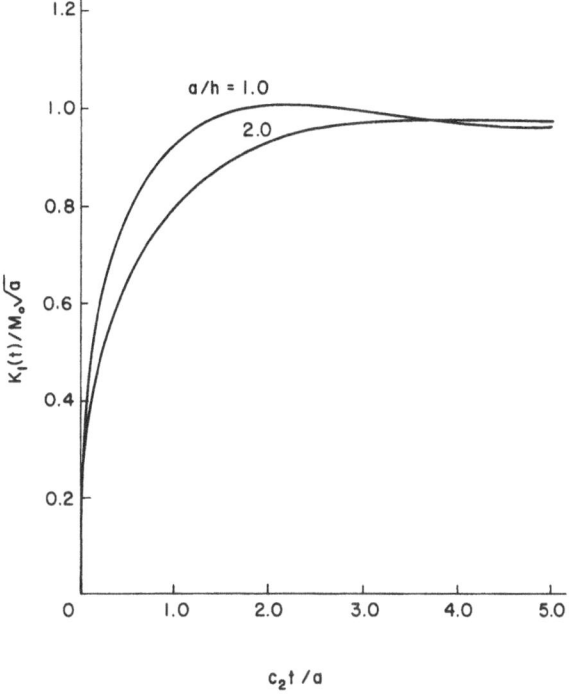

Figure 7.12. Normalized dynamic moment intensity factor versus time for a Classical plate under impact

intensity factor is seen to increase with time and approach the static value. Decreasing plate thickness tends to lower the $K_1(t)$ value for small time t.

Sudden bending of a Mindlin plate. The effect of bending on the transient response of a cracked plate using the Mindlin theory [2] is considered. In this theory, the rectangular components of the displacement vector are given by

$$u_x(x, y, z, t) = z\psi_x(x, y, t) \tag{7.131a}$$

$$u_y(x, y, z, t) = z\psi_y(x, y, t) \tag{7.131b}$$

$$u_z(x, y, z, t) = \psi_z(x, y, t) \tag{7.131c}$$

Refer to Figure 7.2 for the physical meaning of ψ_x, ψ_y and ψ_z. The expressions for the moments and shear forces are the same as those shown in equations (7.30) and (7.31) except that the exponential factor, $\exp(-i\omega t)$, should be removed and ψ_x, ψ_y and ψ_z are to be understood as functions cf x, y and t. Keeping this in mind, the equations of motion of three-dimensional elasticity theory may be written in terms of the displacement functions:

$$6D\left[(1 - v)\nabla^2\psi_x + (1 + v)\frac{\partial}{\partial x}\left(\frac{\partial\psi_x}{\partial x} + \frac{\partial\psi_y}{\partial y}\right)\right]$$
$$- \pi^2\mu h\left(\psi_x + \frac{\partial\psi_z}{\partial x}\right) = \rho h^3\frac{\partial^2\psi_x}{\partial t^2} \tag{7.132a}$$

$$6D\left[(1 - v)\nabla^2\psi_y + (1 + v)\frac{\partial}{\partial y}\left(\frac{\partial\psi_x}{\partial x} + \frac{\partial\psi_y}{\partial y}\right)\right]$$
$$- \pi^2\mu h\left(\psi_y + \frac{\partial\psi_z}{\partial y}\right) = \rho h^3\frac{\partial^2\psi_y}{\partial t^2} \tag{7.132b}$$

$$\pi^2\mu h\left(\nabla^2\psi_z + \frac{\partial\psi_x}{\partial x} + \frac{\partial\psi_y}{\partial y}\right) = 12\rho h\frac{\partial^2\psi_z}{\partial t^2} \tag{7.132c}$$

Applying the Laplace transform to equations (7.132) gives

$$6D\left[(1 - v)\nabla^2\psi_x^* + (1 + v)\frac{\partial}{\partial x}\left(\frac{\partial\psi_x^*}{\partial x} + \frac{\partial\psi_y^*}{\partial y}\right)\right]$$
$$- \pi^2\mu h\left(\psi_x^* + \frac{\partial\psi_z^*}{\partial x}\right) = \rho h^3 p^2\psi_x^* \tag{7.133a}$$

$$6D\left[(1 - v)\nabla^2\psi_y^* + (1 + v)\frac{\partial}{\partial y}\left(\frac{\partial\psi_x^*}{\partial x} + \frac{\partial\psi_y^*}{\partial y}\right)\right]$$
$$- \pi^2\mu h\left(\psi_y^* + \frac{\partial\psi_z^*}{\partial y}\right) = \rho h^3 p^2\psi_y \tag{7.133b}$$

$$\pi^2\mu h\left(\nabla^2\psi_z^* + \frac{\partial\psi_x^*}{\partial x} + \frac{\partial\psi_y^*}{\partial y}\right) = 12\rho h p^2\psi_z^* \tag{7.133c}$$

where p is the Laplace transform variable. The reduction of the preceding equations to a set of homogeneous ordinary differential equations necessitates the usage of the Fourier cosine and sine transforms. In the case of symmetric bending with respect to the crack plane, ψ_y and ψ_z are even in x and ψ_x is odd in x. That is

$$\psi_x(x, y, t) = -\psi_x(-x, y, t) \tag{7.134a}$$

$$\psi_y(x, y, t) = \psi_y(-x, y, t) \tag{7.134b}$$

$$\psi_z(x, y, t) = \psi_z(-x, y, t) \tag{7.134c}$$

It follows that the Fourier sine transform is applied to equation (7.133a) and the Fourier cosine transform to equations (7.133b) and (7.133c). The resulting couplet ordinary differential equation can then be solved together with the Fourier inversion theorem to render

$$\psi_x^*(x, y, p) = \frac{2}{\pi}\int_0^\infty [s(1 - a_2^*)\, B_1^*(s, p)\exp(-\beta_1^* y) + s(1 - a_1^*)\, B_2^*(s, p)$$

$$\times \exp(-\beta_2^* y) - \beta_3^*\, B_3^*(s, p)\exp(-\beta_3^* y)]\sin(sx)\mathrm{d}s, \; y \geq 0 \tag{7.135a}$$

$$\psi_y^*(x, y, p) = \frac{2}{\pi}\int_0^\infty [(1 - a_2^*)\,\beta_1^*\, B_1^*(s, p)\exp(-\beta_1^* y)$$

$$+ (1 - a_1^*)\,\beta_2^*\, B_2^*(s, p)\,(-\beta_2^* y) - sB_3^*(s, p)\exp(-\beta_3^* y)]\cos(sx)\mathrm{d}s, \; y \geq 0 \tag{7.135b}$$

$$\psi_z^*(x, y, p) = \frac{2}{\pi}\int_0^\infty [B_1^*(s, p)\exp(-\beta_1^* y) + B_2^*(s, p)\exp(-\beta_2^* y)]\cos(sx)\mathrm{d}s,$$

$$y \geq 0 \tag{7.135c}$$

The parameters β_j^* and a_j^* are defined as

$$\beta_j^* = (s^2 + \bar\alpha_j^2)^{\frac{1}{2}}, j = 1, 2, 3 \tag{7.136a}$$

$$a_j^* = \frac{2}{1 - v}\left(\frac{\bar\alpha_j}{\bar\alpha_3}\right)^2, j = 1, 2 \tag{7.136b}$$

and the numbers $\bar\alpha_j$ ($j = 1, 2, 3$) are given in terms of the variable p:

$$\bar\alpha_1^2 = \frac{1}{2}\left(\frac{p}{\omega_0}\right)^2\left\{\frac{1}{S} + \frac{1}{R} + \left[\left(\frac{1}{S} - \frac{1}{R}\right)^2 - \frac{4}{RS}\left(\frac{\omega_0}{p}\right)^2\right]^{\frac{1}{2}}\right\}$$

$$\tag{7.137a}$$

$$\bar{\alpha}_2^2 = \frac{1}{2}\left(\frac{p}{\omega_0}\right)^2 \left\{ \frac{1}{S} + \frac{1}{R} - \left[\left(\frac{1}{S} - \frac{1}{R}\right)^2 - \frac{4}{RS}\left(\frac{\omega_0}{p}\right)^2\right]^{\frac{1}{2}}\right\}$$

(7.137b)

$$\bar{\alpha}_3^2 = \left(\frac{\pi}{h}\right)^2 \left[\left(\frac{p}{\omega_0}\right)^2 + 1\right]$$

(7.137c)

in which R and S have already been defined by equation (7.36) and ω_0 is the cut-off frequency $\pi c_2/h$.

The unknowns B_j^* ($j = 1, 2, 3$) in equations (7.135) are to be determined from the following boundary conditions

$$M_y(x,0,t) = -M_0 H(t); \quad Q_y(x,0,t) = M_{xy}(x,0,t) = 0, \quad 0 \le |x| < a$$

(7.138)

on the crack surface and the symmetry conditions

$$\psi_y(x,0,t) = 0; \quad Q_y(x,0,t) = M_{xy}(x,0,t) = 0, \quad |x| \ge a$$

(7.139)

on the x-axis outside the crack. Expressing B_j^* ($j = 1, 2, 3$) in terms of a single unknown B^* given by

$$B_1^*(s,p) = \frac{B^*(s,p)}{\beta_1^*(\bar{\alpha}_1^2 - \bar{\alpha}_2^2)} \frac{[(1-v)s^2 + \bar{\alpha}_1^2]^2}{(1-v)s^2 + \bar{\alpha}_2^2}$$

(7.140a)

$$B_2^*(s,p) = -\frac{B^*(s,p)}{B_2^*(\bar{\alpha}_1^2 - \bar{\alpha}_2^2)}[(1-v)s^2 + \bar{\alpha}_2^2]$$

(7.140b)

$$B_3^*(s,p) = \frac{sB^*(s,p)}{\bar{\alpha}_1^2 - \bar{\alpha}_2^2}(1-v)(a_1^* - a_2^*)$$

(7.140c)

equations (7.138) and (7.139) reduce to the satisfaction of a pair of dual integrations:

$$\frac{2}{\pi}\int_0^\infty B^*(s,p)\cos(sx)ds = 0, \quad x \ge a$$

(7.141a)

$$\frac{2}{\pi}\int_0^\infty f_2^*(s,p) B_*(s,p)\cos(sx)ds = \frac{M_0}{pD}, \quad 0 < x < a$$

(7.141b)

with f_2^* being a known function:

$$f_2^*(s,p) = \frac{1}{\bar{\alpha}_1^2 - \bar{\alpha}_2^2}\left\{\frac{1 - a_2^*}{\beta_1^*}[(1-v)s^2 + \bar{\alpha}_1^2]^2 - \frac{1 - a_1^*}{B_2^*}[(1-v)s^2\right.$$

$$\left. + \bar{\alpha}_2^2]^2 - (1-v)^2(a_1^* - a_2^*)s^2\beta_3^*\right\}$$

(7.142)

A solution that satisfies equations (7.141) can be written in the form [16]

$$B^*(s, p) = \frac{\pi M_0 a^2}{pD(1 - v^2)} \int_0^1 \sqrt{\xi}\, \Phi_2^*(\xi, p)\, J_0(sa\xi)\mathrm{d}\xi \qquad (7.143)$$

such that Φ_2^* can be computed from a standard Fredholm integral equation of the second kind:

$$\Phi_2^*(\xi, p) + \int_0^1 L_2^*(\xi, \eta, p)\, \Phi_2^*(\eta, p)\mathrm{d}\eta = \sqrt{\xi} \qquad (7.144)$$

The kernel function in equation (7.144) is

$$L_2^*(\xi, \eta, p) = (\xi\eta)^{\frac{1}{2}} \int_0^\infty s[F_2^*(s/a, p) - 1]\, J_0(s\xi)\, J_0(s\eta)\mathrm{d}s \qquad (7.145)$$

where F_2^* is defined as

$$F_2^*(s, p) = \frac{2f_2^*(s, p)}{(1 - v^2)s} \qquad (7.146)$$

For $v = 0.25$, equation (7.144) has been solved numerically by Embley and Sih [16] and the results for Φ_2^* plotted against c_2/pa are displayed graphically in Figure 7.13 for $a/h = 1$ and 2.

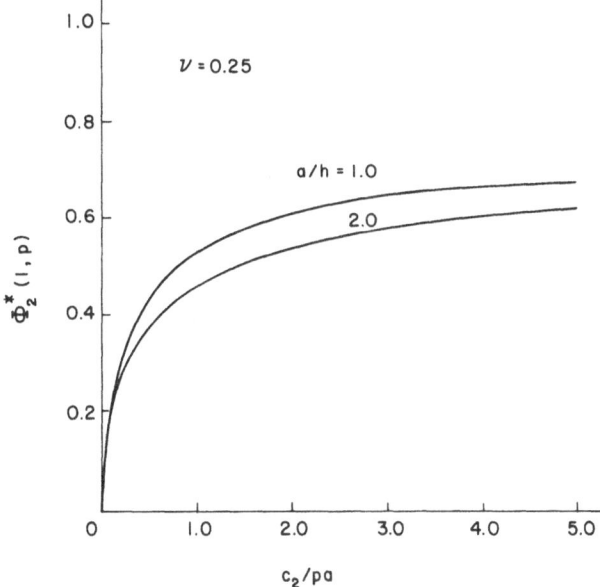

Figure 7.13. Solution of Fredholm integral equation in the Laplace transform domain for a Mindlin plate

To obtain the solution near the crack tip, the Laplace transform of the moments are expressed in terms of infinite integrals which are convergent everywhere except at the singular points corresponding to the crack tips, $x = \pm a$. At these points the moments are unbounded and their behavior can be found by expanding the integrands for large values of the argument s. By retaining only the highest order term in s, using the well-known Bessel integral identities and applying the Laplace inversion theorem, Embley and Sih [16] have found that

$$M_x = \frac{K_1(t)}{(2r)^{\frac{1}{2}}} \cos \frac{\theta}{2} \left[1 - \sin \frac{\theta}{2} \sin \frac{3\theta}{2} \right] + O(1) \tag{7.147a}$$

$$M_y = \frac{K_1(t)}{(2r)^{\frac{1}{2}}} \cos \frac{\theta}{2} \left[1 + \sin \frac{\theta}{2} \sin \frac{3\theta}{2} \right] + O(1) \tag{7.147b}$$

$$M_{xy} = \frac{K_1(t)}{(2r)^{\frac{1}{2}}} \cos \frac{\theta}{2} \sin \frac{\theta}{2} \cos \frac{3\theta}{2} + O(1) \tag{7.147c}$$

and the dynamic moment intensity factor is

$$K_1(t) = M_0 \sqrt{a}\, N_2(t) \tag{7.148}$$

in which $N_2(t)$ is the inverse Laplace transform of $\Phi_2^*(1, p)/p$. The transverse shear forces Q_x and Q_y are non-singular everywhere which is expected

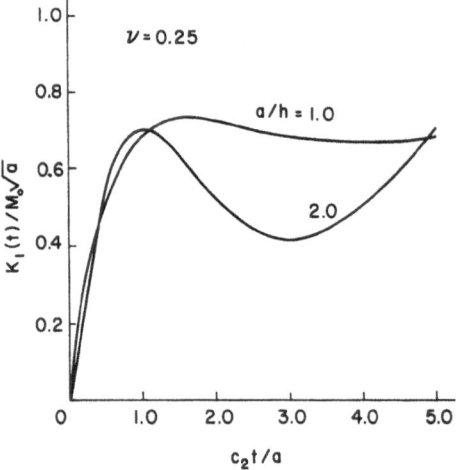

Figure 7.14. Normalized dynamic moment intensity factor versus time for a Mindlin plate under impact

since the applied moment is symmetric with respect to the crack and does not give rise to anti-plane shear deformation of the crack surface.

Figure 7.14 shows a plot of $K_1(t)/M_0\sqrt{a}$ versus the normalized time $c_2 t/a$ for $v = 0.25$. The curves corresponding to $a/h = 1$ and 2 rise rapidly within a short time interval reaching their maximum and then begin to oscillate. For a fixed crack length, the $K_1(t)$-factor for the thicker plate reaches its maximum at a later time. In both cases, the dynamic moment intensity is lower than the static limit.

References

[1] Kirchhoff, G., Über das Gleichgewicht und die Bewegung einer elastischen Scheibe, *Journal reine u. angew. Math.*, 40, pp. 51–88 (1850).

[2] Mindlin, R. D., Influence of rotatory inertia and shear on flexural motions of isotropic elastic plates, *Journal of Applied Mechanics*, 18, pp. 31–38 (1951).

[3] Kane, T. R. and Mindlin, R. D., High-frequency extensional vibrations of plates, *Journal of Applied Mechanics*, 23, pp. 277–283 (1956).

[4] Hartranft, R. J. and Sih, G. C., An approximate three-dimensional theory of plates with application to crack problems, *International Journal of Engineering Science*, 8, pp. 711–729 (1970).

[5] Achenbach, J. D., *Wave propagation in elastic solids*, North-Holland/American Elsevier, Holland (1973).

[6] Copson, E. T., On certain dual integral equations, *Proceedings of Glasgow Mathematical Association*, 5, pp. 19–24 (1961).

[7] Sih, G. C., Paris, P. C. and Erdogan, F., Crack tip stress intensity factors for plane extension and plate bending problems, *Journal of Applied Mechanics*, 29, pp. 306–312 (1962).

[8] Sih, G. C. and Loeber, J. F., Wave propagation in an elastic solid with a line of discontinuity or finite crack, *Quarterly of Applied Mathematics*, 27, 2, pp. 193–213 (1969).

[9] Hartranft, R. J. and Sih, G. C., Effect of plate thickness on the bending stress distribution around through cracks, *Journal of Mathematics and Physics*, 47, 3, pp. 276–291 (1968).

[10] Pao, Y. H., Dynamic stress concentration in an elastic plate, *Journal of Applied Mechanics*, 29, pp. 299–305 (1962).

[11] Sih, G. C., Propagation of elastic waves around a crack, *Proceedings of Third Conference on Dimensioning*, Budapest, pp. 577–588 (1968).

[12] Sih, G. C., Review of three-dimensional stress state in a cracked plate, *International Journal of Fracture Mechanics*, 7, 1, pp. 39–61 (1971).

[13] Macdonald, B. D., *The coupling of in-plane and thickness motions of a cracked plate*, Ph. D. thesis, Lehigh University, Bethlehem (1970).

[14] Loeber, J. F. and Sih, G. C., Diffraction of antiplane shear waves by a finite crack, *Journal of the Acoustical Society of America*, 44, 1, pp. 90–98 (1968).

[15] Sih, G. C., Some elastodynamic problems of cracks, *International Journal of Fracture Mechanics*, 1, 1, pp. 51–68 (1968).

[16] Embley, G. T. and Sih, G. C., Sudden appearance of a crack in a bent plate, *International Journal of Solids and Structures*, 9, pp. 1349–1359 (1973).

Peter D. Hilton

8 | *A specialized finite element approach for three-dimensional crack problems*

8.1 Introduction

The purpose of fracture mechanics is to predict the conditions of failure for structural components containing flaws. Accurate stress analysis is required to determine the stress field in the immediate vicinity of these existing flaws. Typical examples of such flaws in structural members include surface cracks, interior cracks, and through cracks. In each of these cases the near stress field is generally three-dimensional in character. No analytic solution procedures are available for considering the relevent problems and most previous numerical analyses are based on the two-dimensional models of plane stress, plane strain, or plate bending theories. Therefore, it is appropriate to attempt to develop numerical stress analysis procedures applicable to the three-dimensional crack problems of interest. The finite element method is a logical procedure to consider in this regard because of its ability to treat, in a standard manner, a large range of structural geometries and boundary conditions.

This chapter gives a brief review of three-dimensional finite element procedures followed by a presentation of a specialized approach for finite element applications to crack problems. Example problems involving a through crack in a finite thickness plate subjected to symmetric loadings are given to demonstrate this approach. The through crack geometry has been chosen as the primary three-dimensional example for two reasons: first this crack geometry is standard for test specimens and often observed in structural components. Second, literally hundreds of research papers have been written concerning this crack geometry in the past two decades. The vast majority of these papers employed two-dimensional models of plane stress and/or strain to treat the through crack problem for tensile loading and

plating bending theories to analyze the effects of bending moments. Therefore, it is appropriate to study the influence of three-dimensional effects on
this problem and to determine the range of parameters over which the
two-dimensional models yield adequate results for failure prediction.

8.2   Three-dimensional elastic calculations

For purposes of stress analysis a crack is generally modeled as a surface of
displacement discontinuity across which no forces are transferred. Using this
sharp crack edge model of the geometry in conjunction with an elastic,
homogeneous, isotropic model for material response, Hartranft and Sih [2]
have proven that the plane strain singular solution is asymptotically correct
at the crack edge except possibly at corners, i.e., at the points where the
crack edge penetrates the body surface.

The singular portion of the asymptotic solution along the crack edge is
given below for symmetric (mode I) crack problems. A local cylindrical
coordinate system whose axis is tangent to the crack edge is used, Figure 8.1.
The rectangular (local coordinates) components of the stress tensor are

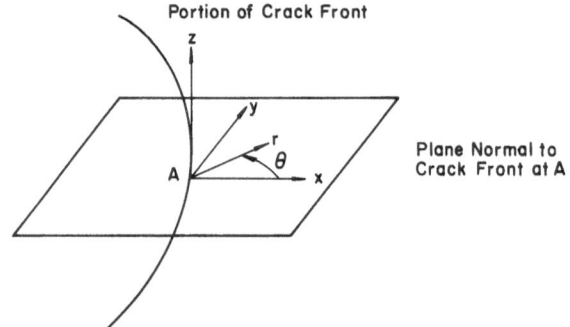

Figure 8.1. Local coordinate system at the crack front

$$\sigma_x = \frac{k_1(z)}{(2r)^{\frac{1}{2}}} \cos \frac{\theta}{2} \left(1 - \sin \frac{\theta}{2} \sin \frac{3\theta}{2} \right) + O(1) \tag{8.1a}$$

$$\sigma_y = \frac{k_1(z)}{(2r)^{\frac{1}{2}}} \cos \frac{\theta}{2} \left(1 + \sin \frac{\theta}{2} \sin \frac{3\theta}{2} \right) + O(1) \tag{8.1b}$$

$$\sigma_z = v(\sigma_x + \sigma_y) + O(1) \tag{8.1c}$$

$$\tau_{xy} = \frac{k_1(z)}{(2r)^{\frac{1}{2}}} \sin \frac{\theta}{2} \cos \frac{\theta}{2} \cos \frac{3\theta}{2} + O(1) \tag{8.1d}$$

$$\tau_{yz} = O(1), \ \tau_{xz} = O(1) \text{ as } r \to 0 \tag{8.1e}$$

The corresponding strain components are

$$\varepsilon_x = \frac{k_1(z)}{(2r)^{\frac{1}{2}}} \frac{(1+v)}{E} \cos \frac{\theta}{2} \left[(1 - 2v) - \sin \frac{\theta}{2} \sin \frac{3\theta}{2} \right] + O(1)$$

$$= k_1(z)\tilde{\varepsilon}_x(r, \theta) + O(1) \tag{8.2a}$$

$$\varepsilon_y = \frac{k_1(z)}{(2r)^{\frac{1}{2}}} \frac{(1+v)}{E} \cos \frac{\theta}{2} \left[(1 - 2v) + \sin \frac{\theta}{2} \sin \frac{3\theta}{2} \right] + O(1)$$

$$= k_1(z)\tilde{\varepsilon}_y(r, \theta) + O(1) \tag{8.2b}$$

$$\varepsilon_z = O(1) \tag{8.2c}$$

$$\gamma_{xy} = \frac{k_1(z)}{(2r)^{\frac{1}{2}}} \frac{2(1+v)}{E} \sin \frac{\theta}{2} \cos \frac{\theta}{2} \cos \frac{3\theta}{2} + O(1)$$

$$= k_1(z)\tilde{\gamma}_{xy}(r, \theta) + O(1) \tag{8.2d}$$

$$\gamma_{yz} = O(1), \ \gamma_{xz} = O(1) \text{ as } r \to 0 \tag{8.2e}$$

and the displacement components relative to the crack edge are

$$u_x = \frac{(1+v)}{4E} k_1(z)(2r)^{\frac{1}{2}} \left[(5 - 8v) \cos \frac{\theta}{2} - \cos \frac{3\theta}{2} \right] + O(r)$$

$$= k_1(z)\tilde{u}_x(r, \theta) + O(r) \tag{8.3a}$$

$$u_y = \frac{(1+v)}{4E} k_1(z)(2r)^{\frac{1}{2}} \left[(7 - 8v) \sin \frac{\theta}{2} - \sin \frac{3\theta}{2} \right] + O(r)$$

$$= k_1(z)\tilde{u}_y(r, \theta) + O(r) \tag{8.3b}$$

$$u_z = O(r) \text{ as } r \to 0 \tag{8.3c}$$

where E is Young's modulus and v is Poisson's ratio.

The amplitude, $k_1(z)$, of this singular solution is not determined by the asymptotic analysis referenced above, but, depends on the global geometry and loading conditions for the particular component under consideration. For the special case of plane strain, $k_1(z)$ is constant along the crack front, i.e., $k_1(z) = k_1$, and is known as the stress intensity factor. It can serve as a parameter to represent the effect of load transmission to the crack tip region. Hence, regardless of the criterion of fracture, it is useful and necessary to calculate the stress intensity factor for a specimen and/or structural component containing a crack in terms of geometry and loading. This need has led to considerable interest and effort in the development of numerical procedures for analysis of crack problems, particularly for three-dimensional problems where most of the analytical methods become unmanageable. The finite element method has received particular attention because of its ability to treat, in a standard manner, complicated geometries and boundary conditions.

8.3   Finite element method—background

The finite element displacement method [3] is based on the variational principle of minimum potential energy, employing a modified Ritz procedure to obtain approximate solutions. The domain under consideration is divided into subregions called elements. The displacement field within each element is then approximated in terms of discrete (nodal) displacement values using interpolating polynominals known as displacement shape functions. Using this assumed form for the displacement field, the strain field, stress field, and finally the strain energy of an element can be expressed in terms of the unknown nodal displacement values. The potential energy for the body is then calculated as the sum of the strain energies of the elements minus the work done by applied forces.

Minimization of this potential energy functional with respect to the nodal displacement values results in a set of governing algebraic equations which is to be solved for these discrete quantities. The stress and strain fields within an element are then expressible in terms of these nodal displacement values.

Two-dimensional finite element procedures for crack problems have been the subject of a number of recent review articles [4, 5]. It has been recognized that the character of the crack tip singularity must be introduced a priori into the finite element method to obtain accurate values for the stress intensity factor (or factors for nonsymmetric situations). Several procedures have been developed for this purpose. The first, presented by W. K. Wilson [6], involved the formulation of a circular crack tip element whose displacement shape function was based on the asymptotic solution for the near tip displacement field, equations (8.3) with $k_1(z) = k_1$; i.e., with the stress intensity factor treated as a generalized nodal displacement component (degree of freedom). The approach was extended to elastic-plastic problems through the introduction of the plastic singularity by Hilton and Hutchinson [7]. Tracey [8] later developed an alternative approach using a sequence of pie-shaped elements to surround the crack tip with displacement shape functions that contain the square root dependence of the near tip displacement field on distance from the crack tip. Pian [9] developed a hybrid crack tip element consistent with his standard elements. Still other approaches have been presented. The one which is of interest for the present work was due to Benzley [1]. This approach employs a rectangular grid pattern with special treatment for the elements adjacent to the crack tip, i.e., their displacement field includes the asymptotic field, equations (8.3), as well as the standard terms. The author is not aware of any analogous finite element treatments for plating bending problems involving cracks.

The only previous three-dimensional treatment of crack problems by specialized finite element procedures of which the author is aware is due to Tracey [10]. Mr. Tracey has extended his two-dimensional approach mentioned above to three-dimensions and treated both through crack and surface crack (semi and quarter circle crack edges were considered) problems.

The Tracey approach enforces a square root singularity in the in-plane strain components and does not allow for non-singular contributions to these components in the immediate vicinity of the crack edge. An alternative approach employing Tracey's element goemetry with an extension of Benzley's displacement shape functions to three-dimensions will be presented here. The new approach is believed to be a significant improvement to the work of Tracey because it allows for nonsingular contributions to the strain components along the crack edge. These contributions, while not particularly important for two-dimensional problems, allow for the large distortional strains with small (or zero) stress intensity factor believed to exist in the vicinity of the intersection of the crack edge with the body surface.

The finite element approach to three-dimensional crack problems to be described here is applicable to curved (smooth) crack fronts; however, the examples considered are for a straight crack front normal to the plate surface. The straight crack front geometry is implicitly assumed in all two-dimensional analyses and also it is employed in manufacture of test specimens. On the other hand, test specimens, after sufficient crack growth, are observed to contain a curved (thumb nail shaped) crack front. Thus the straight crack front assumed here is not believed to be the natural crack geometry. One

aim of the present work is to explain the growth of the initial manufactured
crack front into the thumb nail shape observed in laboratory specimens.

8.4   Specialized elements for the crack edge

The finite element displacement method incorporating isoparametric elements
[3] is the basis of the approach to be presented. In general, an element is
enclosed by six surfaces with nodal points chosen along the element edges
(a five surface element will be introduced later in the text). A coordinate
transformation is employed to map the element into a cube, Figure 8.2. The
inverse mapping function can be expressed in the form

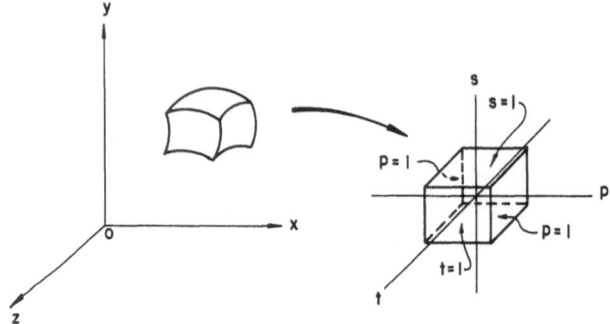

Figure 8.2. Mapping of curvilinear finite elements

$$\begin{bmatrix} x \\ y \\ z \end{bmatrix} = \sum_i N_i(p, s, t) \begin{bmatrix} x_i \\ y_i \\ z_i \end{bmatrix}$$

where $i = 1$, the number of nodes associated with the element, and (x_i, y_i, z_i)
are the coordinates of the ith node. The functions $N_i(p, s, t)$ associated with
the mapping are polynomial functions with the property

$$N_i(p, s, t) = \begin{cases} 1 \text{ for } p, s, t = p_i, s_i. t_i \text{(the point in the mapped shape which} \\ \qquad\qquad\qquad\text{corresponds to } x_i, y_i, z_i) \\ \\ 0 \text{ for } p, s, t = p_j, s_j, t_j\, j \neq i \end{cases} \tag{8.5}$$

The same functions $N_i(p, s, t)$ are used in the approximation for the displacement field within the element, i.e., the interior displacement components

are expressed in terms of the corresponding nodal components as

$$
\begin{bmatrix} u_x \\ u_y \\ u_z \end{bmatrix} = \sum_i N_i(p, s, t) \begin{bmatrix} u_{x_i} \\ u_{y_i} \\ u_{z_i} \end{bmatrix}
\tag{8.6}
$$

It is now appropriate to mention that these functions are known as 'displacement shape functions'.

The strain components are obtained in terms of the nodal displacement components by differentiating equation (8.6), i.e.,

$$
\varepsilon_x = \sum_i \frac{\partial N_i}{\partial x} u_{x_i}, \quad \gamma_{xy} = \sum_i \frac{\partial N_i}{\partial y} u_{x_i} + \sum_i \frac{\partial N_i}{\partial x} u_{y_i}
\tag{8.7a}
$$

$$
\varepsilon_y = \sum_i \frac{\partial N_i}{\partial y} u_{y_i}, \quad \gamma_{yz} = \sum_i \frac{\partial N_i}{\partial z} u_{y_i} + \sum_i \frac{\partial N_i}{\partial y} u_{z_i}
\tag{8.7b}
$$

$$
\varepsilon_z = \sum_i \frac{\partial N_i}{\partial z} u_{z_i}, \quad \gamma_{zx} = \sum_i \frac{\partial N_i}{\partial x} u_{z_i} + \sum_i \frac{\partial N_i}{\partial z} u_{x_i}
\tag{8.7c}
$$

where

$$
\begin{bmatrix} \dfrac{\partial N_i}{\partial p} \\[2ex] \dfrac{\partial N_i}{\partial s} \\[2ex] \dfrac{\partial N_i}{\partial t} \end{bmatrix} = \begin{bmatrix} \dfrac{\partial x}{\partial p} & \dfrac{\partial y}{\partial p} & \dfrac{\partial z}{\partial p} \\[2ex] \dfrac{\partial x}{\partial s} & \dfrac{\partial y}{\partial s} & \dfrac{\partial z}{\partial s} \\[2ex] \dfrac{\partial x}{\partial t} & \dfrac{\partial y}{\partial t} & \dfrac{\partial z}{\partial t} \end{bmatrix} \begin{bmatrix} \dfrac{\partial N_i}{\partial x} \\[2ex] \dfrac{\partial N_i}{\partial y} \\[2ex] \dfrac{\partial N_i}{\partial z} \end{bmatrix}
\tag{8.8}
$$

is inverted to obtain the needed partial derivatives. This calculation is expressed formally as

$$
\varepsilon = [B] \delta
\tag{8.9}
$$

where

$$
\varepsilon = (\varepsilon_x, \varepsilon_y, \varepsilon_z, \gamma_{xy}, \gamma_{yz}, \gamma_{zx})
$$

$$
\delta = (u_{x_1}, u_{y_1}, u_{z_1}, u_{x_2}, \ldots u_{z_n})
\tag{8.10}
$$

n = number of nodes per element

The stress components are related to the strain components through the generalized Hooke's law

$$\sigma = [D]\, \varepsilon \tag{8.11}$$

The strain energy in an element can then be expressed as

$$W = \tfrac{1}{2}\, \delta^T [k]\, \delta \tag{8.12}$$

where $[k]$ is the element stiffnes matrix given by

$$[k] = \int_{-1}^{1} \int_{-1}^{1} \int_{-1}^{1} [B]^T [D][B] \det J \, dp \, ds \, dt \tag{8.13}$$

and where $\det J$ is the determinate of the Jacobian for the transformations given in equations (8.8).

The potential energy of the body, calculated as the sum of the element strain energy contributions minus the work done by applied forces, is minimized with respect to nodal displacement components to obtain a set of linear algebraic equations for the determination of their values.

Elements which have an edge that contacts the crack edge are treated specially. The approximate form for their displacement field is chosen so as to include the crack edge singularity, equations (8.3), with unknown amplitude $k_1(z)$ plus the standard nonsingular terms for the corresponding isoparametric element. The variation of the stress intensity factor along the crack edge is approximated in a piece-wise fashion consistent with the finite element procedure. Letting s be the variable in the mapped space corresponding to the crack edge, the stress intensity factor is assumed in the form

$$k_1(s) = \sum_i N_i^*(s) k_i$$

where the summation is for values of i from one to the number of nodes along the edge. The displacement field in an element adjacent to the crack edge is then approximated in the local coordinate system shown in Figure 8.1 as

$$u_x = \sum_i N_i(p,s,t)\left[u_{x_i} - \tilde{u}_{x_i} \sum_j k_j N_j^*(s)\right] + \tilde{u}_x(r,\theta) \sum_j k_j N_j^*(s) \tag{8.14a}$$

$$u_y = \sum_i N_i(p,s,t)\left[u_{y_i} - \tilde{u}_{y_i} \sum_j k_j N_j^*(s)\right] + \tilde{u}_y(r,\theta) \sum_j k_j N_j^*(s) \tag{8.14b}$$

$$u_z = \sum_i N_i(p,s,t) u_{z_i} \tag{8.14c}$$

where $\tilde{u}_x(r,\theta)$ and $\tilde{u}_y(r,\theta)$ are given by equations (8.3) and $\tilde{u}_{x_i}$, $\tilde{u}_{y_i}$ are the values of these functions at the ith node of the element.

The strain components are obtained by differentiating equations (8.14) as

$$\varepsilon_x = \sum_i N_{i,x} \left[u_{x_i} - \tilde{u}_{x_i} \sum_j N_j^* k_j \right] + \sum_j N_j^* \tilde{\varepsilon}_x k_j \tag{8.15a}$$

$$\varepsilon_y = \sum_i N_{i,y} \left[u_{y_i} - \tilde{u}_{y_i} \sum_j N_j^* k_j \right] + \sum_j N_j^* \tilde{\varepsilon}_y k_j \tag{8.15b}$$

$$\varepsilon_z = \sum_i N_{i,z} \, u_{z_i} \tag{8.15c}$$

$$\gamma_{xy} = \sum_i N_{i,y} \left[u_{x_i} - \tilde{u}_{x_i} \sum_j N_j^* k_j \right] + \sum_i N_{i,x} \left[u_{y_i} - \tilde{u}_{y_i} \right.$$

$$\left. \times \sum_j N_j^* k_j \right] + \sum_j N_j^* \tilde{\gamma}_{xy} k_j \tag{8.15d}$$

$$\gamma_{yz} = \sum_i N_{i,z} \left[u_{y_i} - \tilde{u}_{y_i} \sum_j N_j^* k_j \right] + \sum_i N_{i,y} \, u_{z_i}$$

$$+ \sum_j N_{j,z}^* \left(\tilde{u}_{y_i} - \sum_i N_i \, \tilde{u}_{y_i} \right) k_j \tag{8.15e}$$

$$\gamma_{zx} = \sum_i N_{i,z} \left[u_{z_i} - \tilde{u}_{z_i} \sum_j N_j^* k_j \right] + \sum_i N_{i,z} \, u_{z_i}$$

$$+ \sum_j N_{j,z}^* \left(\tilde{u}_{x_i} - \sum_i N_i \, \tilde{u}_{x_i} \right) k_j \tag{8.15f}$$

The last summations in the equations for ε_x, ε_y, γ_{xy} contain the strain singularity while all other terms remain finite at the crack tip.

Note that the nodal values of the stress intensity factor are treated as generalized nodal displacement components, i.e., let

$$\delta^* = u_{x_1}, u_{y_i}, u_{z_i}, u_{x_2}, \ldots, u_{z_m}, k_1, k_2, \ldots k_n$$

where n is the number of nodes along the crack front and m is the number of nodes of the elements adjacent to the crack front. Then the strains can be expressed in the form

$$\varepsilon = [B^*] \delta^*$$

and the element stiffness matrix for elements adjacent to the crack front is given by

$$[k] = \int_{-1}^{1} \int_{-1}^{1} \int_{-1}^{1} [B^*]^T [D][B^*] \det J \, dp \, ds \, dt \tag{8.16}$$

It needs to be pointed out that in cases where the local coordinate system (x, y, z) along the crack edge differs from the global coordinate system (X, Y, Z), a coordinate transform [3] for the nodal displacement vector δ is required in the form

$$\delta_{x,y,z} = [\lambda] \delta_{X,Y,Z} \tag{8.17}$$

The element stiffness matrix then becomes

$$[k] = \int_{-1}^{1} \int_{-1}^{1} \int_{-1}^{1} [\lambda]^T [B^*]^T [D][B][\lambda] \det J \, dp \, ds \, dt \qquad (8.18)$$

8.5 Applications to crack problems

The results to be presented are based on use of 20 node isoparametric elements with nodes at each element corner and an intermediate node along each edge. This element yields a quadratic approximation in the mapped space for the displacement field along each edge with some additional cubic terms in the interior. Specialized elements with rectangular cross section (the logical extension of Benzley's element) and with pie-shaped cross section (similar to Tracey's element) have been tested. The pie-shaped configuration was found preferable because it enabled more accurate modeling of circumferential variations for the displacement field in the vicinity of the crack edge. Typical grid patterns are shown in Figures 8.3 and 8.4.

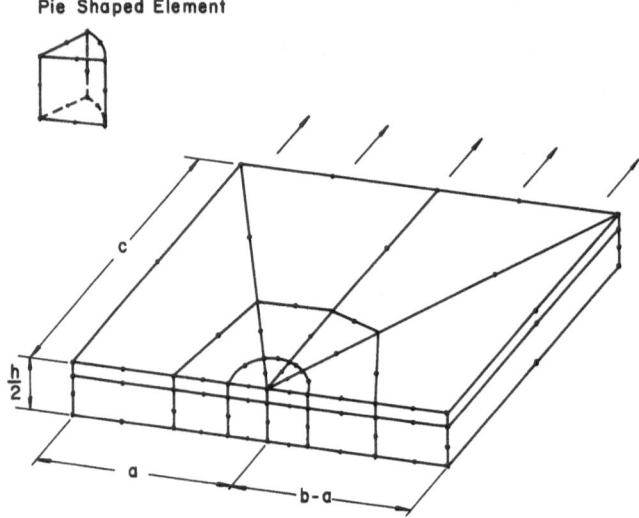

Figure 8.3. First octant for center cracked plate with two layer grid pattern

The variation of the stress intensity factor along the portion of the crack edge corresponding to a particular element was taken as quadratic in the mapped space for consistency with the element chosen, i.e.,

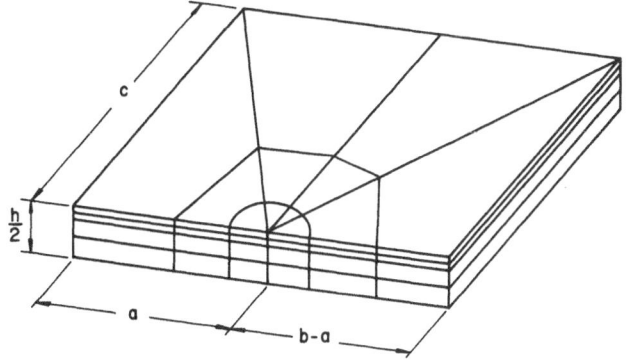

Figure 8.4. Refined (four layer) grid pattern

$$k_1(s) = \sum_1^3 N_j^*(s)k_j$$

where

$$N_1^* = s(s-1)/2$$

$$N_2^* = (1 - s^2)$$

$$N_3^* = s(s+1)/2$$

8.6 Details of the analysis

The specialized finite element procedure described above was employed to consider the problem of a through crack in a plate subjected to (a) tensile and (b) bending loads. The in-plane dimensions of the plate were kept constant at the values shown in Figure 8.5 and the loading was maintained at either $\sigma_0 = 1$ psi for the tension case or $\sigma_0 = 2y/h$ for the bending problem. The influence of plate thickness was studied by considering two values for h at 1.0 inch and 0.6 inch.

Note that the geometry chosen for consideration makes both the tension and bending problems symmetric about the xz and yz planes. Further, the tension case is symmetric about the xy plane while the bending problem is skew-symmetric about the same xy plane.

In this analysis it is assumed that no contact exists between the crack faces; yet, the results for the bending case will indicate a negative stress

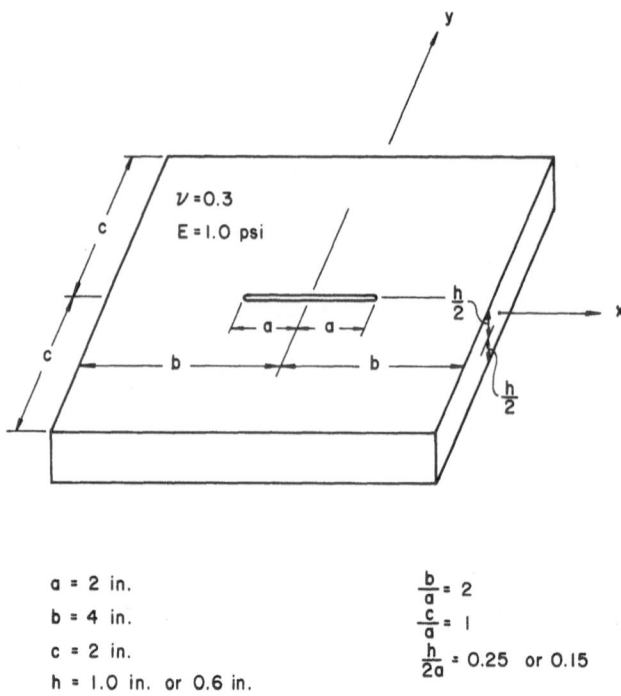

a = 2 in. $\frac{b}{a}$ = 2

b = 4 in. $\frac{c}{a}$ = 1

c = 2 in. $\frac{h}{2a}$ = 0.25 or 0.15

h = 1.0 in. or 0.6 in.

Figure 8.5. Plate geometry

intensity factor and material overlap in the compressive side of the plate. This inconsistency is avoided by considering plates subjected to simultaneous tension and bending with sufficient tension applied to prevent the overlap. Three-dimensional finite element calculations for crack closure problems associated with bending loads can be carried out using an iterative procedure to determine the region of crack face contact. While this approach is not conceptually difficult its application to three-dimensional crack problems requires excessive computation time and is therefore postponed in the hope of determining methods which will substantially reduce computational costs. The linear character of elasticity solutions allows for the use of the superposition principle to add tensile and bending solutions in appropriate proportions such that the result satisfies the consistency argument even though the bending solution itself is not meaningful. This is the approach taken here.

Finite element analyses were carried out for two grid patterns: the first shown in Figure 8.3 contains 24 elements and 161 nodes: The second grid is a refinement of the first developed by dividing each of the two layers of elements in half in the thickness direction to form a four layer grid with 48 elements and 279 nodes, Figure 8.4.

One of the objectives of this work is to determine the variation of the stress intensity factor through the specimen thickness and the variations of the near field stress components and energies in the thickness direction and with distance from the crack edge. Graphs illustrating these variations are given in Figures 8.6–8.11 for the tensile case and Figures 8.12–8.14 for the bending problem.

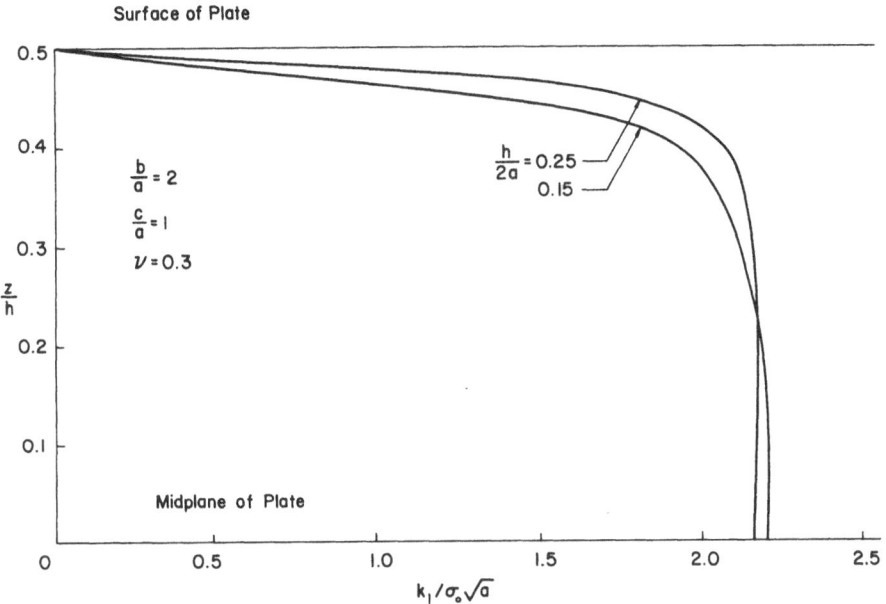

Figure 8.6. Variation of stress intensity factor along crack front, tensile loading

Notice that for the case of tensile loading the stress intensity factor is almost constant along most of the crack edge but decreases sharply near the plate surface. The slope or rate of decrease is increased as the grid pattern is refined; however, the numerical values for the stress intensity factor at the free surface are far from zero. Considerable research efforts concerning this behavior of the stress intensity in the vicinity of the free surface have been reported by Hartranft and Sih [2, 11, 12]. Their results indicate that a thin

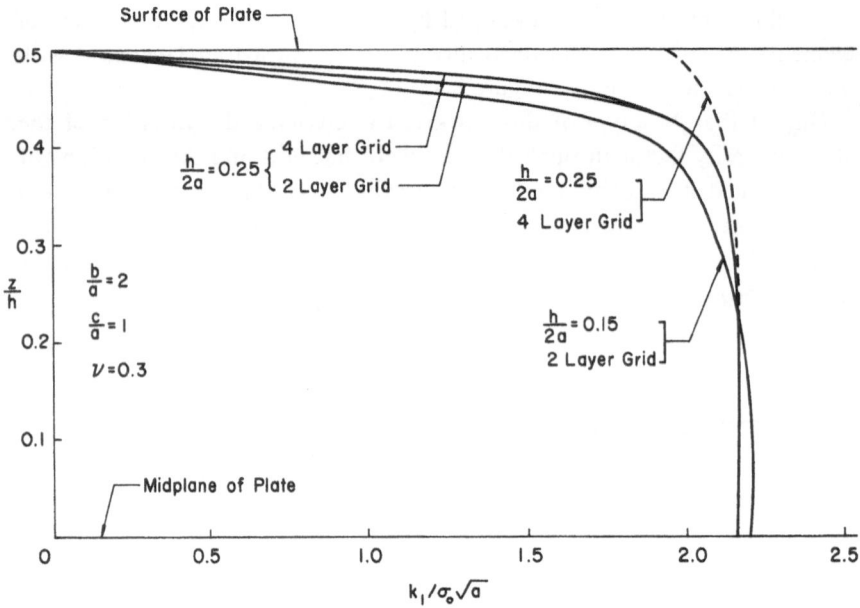

Figure 8.7. Influence of procedure on the variation of the stress intensity factor along the crack front, tensile loading

boundary layer is expected at the free surface across which the stress intensity factor varies very rapidly and that it should approach zero at the free surface. Their calculations further clearly indicate that a numerical procedure, as used here, without special treatment of the boundary layer region is not adequate to model these rapid variations and should not be expected to result in zero stress intensity at the free surface. On the other hand, the finite element procedure used here enables one to enforce $k_1(h/2) = 0$. That condition has been imposed on some of the runs reported here and comparisons with and without enforcing $k_1(h/2) = 0$ are shown in Figures 8.7–8.10 for the tensile loading situation and in Figures 8.12–8.14 for the bending case. Notice that, to the accuracy of the results obtained and reported in graphical form by the finite element procedure, the interior fields are not substantially altered by the introduction of the surface constraint, $k_1(h/2) = 0$.

It is appropriate to make some estimates of the accuracy of the results presented here. For that purpose, a two-dimensional finite element calcula-

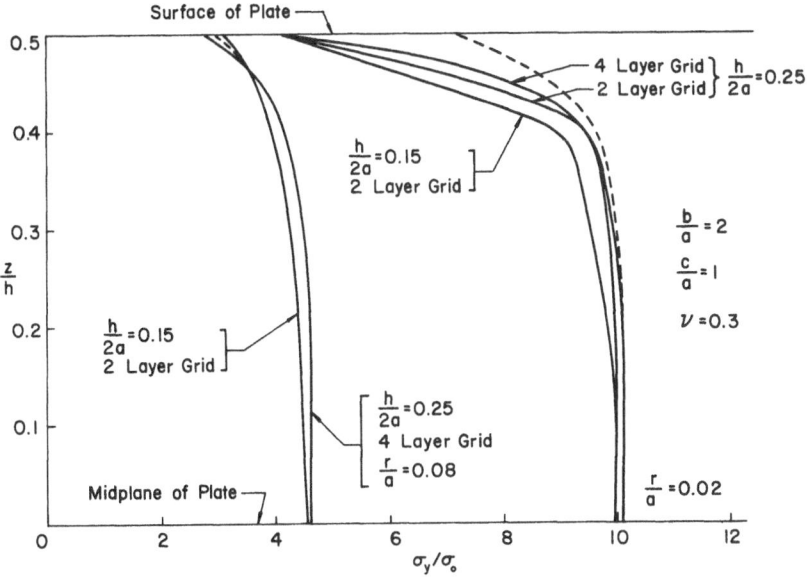

Figure 8.8. Variation of the in-plane normal stress through the plate thickness

tion (plane strain) was performed using the NSRDC [13] twelve node isoparametric element program with a circular singularity element at the crack
tip. Results from that program with appropriate grid patterns are believed
to be accurate to 2%. The value of the stress intensity factor obtained from
the two-dimensional finite element analysis is $k_1 = 1.96\ \sigma_0\sqrt{a}$. This result
is within 10% of the interior model values obtained for the stress intensity
factor using the three-dimensional finite element solution procedure for the
case of tensile loading. It is interesting to note that the three-dimensional
results for the stress intensity in the interior of the plate are actually larger
in magnitude than that obtained from the two-dimensional analysis. As the
finite element method generally underestimates the primary unknowns (in
this case, the nodal values for the stress intensity and for the displacement
components), this observation is an indication that the numerical results
for the stress intensity factor obtained from the three-dimensional analysis
are at least as accurate as is implied by the 10% maximum deviation from
the two-dimensional predictions. Tracey [10] also observed that comparable

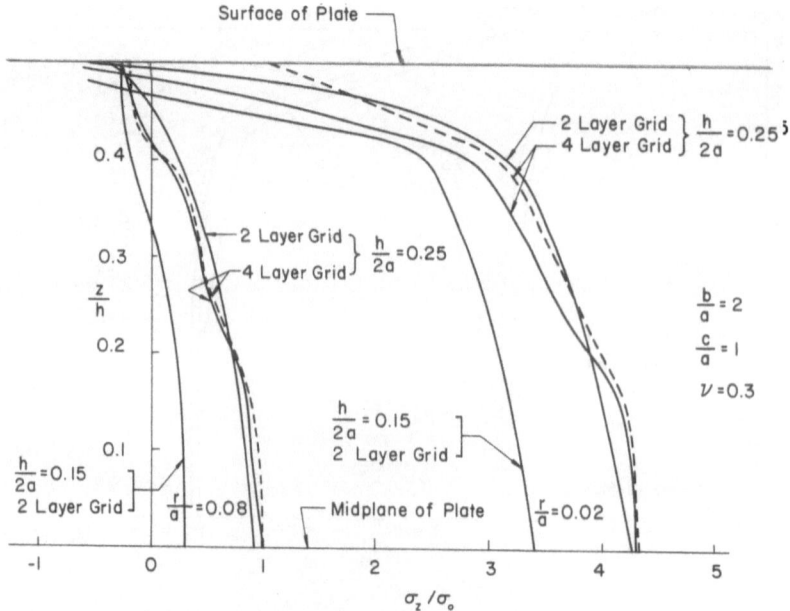

Figure 8.9. Variation of transverse normal stress at $\theta = 0$ through the plate thickness, tensile loading

grid patterns yielded more accurate results for three-dimensional analysis of crack problems than for corresponding two-dimensional calculations.

Comparative results required to estimate the accuracy of the bending solutions are more difficult to obtain. Solutions for an infinite plate containing a through crack and subjected to a uniform far field bending moment distribution have been obtained using classical plate bending theory in 1962 by Sih et al [14]. The solution yields $k_1 = (12z/h^3)M_0\sqrt{a}$. Hartranft and Sih [15] solved the same problem employing the Reissner plate bending theory in 1968. The latter solution results in a thickness effect prediction, i.e.,

$$k_1 = F(h/a)\ \frac{12z}{h^3}\ M_0\ \sqrt{a}$$

where the function $F(h/a)$ has been determined numerically and presented a graphical form. The author is not aware of analogous solutions for finite size plates which would predict the influence of the in-plane dimensions (b/a and c/a) on the stress intensity amplitude as well as distribution. A crude

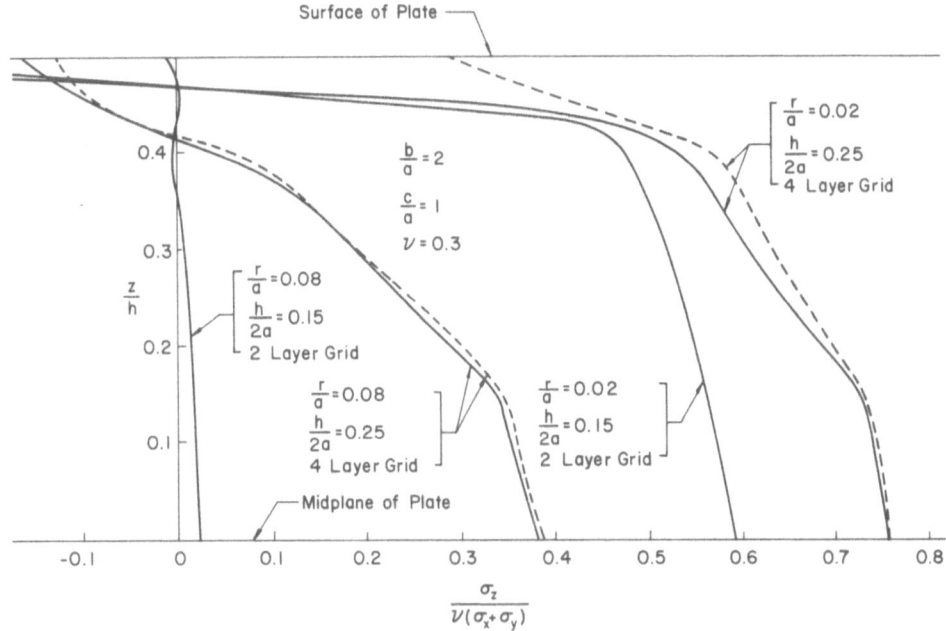

Figure 8.10. Degree of plane strain constraint at $\theta = 0$, tensile loading

estimate of this effect can be obtained from the plane strain solution discussed above. That solution indicates that the stress intensity factor for the in-plane geometry considered here is approximately 1.96 times the corresponding value for an infinite plate. If a similar influence of in-plane dimensions is anticipated for the bending problem, the Sih et al and Hartranft – Sih solutions given above should be multiplied by 1.96 before comparison with the finite element results reported here. The normalized slopes for the stress intensity factor against distance from the mid-plane $[(k_1/\sigma_0\sqrt{a})/(2z/h)]$ reported by Hartranft and Sih and corrected for the in-plane plate dimensions as discussed above are 1.17 and 1.37 for $h = 0.6$ and 1.0 respectively.

Three dimensional finite element predictions for the transverse variation of the stress intensity factor in the bending case are reported in Figure 8.12 for the two and four layer grid patterns employed. The two layer grid pattern calculations have been performed for both the cases of k_1 treated as a free parameter at the plate surface and k_1 constrained to be zero at that surface. (Results are shown in Figures 8.12–8.13 for $h/2a = 0.25$). The four layer grid

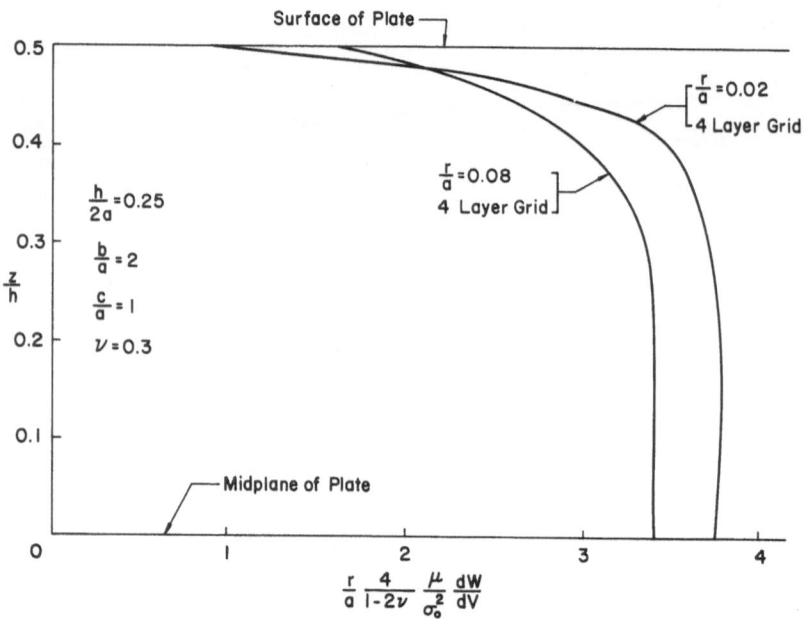

Figure 8.11a. Variation of strain energy density minimum ($\theta = 0$) across plate thickness, tensile loading, $h = 0.5a$

pattern calculations have only been carried out for the constrained situation. Notice that for the bending problem, unlike the tensile case, the four layer grid pattern calculation yields larger values for the stress intensity factor in the plate interior than those predicted by the two layer grid results. This is an indication that the finer grid pattern is necessary to obtain accurate numerical results for the plate bending problem. These fine grid calculations result in normalized slopes for the stress intensity factor $(k_1/\sigma_0\sqrt{a})/(2z/h)$ in the specimen interior of 1.23 and 1.25 for $h = 0.6$ and 1.0, respectively.

The trend of increasing normalized slope for the stress intensity factor predicted by Hartranft and Sih is confirmed by the present results. Further, the magnitudes for this slope, reported here, are in reasonable agreement with those given by the Hartranft – Sih calculations.

8.7   Results of the finite element analysis

The finite element solution was obtained for the geometry shown in Figure

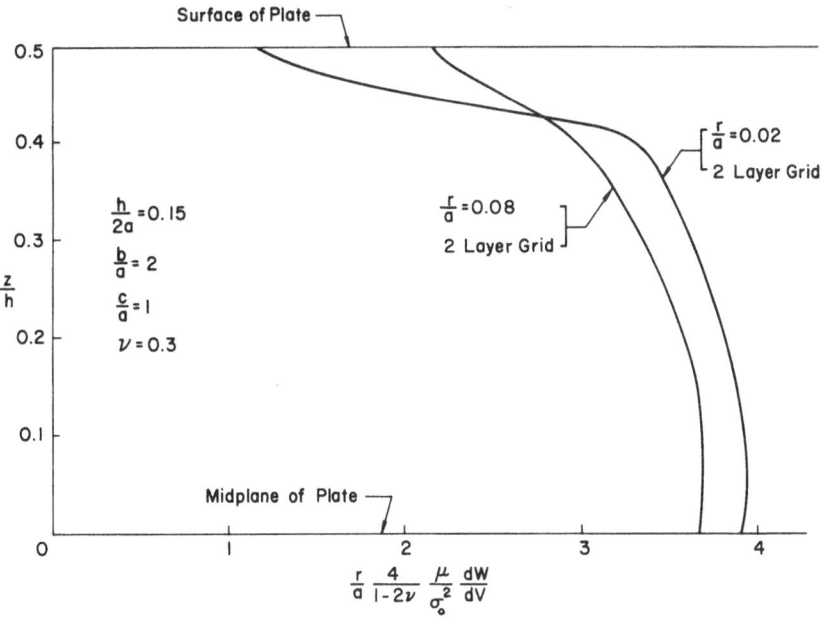

Figure 8.11b. Variation of strain energy density minimum ($\theta = 0$) across plate thickness, tensile loading, $h = 0.3a$.

8.5. From dimensional considerations, there are three basic geometrical parameters: the ratios of plate width, plate length, and plate thickness to crack length. These ratios, b/a, c/a, $h/2a$, take the respective values, 2.0, 1.0, 0.25 in one case, and 2.0, 1.0, 0.15 in another. Because of the linear nature of the elastic analysis, the stress state may be normalized by the applied stress, σ_0 which is uniform for the tensile case.

The basic result, i.e., the variation of the computed stress intensity factor along the crack front from midplane to the surface, is shown for the case of tensile loading in Figure 8.6. For the two plate thicknesses used the values for the stress intensity factor are fairly close along the center portion of the crack front, but deviate significantly near the surfaces. It is expected that a wider range of plate thicknesses would produce larger differences throughout the plate. The zero value of the stress intensity factor at the surface of the plate has been imposed on the problem. The effect of allowing the finite element method to determine the value at the surface, and the accuracy of the results can be judged by referring to Figure 8.7.

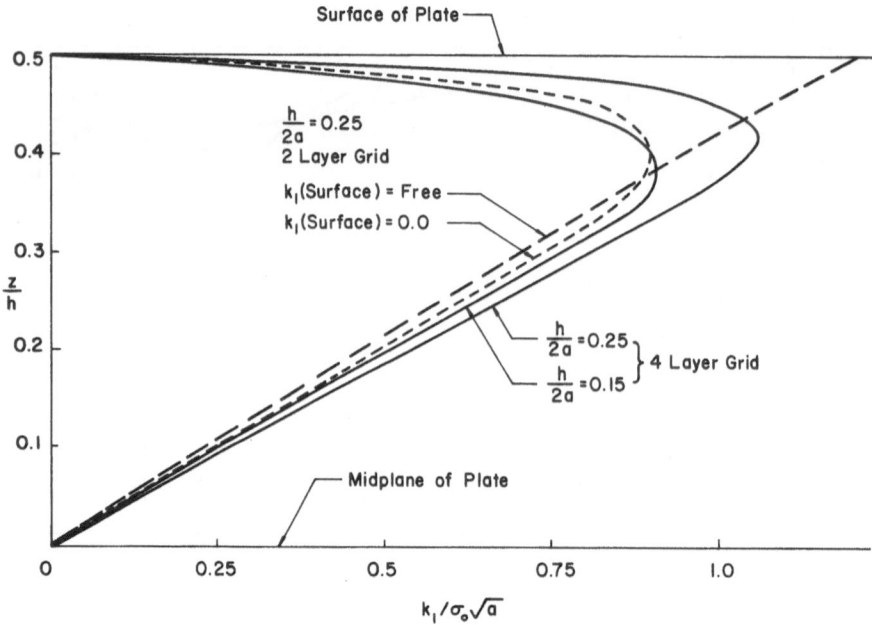

Figure 8.12. Variation of stress intensity factor across plate thickness for bending case

The variations in Figure 8.7 of the stress intensity factor show the effects of increasing the number of elements modeling the plate. For $h/2a = 0.25$, there are two curves which both result in $k_1 = 0$ at the surfaces. Comparison of them will reveal that they differ only slightly near the surface of the plate. The curve corresponding to four layers (48 elements) gives higher values than that for two layers (24 elements) of elements. This indicates a more rapid decrease of stress intensity factor in the boundary layer for the more accurate solution.

A comparison of the curves in Figure 8.7 for $h/2a = 0.25$ which result from the four layer grid pattern may be made. For one of the curves, the value of k_1 at the surface was determined from the finite element solution in the same way as at all other points. For the other, this value was chosen to be zero. The curves may be seen to be indistinguishable for $z/h < 0.3$, and quite distinct at the surfaces. Nevertheless, the decrease shown at the surface by the curve in which k_1 at the surface was left free reinforces the suspicion [2, 16, 17] that in the boundary layer, the stress singularity is relatively less important.

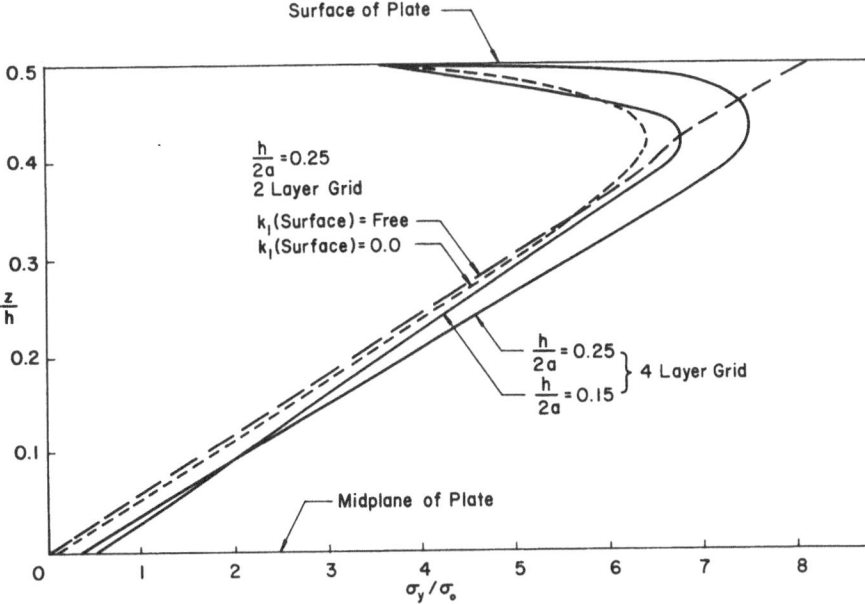

Figure 8.13. Variation of σ_y/σ_0 across plate thickness for bending problem

To further investigate the condition of the neighborhood of the crack front, stresses were computed. These stresses include the nonsingular as well as the singular parts. For the positions considered, the nonsingular portion is seen to be considerable. The normal stress component, σ_y, acting on the plane directly ahead of the crack front is shown in Figure 8.8. The two families of curves are for two distances ahead of the crack front. The larger stresses act at the closer points a distance of 0.02a, and the smaller at 0.08a. The decrease in stress as the surface is approached is primarily due to the decrease in the singular terms (compare Figures 8.6 and 8.7). However, since the singular terms are zero at the surface for all but two curves (shown dashed), the surface values of σ_y must be due to nonsingular effects. At the larger distance from the crack, the singular part is relatively small.

The transverse normal stress, σ_z, acting on planes parallel to the plate surfaces is shown in Figure 8.9 for the same thicknesses and distances from the crack front used for σ_y. Again, as for Figure 8.8, the dashed curves were obtained without constraining k_1 to be zero at the surfaces. This stress should be zero at the surface, but because of the nature of finite element

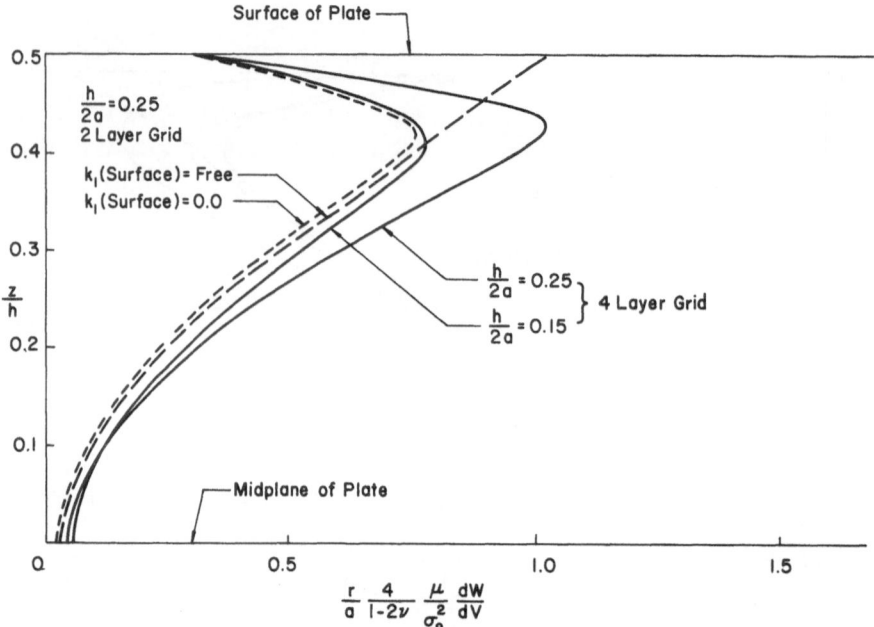

Figure 8.14. Transverse variation of normalized strain energy density for bending case

methods imposing displacement fields, it is not. The surface values are low compared to other values, and so the small discrepancy is tolerable. The element interfaces for the four layer grid pattern occur at $2z/h = 0.2$, 0.4, and 0.45. Rapid changes in magnitude observed for σ_z/σ_0 at these points (Figure 8.9) are numerical and, as such, indicate the order of inaccuracies associated with the finite element calculations employing this grid pattern.

The singular portion of the crack tip stress state is known [2] to satisfy the plane strain condition

$$\sigma_z = \nu(\sigma_x + \sigma_y)$$

Figure 8.10 illustrates the effect of the nonsingular terms on the validity of this condition for points on the plane directly ahead of the crack. The ratio of σ_z to the plane strain value is plotted as a function of distance from the midplane of the plate for the various parameters considered. The closer to the crack front that the ratio is computed, the closer to the plane strain value of one it becomes. In addition, for a given distance from the crack front, the deviation from the plane strain value increases as the surface is approached. Thus, as the model of reference [17] proposed, the stress singularity may be taken to be a state of plane strain except in a boundary layer near the surface. Notice, finally, that the thinner plate deviates significantly more from plane strain than the thicker plate.

The plane strain value of the strain energy density for an infinite, uniformly stressed, cracked body has been discussed in [18]. The minimum value of the limit

$$S_{min} = \frac{1 - 2v}{4\mu}\sigma_0^2 a, \quad S = \lim_{r \to 0} r\frac{dW}{dV}$$

is attained in the plane directly ahead of the plane of the crack. The fracture criterion proposed in [18]* is equivalent to several others in cases such as this in which only Mode I loading is present.

In this three-dimensional case, the strain energy density, dW/dV, was computed for several values of distance, r, from the crack front and several positions through the thickness. The results are normalized by the plane strain value given above. As shown in Figure 8.11, this value,

$$\frac{r}{a} \quad \frac{4}{1 - 2v} \quad \frac{\mu}{\sigma_0^2} \quad \frac{dW}{dV}$$

is large at midplane, reflecting the large stress intensity factors there. The values decrease as the surface is approached but, because of the nonsingular terms included, not to zero. The points closer to the crack front show the sharpest decrease at the surface. For values of r close to zero, the curves in Figure 8.11 approach the variation of the normalized stress intensity factor squared as obtained from Figures 8.6 and 8.7, i.e.,

$$\frac{r}{a} \quad \frac{4}{1 - 2v} \quad \frac{\mu}{\sigma_0^2} \quad \frac{dW}{dV} \quad \frac{k_1^2}{\sigma_0^2 a} \quad \text{as } r \to 0$$

The finite element solutions for bending were obtained using the same plate geometries (Figure 8.5) and grid patterns (Figures 8.3 and 8.4) as considered in the tensile loading case. The tractions applied to the surfaces $y = \pm c$ for this case are

---

* Fracture occurs in the direction of minimum S at a load which causes S_{min} to reach a critical value, S_{cr}, characteristic of the material.

$$\sigma = 2 \frac{\sigma_0 z}{h}$$

with $\sigma_0 = 1$ psi
leading to a bending moment per unit length of

$$M_0 = \frac{\sigma_0 h^2}{6}$$

Numerical results for the variations of the stress intensity factor across the tensile portion of the plate thickness are given in Figure 8.12. Calculations were performed for two values of the normalized plate thickness, $h/2a = 0.15$ and 0.25. For the thicker plate computations were carried out with the stress intensity at the plate surface unconstrained and with the zero values of the stress intensity factor imposed at the plate surface. Unlike the tensile case, the results for the stress intensity factor with no surface constraint imposed show little tendency to drop off at the free surface. In this case the tendency of the numerical results for the stress intensity factor to approach zero at the free surface is opposed by the behavior interior to the plate for which the stress intensity factor is expected to increase with distance from the mid-plane.

Notice further that, unlike the tensile loading case, refinement of the grid pattern in the transverse direction for the bending problem results in significantly improved predictions for the stress intensity factor in the plate interior. This observation indicates that the four layer grid pattern is needed to obtain accurate solutions for the plate bending problem considered here.

In comparing the results presented in Figure 8.12 for the fine grid calculations with the stress intensity factor constrained at the plate surface, it is of interest to note that the relative 'boundary layer' thickness increases as plate thickness decreases. Similar observations can be made for the tensile loading case. They indicate a more dominate influence of the plate surfaces on the interior solution for thinner plates and, as such, suggest that two-dimensional models become less applicable as plate thickness is decreased.

Figure 8.13 contains curves for the variation of the normalized stress component (σ_y/σ_0) normal to the crack plane along the line $r = .01a$ directly ahead of and parallel to the crack edge. These curves also show the influence of the conditions imposed at the plate surface and the thickness effect on the near field transverse distributions.

Figure 8.14 contains plots of the normalized strain energy density directly ahead of the crack edge ($r = .01a$) as a function of the transverse variable $2z/h$.

8.8 Summary

A three-dimensional finite element procedure has been developed for treating crack problems. Results were obtained for problems involving a plate containing a through crack and subjected to tensile or bending loads. Transverse variations for the stress intensity factor and stress components in the vicinity of the crack edge were reported. These results indicate the influence of plate thickness and demonstrate the limitations of previously reported solutions based on the two-dimensional models of plane stress, plane strain, and plate bending theories.

References

[1] S. E. Benzley, Representation of Singularities with Isoparametric Finite Elements, *Int. J. Num. Meth. Eng.*, Vol. 8, pp. 537–545 (1974).

[2] R. J. Hartranft and G. C. Sih, The Use of Eigenfunction Expansions in the General Solution of Three-Dimensional Crack Problems, *J. Math. Mech.*, Vol. 19, pp. 711–729 (1970).

[3] O. C. Zienkiewicz, *The Finite Element Method in Engineering Science*, McGraw-Hill, London (1971).

[4] P. D. Hilton and G. C. Sih, Application of the Finite Element Method to the Calculations of Stress Intensity Factors, *Methods of Analysis and Solutions of Crack Problems*, pp. 426–483, (G. C. Sih, editor), Noordhoff International Publishing, Leyden, The Netherlands (1973).

[5] J. R. Rice and D. M. Tracey, Computational Fracture Mechanics, *Numerical and Computer Methods in Structural Mechanics*, (S. J. Fenves, editor), Academic Press, New York, pp. 585–623 (1973).

[6] W. K. Wilson, *Combined Mode Fracture Mechanics*, Ph. D. Dissertation, University of Pittsburgh (1969).

[7] P. D. Hilton and J. W. Hutchinson, Plastic Intensity Factors for Crack Problems, *Eng. Fract. Mech.*, Vol. 3, pp. 435–451 (1971).

[8] D. M. Tracey, Finite Elements for Determination of Crack Tip Stress Intensity Factors, *Eng. Fract. Mech.*, Vol. 3, pp. 255–265 (1971).

[9] T. H. Pian, P. Tong and C. H. Luk, Elastic Crack Analysis by a Finite Element Hybrid Method, *Proc 3rd Int. Conf. on Matrix Methods in Struct. Mech.*, Wright-Patterson Air Force Base, Ohio, (November 1971).

[10] D. M. Tracey, Finite Elements for Three-dimensional Elastic Crack Analysis, *Nuclear Eng. and Design*, Vol. 26, pp. 282–290 (1973).

[11] R. J. Hartranft and G. C. Sih, Alternating Method Applied to Edge and Surface Crack Problems, *Methods of Analysis and Solutions of Crack Problems*, pp. 179–238, (G. C. Sih, editor), Noordhoff International Publishing, Leyden, The Netherlands (1973).

[12] G. C. Sih, Three-Dimensional Stress State in a Cracked Plate, *Int. J. Fract.*, Vol. 7, pp. 39–61 (1971).

[13] P. D. Hilton, L. N. Gifford Jr. and O. Lomacky, Finite Element Fracture Mechanics Analysis of Two-Dimensional and Axisymmetric Elastic and Elastic-Plastic Cracked Structures, NSRDC Report 4493, (November 1974).

[14] G. C. Sih, P. C. Paris, and F. Erdogan, Crack Tip Stress Intensity Factors for Plane Criterion and Plate Bending Problems, Transactions, *Am. Soc. Mechanical Engineers*, *J. Appl. Mech.*, Vol. 29, pp. 307–312 (1962).

[15] R. J. Hartranft and G. C. Sih, Effect of Plate Thickness on the Bending Stress Distribution Around Through Cracks, *J. of Maths.* and *Physics*, Vol. 47, No. 3, pp. 276–291 (1908).

[16] G. Villarreal, G. C. Sih and R. J. Hartranft, Photoelastic Investigation of a Thick Plate with a Transverse Crack, *J. Appl. Mech.*, Vol. 42, pp. 9–14 (1975).

[17] R. J. Hartranft and G. C. Sih, An Approximate Three-Dimensional Theory of Plates with Applications to Crack Problems, *Int. J. Engr. Sci.*, Vol. 8, pp. 711–729 (1970).

[18] G. C. Sih, A Special Theory of Crack Propagation, *Methods of Analysis and Solutions of Crack Problems*, pp. XXI-XLV, (G. C. Sih, editor), Noordhoff International Publishing, Leyden, The Netherlands (1973).

Author's index

Subject index